Nature-based Tourism and Wellbeing

In Memoriam

In 2018, three editors of this work (Iride Azara, Eleni Michopoulou and Federico Niccolini) published *Tourism, Health, Wellbeing and Protected Areas*. It was co-edited by our esteemed colleague Alan Clarke, who passed away in 2021 at the age of 65. Alan had been a friend, and a mentor for Alessio Cavicchi as well. We will never forget his openness, enthusiasm, academic knowledge, kindness, generosity and that attitude of spurring one on to always do better. This book is dedicated to his memory with our love.

Nature-based Tourism and Wellbeing

Impacts and Future Outlook

Edited by

Federico Niccolini

Department of Political Sciences, University of Pisa, Italy

James R. Barborak

Center for Protected Area Management, Colorado State University, USA

Iride Azara

Centre for Contemporary Hospitality and Tourism, College of Business, Law and Social Sciences, University of Derby, UK

Eleni Michopoulou

Department of Innovation and Research, College of Business, Law and Social Sciences, University of Derby, UK

Alessio Cavicchi

Department of Agriculture, Food and Environment, University of Pisa, Italy

CABI

CABI is a trading name of CAB International

CABI
Nosworthy Way
Wallingford
Oxfordshire OX10 8DE
UK

Tel: +44 (0)1491 832111
E-mail: info@cabi.org
Website: www.cabi.org

CABI
200 Portland Street
Boston
MA 02114
USA

T: +1 (617)682-9015
E-mail: cabi-nao@cabi.org

A catalogue record for this book is available from the British Library, London, UK.

ISBN-13: 9781800621398 (hardback)
9781800621404 (ePDF)
9781800621411 (ePub)

DOI: 10.1079/9781800621411.0000

Commissioning Editor: Claire Parfitt
Editorial Assistant: Lauren Davies
Production Editor: Shankari Wilford

Typeset by Exeter Premedia Services Pvt Ltd, Chennai, India
Printed and bound in the USA by Integrated Books International, Dulles, Virginia

Contents

Contributors

Isabelle Adler, University of Lleida, Lleida, Spain. Email: adler.isi@gmx.de

Monika Bachinger, University of Applied Forest Sciences, Rottenburg, Germany. Email: bachinger@hs-rottenburg.de

Thomas Bausch, Competence Center Tourism and Mobility, Free University of Bozen-Bolzano, Italy; Department of Tourism, Munich University of Applied Sciences, Germany. Email: thomas.bausch@unibz.it and thomas.bausch@hm.edu

Kabindra Bhatta, Nepal Tourism Board, Kathmandu, Nepal

Michael Bischof, Institute of Ecomedicine, Paracelsus Medical University Salzburg, Austria. Email: michael.bischof@pmu.ac.at

Clara-Jane Blye, Faculty of Kinesiology, Sport and Recreation, University of Alberta, Edmonton, Canada. Email: clarajan@ualberta.ca

Myroslava Bosovska, Hotel and Restaurant Business Department, Kyiv National University of Trade and Economics, Kyiv, Ukraine. Email: m.bosovskaya@knute.edu.ua

Andrea Zita Botelho, Faculdade de Ciências e Tecnologia, Universidade dos Açores; CIBIO, Centro de Investigação em Biodiversidade e Recursos Genéticos, InBIO Laboratório Associado, BIOPOLIS Program in Genomics, Biodiversity and Land Planning; UNESCO Chair – Land Within Sea: Biodiversity & Sustainability in Atlantic Islands, Ponta Delgada, Portugal. Email: andrea.zc.botelho@uac.pt

Margarita Capdepón Frías, Department of Regional Geographic Analysis and Physical Geography, University of Alicante, Spain. Email: margarita.capdepon@ua.es

Iacopo Cavallini, Department of Economics and Management, University of Pisa, Italy. Email: iacopo.cavallini@unipi.it

Ana Cristina Costa, Faculdade de Ciências e Tecnologia, Universidade dos Açores; CIBIO, Centro de Investigação em Biodiversidade e Recursos Genéticos, InBIO Laboratório Associado, BIOPOLIS Program in Genomics, Biodiversity and Land Planning; UNESCO Chair – Land Within Sea: Biodiversity & Sustainability in Atlantic Islands, Ponta Delgada, Portugal. Email: ana.cm.costa@uac.pt

Sitaram Dahal, Nepal Tourism Board, Kathmandu, Nepal. Email: dahalsram@gmail.com

Sara Duarte, IGOT Institute of Geography and Spatial Planning, University of Lisbon, Portugal. Email: saramduarte@campus.ul.pt

Sonia Ferrari, Department of Business and Legal Science, University of Calabria, Italy. Email: sonia.ferrari@unical.it

Carlos Ferreira, University of Coimbra, CEGOT, Faculty of Arts and Humanities, Coimbra, Portugal and CiTUR (Centre for Tourism Research, Development and Innovation), Estoril, Portugal. Email: carlos.ferreira@uc.pt

Akke Folmer, European Tourism Futures Institute, NHL Stenden University of Applied Sciences, Leeuwarden, The Netherlands. Email: akke.folmer@nhlstenden.com

Martin Fontanari, International School of Management (ISM), Im Mediapark 5C, 50670 Cologne, Germany. Email: martin.fontanari@ism.de

Adriana Ghersi, Department of Architecture and Design, University of Genoa, Italy. Email: adriana.ghersi@unige.it

Maria Goutou, Hellenic Open University, Patras, Greece. Email: mariagoutou@hotmail.com

Yulei Guo, Department of Tourism, Chengdu Research Base of Giant Panda Breeding, Chengdu, China. Email: yulei.guo@panda.org.cn

Elizabeth Halpenny, Faculty of Kinesiology, Sport and Recreation, University of Alberta, Edmonton, Canada. Email: halpenny@ualberta.ca

Erica J. Hando, Nomad Tanzania Tours, Arusha, Tanzania

Stefan Hartman, European Tourism Futures Institute, NHL Stenden University of Applied Sciences, Leeuwarden, The Netherlands. Email: stefan.hartman@nhlstenden.com

Kim Hartmann, University of Strathclyde, Strathclyde Business School, Glasgow, UK. Email: kim.hartmann@strath.ac.uk

Philipp Hilmer, University of Applied Forest Sciences, Rottenburg, Germany

Glen Hvenegaard, University of Alberta, Augustana Campus, Camrose, Canada. Email: gth@ualberta.ca

Nicolaia Iaffaldano, Ionic Department, University of Bari, Italy. Email: nicolaia.iaffaldano@uniba.it

Simon Jones, NatureScapes, Raleigh, USA. Email: sjones@nature-tourism.com

Thomas Jones, College of Sustainability and Tourism, Ritsumeikan Asia Pacific University, Beppu, Japan. Email: 110054tj@apu.ac.jp

Nicholas Karachalis, University of the Aegean, Mytilene, Greece. Email: nkarachalis@aegean.gr

Ahmed A. Karim, University Clinic of Psychiatry and Psychotherapy, University of Tübingen, Germany and Department of Health Psychology and Neurorehabilitation, SRH Mobile University, Riedlingen, Germany

Nataliia Kochkina, Marketing and Business Administration Department, Taras Shevchenko National University of Kyiv, Ukraine. Email: nataliakochkina@knu.ua

Andrea Králiková, Department of Marketing and Trade, Faculty of Business and Economics, Mendel University, Brno, Czech Republic. Email: andrea.kralikova@mendelu.cz

Patrik Kubát, Department of Marketing and Trade, Faculty of Business and Economics, Mendel University, Brno, Czech Republic. Email: patrik.kubat@mendelu.cz

Simona Lo Bianco, Department of Business and Legal Science, University of Calabria, Italy. Email: s.lobianco@fondoambiente.it

Erik Luvaas, Center on Disabilities and Human Development, Department of Movement Sciences, University of Idaho, USA. Email: eluvaas@uidaho.edu

Sebastian Markov, Faculty European Campus Rottal-Inn, Deggendorf Institute of Technology, Germany. Email: sebastian.markov@th-deg.de

Daniela Marzo, Department of Economics and Management, University of Pisa, Italy. Email: daniela.marzo@ec.unipi.it

Angelina Mattivi, University of Applied Forest Sciences, Rottenburg, Germany

Milena Nikolova, BehaviorSMART, Sofia, Bulgaria. Email: milena@behavior-smart.com

Corinna Pippirs, Faculty European Campus Rottal-Inn, Deggendorf Institute of Technology, Germany. Email: corinna.pippirs@th-deg.de

Kateřina Ryglová, Department of Marketing and Trade, Faculty of Business and Economics, Mendel University, Brno, Czech Republic. Email: katerina.ryglova@mendelu.cz

Rita Salvatore, Department of Political Science, University of Teramo, Italy. Email: rsalvatore@unite.it

Ulrich Schmudde, Mid Sweden University, Östersund, Sweden. Email: ulrich.schmudde@miun.se

Rita Schulz, School of Tourism, Hospitality and Events Management, Moi University, Eldoret, Kenya

Abhijeet Shirsat, Department of Recreation, Parks and Tourism Administration, California State University, Sacramento, California. Email: shirsat@csus.edu

Francisco Silva, Escola Superior de Hotelaria e Turismo do Estoril, Centro de Estudos Geográficos, IGOT, Universidade de Lisboa, CiTUR, Estoril, Portugal. Email: francisco.silva@eshte.pt

Anna Sörensson, Mid Sweden University, Östersund, Sweden and Inland Norway University of Applied Science, Norway. Email: anna.sorensson@miun.se and anna.sorensson@inn.no

Georg Christian Steckenbauer, Faculty European Campus Rottal-Inn, Deggendorf Institute of Technology, Deggendorf, Germany. Email: georg.steckenbauer@th-deg.de

Olena Sushchenko, Tourism Department, Simon Kuznets Kharkiv National University of Economics, Kharkiv, Ukraine. Email: olena.sushchenko@hneu.net

Verena Tauber, Competence Center Tourism and Mobility, Free University of Bozen-Bolzano, Italy. Email: Verena.Tauber@unibz.it

Tim Taylor, European Centre for Environment and Human Health, Department of Public Health and Sport Sciences, University of Exeter Medical School, Truro, UK. Email: Timothy.J.Taylor@exeter.ac.uk

Simon M. Thiong'o, School of Events, Hospitality and Tourism Management, Leeds Beckett University, UK. Email: tm_simon@mu.ac.ke and S.Thiongo2234@student.leedsbeckett.ac.uk

Paulo Torres, CIBIO, Centro de Investigação em Biodiversidade e Recursos Genéticos, InBIO Laboratório Associado, BIOPOLIS Program in Genomics, Biodiversity and Land Planning;

UNESCO Chair – Land Within Sea: Biodiversity & Sustainability in Atlantic Islands, Ponta Delgada, Portugal. Email: biol.paulo@gmail.com

Anastasia Traskevich (Анастасия Траскевич), Department of Tourism Management and Economics, Belarus State Economic University, Minsk, Belarus. Email: anastasia.traskevich@akalem.de

Iryna Trunina, Business Administration, Marketing and Tourism Department, Kremenchuk Mykhailo Ostrohradskyi National University, Kremenchuk, Ukraine. Email: irinatrunina@kdu.edu.ua

Brendah N. Uluma, School of Tourism, Hospitality and Events Management, Moi University, Eldoret, Kenya

Rie Usui, College of Sustainability and Tourism, Ritsumeikan Asia Pacific University, Beppu, Japan. Email: rieusui@apu.ac.jp

Maria Anunciação Ventura, Faculdade de Ciências e Tecnologia, Universidade dos Açores; CIBIO, Centro de Investigação em Biodiversidade e Recursos Genéticos, InBIO Laboratório Associado, BIOPOLIS Program in Genomics, Biodiversity and Land Planning; UNESCO Chair – Land Within Sea: Biodiversity & Sustainability in Atlantic Islands, Ponta Delgada, Portugal. Email: maria.am.ventura@uac.pt

Eva Vroegop, Academy of Tourism, Fondazione Campus Lucca, Tuscany, Italy. Email: eva.vroegop@campuslucca.it

Ben Wielenga, European Tourism Futures Institute, NHL Stenden University of Applied Sciences, Leeuwarden, The Netherlands. Email: ben.wielenga@nhlstenden.com

Acknowledgements

The editors would like to thank Dr Michael Manfredo, Head of the Department of Human Dimensions of Natural Resources in the Warner College of Natural Resources at Colorado State University, for his critical role in this book project. Along with other members of the Foundational Consortium from around the world, Dr Manfredo has played a critical role in the Tourism Naturally conference series, begun in 2016, to promote increased professionalism and effectiveness in the global travel and tourism field. Dr Manfredo has been instrumental in bringing people together from around the world and providing leadership and support to various projects that ruminate humans' encounters with nature. Through his highly cited publications on the evolution of societal views of nature, Dr Manfredo is one of the principal sources of inspiration for this volume and the editors.

Even more, the authors would like to thank their dear friend Mike for the vision, passion, commitment, tireless support and generosity with which he has accompanied the 'travel' of some of them in their scientific growth and academic maturation.

1 Diving with Turtles: in Search of Nature in Recalled Vacation Trip Experiences and their Influence on Subjective Wellbeing

Eva Vroegop*
Academy of Tourism, Fondazione Campus Lucca, Tuscany, Italy

Abstract

As nature is ubiquitous in tourism marketing and personal pictures of vacation trips, the question asked in this contribution is to what extent nature is present in people's recalled trip experiences and how this relates to subjective wellbeing (SWB) in the form of emotions and satisfaction with life (SWL). To address this question, a dataset not collected under the theme of nature-based tourism (NBT) was employed. Based on a very broad definition of nature, findings indicate that (i) NBT represents more frequently recalled vacation trip experiences than urban experiences; (ii) a positive relationship exists between recalled active nature-based vacation experiences (NBVE) and the intensity of emotions elicited; (iii) recalling passive experiences elicits different emotions as compared to active ones; (iv) recalled NBVE and emotions elicited have a positive influence on SWL; and (v) the relationship between nature-based experiences and SWL depends on duration and intensity of the recalled trip experience.

Introduction

Since the beginning of tourism as a relevant social and economic phenomenon, the natural world has served as an inspiration for travellers used in tourism-related discourse, be it in marketing and promotion, scientific research, travel writing or simply personal narration of vacation trips. Moreover, the rapid urbanization of the planet, resulting in a majority of the world's population living an urban reality on a daily basis, has continuously reinforced the quest for nature experiences in leisure time and during vacations. In spite of the importance of urban tourism, reference to nature is inevitably present in interpersonal communication about vacation trips. In Italy, for example, when talking about a vacation with someone you just met, the question very often is, 'Are you a mountain or a sea person?'

The motivation for this chapter is linked to this ubiquity of nature in tourism, and arising from a curiosity: Will nature also be omnipresent in recalled vacation trip experiences in a dataset not collected under the theme of nature-based tourism (NBT)?

One originality of the current contribution is that it does not depart from any of the definitions proposed in NBT literature, but uses the data to identify the 'based' in 'nature-based' along the dimensions of intensity and duration of an experience in nature. The reason is not a lack of existing sound and operable definitions of NBT, but the fact that the data used in this study have not been collected for the theme of nature-based tourism. Hence, it does not allow all reported experiences to be identified unequivocally with reference to one of these NBT definitions. Not having the possibility to depart from an *a priori*

*eva.vroegop@campuslucca.it

 Nature-based Tourism and Wellbeing: Impacts and Future Outlook (eds F. Niccolini *et al.*)
DOI: 10.1079/9781800621411.0001

definition of NBT, the current approach – based on a survey on any kind of vacation trip recall – consists in coding verbal references to nature-related experiences and then empirically checking for their influence on subjective wellbeing (SWB). If there is a positive impact of a nature-related vacation trip recall on SWB, this kind of nature-related experience will be considered as being nature-based. In this sense, nature-based is what creates wellbeing.

Recalling positive vacation trip experiences generally makes people feel well. As memory carries our experiences forward in time, this permits memories from past experiences to influence people's subjective wellbeing today. This is what Elster and Loewenstein (1992) refer to as 'memory utility', which originates from reflecting about our personal past, as in the case of vacation trip experiences. Even though experiences are individual constructs, our tourism experiences are intrinsically linked to the surrounding environment (Fossgard and Fredman, 2019). Unsurprisingly, among the freely recalled trip experiences in the present dataset numerous entries relate to nature, which encompasses the Earth's natural environment, its animals and plants (Newsome *et al.*, 2012).

Given the set-up just described, the research presented here consists in searching statistical evidence for NBT with regard to the research questions and hypotheses in Table 1.1.

This chapter contributes to the literature on nature-based tourism experiences in two ways. Firstly, presenting a new take on vacation trip memories, providing evidence that vacation trip experiences will create emotions whenever recalling; these will be called 'long-lasting emotional memories'. Secondly, the current research is not about choosing or having chosen nature-based vacations, but about choosing whether to recall nature-based experiences among many past vacation experiences, thus verifying the relevance of nature-based experiences in any form of tourism experience. Thirdly, in terms of method, this chapter presents a first operationalization in tourism of the idea proposed by Mannell (1980, as cited in Mannell and Iso-Ahola, 1987) to identify the contact with nature in terms of duration and intensity.

Nature, Recalled Vacation Experiences and Subjective Wellbeing

To investigate recalled tourism experiences, and especially nature-based ones, the concept of

Table 1.1. Research questions and hypotheses.

Research questions		Hypotheses	
1	Are nature-based experiences frequent elements of a freely recalled vacation trip?	H_1	Nature-based experiences are expected to represent a more frequently recalled vacation trip experience than, e.g. urban experiences.
2	Do nature-based vacation experiences create long-lasting memories in terms of emotions elicited when recalling them?	H_2	The relationship between recalled nature-based experiences and the intensity of emotions elicited is expected to be positive.
3	Do different kinds of nature-based experiences elicit different emotions?	H_3	The recall of passive experiences is expected to elicit different emotions as compared to recalling active ones.
4	Do recalled vacation experiences that are nature-based and emotions elicited while recalling them influence satisfaction with life?	H_4	Recalled vacation experiences that are nature-based and emotions elicited while recalling them are expected to have a positive influence on satisfaction with life.
5	Do duration and intensity of a recalled nature-based vacation experience influence satisfaction with life?	H_5	Both the duration and intensity of a recalled nature-based vacation experience are expected to positively influence satisfaction with life.

the tourist experience provides a natural point of departure. The tourist experience involves intricate psychological processes within individuals. In a review of tourism psychology studies, Larsen (2007, p. 12) concludes that 'tourist experiences are well conceptualized as cognitive and particularly memory processes'. Memories are created within the individual tourist by means of a reconstructive process. Building on this notion, Larsen defines a tourist experience as 'a past personal travel-related event strong enough to have entered long-term memory' (2007, p. 15). Memory is a key theme in cognitive psychology, and tourism research can benefit from recent developments in this field (Pearce and Packer, 2013; Scott, 2020). Namely, cognitive psychology studies how people think, aiming to grasp the mechanisms that determine how the mind processes internal and external stimuli and makes sense of our experiences (Scott, 2020; Skavronskaya *et al.*, 2020). Although experiences are unique mental creations, the environment in which we travel plays a fundamental role in our tourist experiences (Fossgard and Fredman, 2019). Hence, the present study builds on memory research to investigate recalled nature-based tourism experiences and their relationship with SWB.

When people spend time in nature, this is found to enhance their SWB, including a greater satisfaction with life (SWL) (Capaldi *et al.*, 2015). The natural environment, for instance wildlife and scenery, has been found to elicit different emotions in tourists (Chhetri *et al.*, 2004; Farber and Hall, 2007). Moreover, emotions are aroused also when people think back on a past vacation trip experience. An individual's emotions when recalling a past vacation experience can differ from the emotions lived during the trip itself. For instance, when the individual's on-site emotions were negative, but their emotions while thinking back on the same trip event are positive (Tung *et al.*, 2017). Hence, 'mental time travel' has the ability to influence our cognitions, emotions and decisions in the here and now (Van Boven and Ashworth, 2007). Furthermore, emotions are at the root of tourists' most important experiences and may be the primary driver of why some vacation experiences become more memorable than others (Chandralal and Valenzuela, 2013; Knobloch *et al.*, 2014; Chandralal and Valenzuela, 2015; Tung *et al.*, 2017). The psychology literature discerns between individuals' 'remembered emotions' on the one hand and their emotional reactions while recalling past experiences on the other. Regarding the latter, psychologists have acknowledged the influence of retrospection on people's emotions in the here and now beyond the information that memories transmit (Mather and Carstensen, 2005; Van Boven and Ashworth, 2007). Although extensive literature exists on the influence of emotions on remembered holiday experiences, emotions elicited when recalling vacation trip experiences have been neglected in tourism research (Filep *et al.*, 2013). This is surprising, as emotions while recalling past personal events are omnipresent in people's everyday lives (Van Boven and Ashworth, 2007). Moreover, thinking about positive emotional experiences is an important element of everyday life as it can enhance subjective wellbeing and psychological health (Denkova *et al.*, 2013; Speer *et al.*, 2014).

The relationship between taking vacations and people's SWB has been studied by occupational psychologists and tourism scholars alike. A meta-analysis showed that vacations have a small but positive effect, although the effects on SWB are rather short-lived (De Bloom *et al.*, 2009) and life satisfaction declines upon return home (Chen and Petrick, 2013). A lack of evidence exists when it comes to the long-term influence of vacations on people's wellbeing (Chen and Petrick, 2013; Knobloch *et al.*, 2017). Furthermore, research testing whether specific vacation experiences and activities have an influence on SWB remains scarce.

In the tourism literature, different authors have suggested a connection between nature-based tourism and SWB. In the context of a most recent vacation, Lehto (2013) asked university students and staff to evaluate the restorative value of their holiday destination just visited, presenting initial evidence that nature-based vacations facilitate recovery from mental fatigue better than city-based destinations. This finding held for the elements of the destination's 'compatibility' with the individual's personality, feeling physically as well as mentally away from their daily environment, and for confusion and chaos at the destination ('discord'). No differences between natural and urban environments were found for the variety of activities offered at the holiday destination ('extent') and whether holidaymakers perceived the destination as 'fascinating'. In the context of hiking holidays, the tourist motivation 'enjoying the natural environment and escaping from daily life' was the dominant factor influencing the SWB of hiking tourists during their vacation (Kim *et al.*, 2015). In a systematic review of previous research on the link between nature-based tourism and SWB, Qiu *et al.* (2021, p. 15) conclude that one can view NBT as a 'public-wellness product to improve the health and wellbeing of visitors'.

For obvious reasons, the results obtained in the various studies heavily depend on the definition of nature-based. From the abundance of NBT definitions the author mentions just a few here, which are considered sound and operable, like, for example, the following one coined by Fredman and Tyrväinen (2010, p. 179) for Finland: 'Typically this means travelling to and staying overnight in locations close to protected areas, forests, lakes or the sea or the countryside and participating in activities compatible with the location's natural qualities'. In a global analysis of NBT definitions, Fredman *et al.* (2009, as cited in Fredman and Tyrväinen, 2010, p. 180) have determined four common themes in nature-based tourism: '(i) visitors to a nature area; (ii) experiences of a natural environment; (iii) participation in an activity; and (iv) normative components related to, e.g., sustainable development and local impacts'. Based on tourism's general definition by the World Tourism Organization (UNWTO, 2010), an instance of a broad definition may be 'activities by humans occurring when visiting nature areas outside one's ordinary residence' (Fredman and Haukeland, 2021, p. xiv). This minimalistic, non-normative definition of NBT is inclusive in terms of environment, activity, and enjoyment level (Fredman and Haukeland, 2021). Other authors, instead of defining nature-based tourism, have classified NBT by distinguishing between travel motivations (i.e. different levels of *nature as point of attraction* from 'nature protection' to 'nature experience', 'sports and adventure', and 'hedonistic') and the extent to which service arrangements are booked (i.e. from 'independent' to *à la carte*, 'customized' and 'fully standardized'), resulting in a product-based typology of 16 different types (Arnegger *et al.*, 2010).

For the reasons exposed above, this chapter does not and cannot depart from one of these definitions, searching instead for the relevance of 'based' in 'nature-based' in terms of intensity and duration of the individual's exposure to nature in a recalled vacation experience following Mannell (1980, as cited in Mannell and Iso-Ahola, 1987). In discussing the psychological 'nature' of tourism experiences, they observed that the intensity and duration of a vacation experience have been relatively ignored in tourism research. In terms of intensity, specifically regarding SWB, De Bloom *et al.* (2009) recommend examining active versus passive vacation activities.

Research Method

This contribution exploits data on vacation trip memories collected in Switzerland in 2018 asking participants to freely recall a past holiday trip that comes to mind. The questionnaire was conducted among a sample of 1003 Swiss adults stratified by age, gender and language (German-speaking and French-speaking). The following analyses are based on 997 observations as six had to be excluded due to missing data.

This is the first time, to the author's knowledge, that verbal memories are collected without reference to a specific type of vacation, destination or point in time undertaken (e.g. 'most recent'), in the context of a large survey. Note that in memory research the dominant methods are in-depth interviews and small-group experiments. The survey participants, not knowing beforehand what the survey theme would be, were, at the start of the survey, confronted with the request: 'Think of a past vacation trip that spontaneously comes to your mind'. Subsequently, participants were asked to describe this trip in three words. Later in the questionnaire, participants were invited to think of one moment, experience or event during the recalled vacation trip and describe it in one sentence (not prompting respondents for 'special', 'best', 'worst', etc.). This second statement was asked after collecting the usual vacation trip characteristics such as the destination, duration, travel party and repeat visits, of which the destination and duration were used to complement the verbal memory descriptors.

Both the words and sentence are used to examine the connection between nature-related tourism experiences and SWL. This relation is expected to depend on the coding of the verbal statements as described below. The qualitative information represents the main material on which this study is based, because it contains the essence of the vacation memory. It is this text-based information that is used here to inquire about the importance of NBT experiences in people's memories, when compared to other forms of tourism and its impact on their SWL today. In contrast to most of the vacation memory literature on the issue, especially on memorable tourism experiences, this survey did not ask for memorable experiences, but for a vacation trip that freely comes to the mind of the survey participant. The memory content, in the present case with a special focus on nature-based experience recalls, and its relation with emotions elicited while recalling and SWL represent the research interest in this chapter.

Emotions were measured following Kim and Fesenmaier's (2017) work on post-trip experiences for a hypothetical vacation, adapting their Likert scale item 'At this moment, I feel ...' into 'Remembering this vacation trip, I feel ...'. The final emotion scale is composed of ten subjective feeling pairs (e.g. happy–sad, excited–indifferent).

Subjective wellbeing is assessed using two different scales, namely at the start of the survey by the ten-item Brief Inventory of Thriving (BIT) scale (Su *et al.*, 2014), and at the end of the survey along the five-item Satisfaction with Life scale (SWLS) (Diener *et al.*, 1985). The BIT consists of six (out of seven) fundamental dimensions that measure the intricate concept of psychological wellbeing: (i) subjective wellbeing (SWB), which is characterized by high levels of positive emotions and SWL; (ii) sense of support and belonging in relationships; (iii) interest and engagement in daily activities; (iv) purpose and meaning in life; (v) mastery (i.e. a sense of self-worth, self-efficacy and accomplishment); and (vi) optimism – not including the seventh dimension of autonomy (Su *et al.*, 2014). The SWLS measures the life satisfaction dimension of SWB and has been widely used in tourism studies, for instance to investigate the effect of tourism recovery experiences on SWL (Chen *et al.*, 2016). To the author's knowledge, this is the first time that the BIT is employed in a tourism context.

In empirical terms, the connection between memories, emotions and SWL will be analysed in two steps. Namely, first in a descriptive analysis, and a subsequent econometric analysis in the form of a multiple regression with SWL as a dependent variable. Another option considered was to use structural equation modelling (SEM) for understanding the complexity of these relations, but using a complex behavioural model like SEM on a dataset not created for the research questions in this study did not seem appropriate. Therefore, the author opted for a simple impact analysis employing a linear regression model in order to test for systematic effects of nature-based recalls and emotions on SWL. First, using descriptive statistics, the kind of vacation experiences that create long-lasting emotional memories will be discussed, with a focus on nature-based experiences.

The main contribution of this research lies in the coding of the verbal information. After careful consideration, the author opted for a manual coding emerging from the need to combine information contained in four text variables (three words and one sentence) with information on trip duration and destination. Moreover, the text in the dataset stemming from responses in the French and German language would have implied performing separate computer coding by language and then combining them with the information on the destination and trip duration. Finally, electronic coding of the words written by participants only, translated into English, would not have provided a base for the analysis intended here, given that the words are mostly very generic (for an illustration see Fig. 1.1).

For a coherent coding of nature-based instances in the data, this study opts for a new conceptual but pragmatic definition. Considering the term 'nature-based', a definition should be clear on what is 'nature' and what is 'based' on nature. Moreover, distinguishing between, for instance, pristine nature, preserved nature, transformed nature or 'arranged' nature is conceptually difficult and in the current data practically impossible. Therefore, this investigation adopts a pragmatic approach in accordance

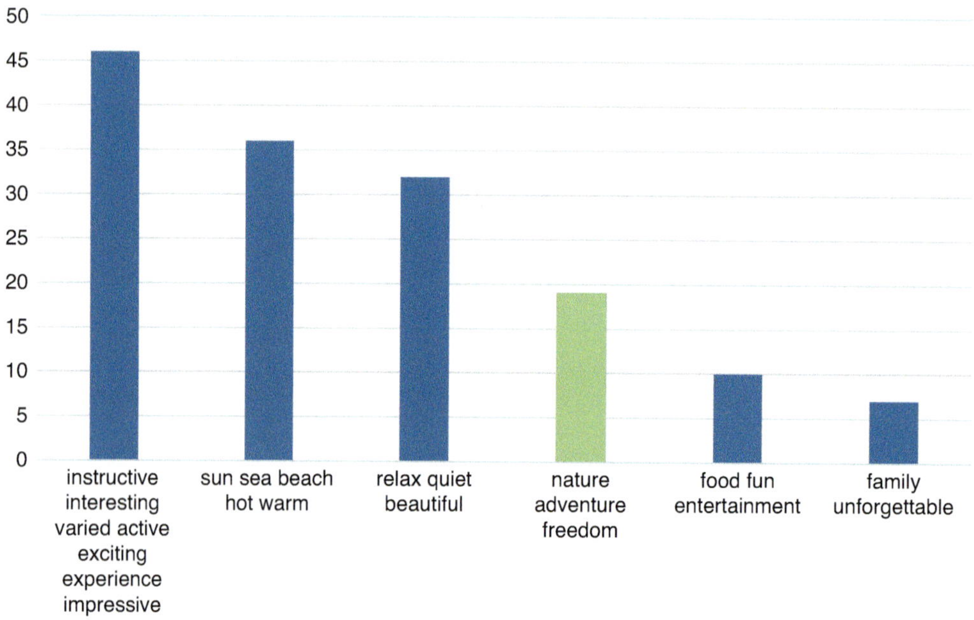

Fig. 1.1. Ad hoc grouping of most frequent words in first 5% of observations.

with Newsome *et al.* (2012), considering all forms of flora and fauna mentioned by respondents as 'nature', with two exceptions. Firstly, indoor activities linked to water (wellness) or animals (zoo) are not considered. Secondly, anything mentioned in relation to beach holidays is put in a separate category.

While the first restriction may be debatable, as it implies considering only outdoor activities, it has the advantage of being rather clear.

The special treatment of beach holidays is motivated as follows. Traditional beach holidays do not fit into the coding scheme developed. Sun, sand, sea, waves, wind and sunsets are all natural elements, and jumping in the waves of the sea for the first time undoubtedly creates lasting nature-based memories for a Swiss child. As all beach vacations will necessarily imply continuous active immersion with the element, they would all fall into the category 'nature-based vacation'. Yet, in this category one would expect to find completely different vacation trips, such as a Himalaya trek or a trek through Nordic wildernesses. Hence, to consider a traditional beach vacation as nature-based seems to stretch the concept unduly. The proposed solution to this problem is to code a separate category called 'beach' and then check empirically whether or not there are differences with respect to the various categories of 'nature-based'.

As the present concept of 'nature' is highly generic, in order to identify different kinds of nature-based tourism experiences, this study concentrates on 'based'. Following the suggestions by Mannell (1980, as cited in Mannell and Iso-Ahola, 1987) in a tourism psychology context, the current study distinguishes between two dimensions of a trip experience based on nature: (i) duration; and (ii) intensity of the contact with nature. To simplify the coding task, the present inquiry employs binary classifications for both experience dimensions:

- duration: 'excursion, visit, moment' versus 'vacation, multi-day trip/trek'; and
- intensity: 'passive, consuming' versus 'active, engaged'.

Regarding duration, the distinction between moment and vacation is based on whether the moment described in one sentence is part of a vacation trip that is nature-based as a whole (coded as 'vacation'), or whether this moment

		Intensity	
		Passive	Active
Duration	Moment	• Whale/dolphin watching during a Maldives vacation • Enjoying/admiring the landscape/view during a desert excursion while on vacation in Marrakesh	• Diving with dolphins, sharks, turtles • Day excursion to mountains, natural parks etc.
	Vacation	• Guided safari • Arctic cruise	• Himalaya trek • Vacation/trek in a natural reserve

Fig. 1.2. Coding examples differentiated by experience intensity and duration.

is part of a non-nature-based vacation (coded as 'moment'). An example of a nature-based vacation trip is sighting wild animals during a safari, while the latter often takes the form of an excursion during a beach holiday, a cruise or round trip. The examples provided in Fig. 1.2 illustrate this further.

The distinction between active and passive vacation activities (De Bloom *et al.*, 2009) is a traditional one, classifying physical activities (in nature) in the active category. Selected kinds of motorized experiences, where the participant was exploring nature using their own means of transport (e.g. driving a van through northern Lapland), were included in the active category as well.

As the current data were not collected for the purpose of analysing nature-based tourism, it permits putting 'freely recalled trip experiences that are nature-based' in context, comparing it with other forms of vacation experiences. Therefore, coding was also applied to the verbal statements to identify three other kinds of experiences, namely (i) beach-based vacation experiences, as mentioned earlier; (ii) urban-based touristic experiences; and (iii) reference to interactions with the local population. The coding of beach vacation experiences was straightforward, because destination and duration together with the stereotypical words 'sand', 'sea', 'beach', 'waves' and 'sunset' allowed for a reliable classification. Interaction with locals was again not representing any challenge, given that most of the sentences were referring generically to the friendliness of the people, and statements of concrete interaction did not go beyond two mentions of a dinner with the host family. Therefore, any mention of the local population, be it in the words or sentences written by the survey participants, were treated as an indication of an interaction. This very 'generous' definition still resulted in a low number of 40 observations, and, not surprisingly, the respective variable does not return significant results. Determining urban experiences was simple in cases where participants were referring to an urban experience or a city destination in their description of a moment they recalled. More challenging was the very frequent mention of cities during long round-trip vacations, typically in overseas destinations like the USA, Canada and Australia, but no urban experience in the description of the moment. In these cases, the experience was only coded as urban if at least one city was also mentioned in the description of the moment.

Findings

The findings are presented and discussed in three steps. Firstly, the relative importance of nature-based experiences in descriptions of a freely recalled vacation trip as compared to other memories in terms of frequencies (H_1). Secondly, performing a series of independent samples t-tests,

Table 1.2. Frequencies of different recalled vacation experiences.

Nature-based							Other coded		
Passive moment	Passive vacation	Passive total	Active moment	Active vacation	Active total	Nature-based	Beach-based	Urban-based	Locals-based
94	30	124	105	30	135	259	138	160	40

checking for significant differences between presence or absence of a specific type of memory on emotions and SWL. This in order to assess the importance of the various categories and items in the two constructs of interest, namely emotions and SWL (H_2, H_3, H_4). In a third step, the results of a simple linear regression model are presented. This model contains selected socio-demographic characteristics, vacation descriptors, emotions, as well as four types of recalled vacation trip experiences, identified with the preceding independent samples t-tests as covariates, and SWL as a dependent variable (H_5).

Table 1.2 contains the frequencies of different vacation trip recalls as defined by the present coding. In the first instance, this permits the examination of the share of NBT experiences in all recalled experiences. It is important to note that when respondents are posed the question to recall any of past vacation trips, they will intuitively screen a smaller to larger sample (from one to all past vacations) of their vacation history. This will evoke free recalls of various kinds – places, experiences, people, moments, etc. – and these flashbacks will simultaneously elicit emotions during recall. As a consequence, the frequencies do not reflect the shares in vacations of the Swiss population, but the share of those vacation trips they want to recall. Deviations from the vacation shares contain interesting clues as to the relative importance of certain vacation experiences. Unfortunately, however, no Swiss statistics on the importance of nature-based vacations exist to date.

A first observation is that, according to the coding, nature-based memories with a share of 26% play a rather important role in a free recall experiment of vacation experiences, being the most frequent of the four observed categories. Note that the not-coded 40% of the sample (394 observations) does not fall into any of the four categories of nature-, beach-, urban- and locals-based. It is composed on the one hand of round trips, cruises and probably other forms of tourism experiences. On the other hand, and more importantly, this group is formed by verbal statements that do not permit concluding on a specific type of tourism experience. Often, this group consists of exclamations, adjectives in superlatives or expressions of positive and negative evaluations (e.g. 'best trip ever' or 'worst trip ever'). For this reason, it was decided not to classify these experiences solely based on information about the destination and trip duration. Regarding research question 1 on the relative importance of nature-based experiences in recalled vacation trips, hypothesis 1 is confirmed. Freely recalled trip experiences that were nature-based are significantly more frequent than the other ones identified (beach-, urban- and locals-based).

While beach- and urban-based experiences are important elements of participants' freely recalled vacation experiences as well, interaction with locals is – in spite of the large definition employed – rather marginally present in the reported trip experiences. Of course, it cannot be known whether this is due to the underlying vacation choices of the respondents or the decision on what kind of recalled vacation trip experience to report. In this context, it is relevant to mention that the sample is representative of the Swiss population in terms of basic socio-demographics (gender, age, language), but not necessarily in terms of vacation behaviour. A substantial and interesting bias in the dataset is that of the reported vacations, the domestic ones (8%) are dramatically under-represented when compared to official statistics of Swiss travel behaviour (33%). This while overseas holiday trips (37% of recalled vacation trips) are heavily over-represented as compared to the Swiss travel statistics (7%). Continental tourism in Europe (55%) has a share similar to the statistics (60%) (Swiss Federal Statistical Office, 2019).

Taking a closer look at the composition of nature-based tourism, it emerges that while memories of passive and active nature-based experiences are of equal relevance, nature-based moments are much more present than whole-vacation experiences. The following objective is to assess the relevance of the four types of nature-based tourism experiences, comparing their relation to and impact on emotions and SWL. These considerations start with reporting the results of a series of independent samples t-tests for each of the categories. The null hypothesis is that membership of a category does not make a difference in terms of the dependent variables' mean (emotions and SWL) with respect to non-membership.

Table 1.3 presents statistically significant results regarding emotions. The table reports only the sign and the level of significance (signs inside parentheses are significance levels based on the Welch test for unequal variances; $^{*}p<0.10$, $^{**}p<0.05$, $^{***}p<0.01$).

Examining the results in Table 1.3 in relation to research question 2 and 3, it is observed that a relevant and differentiated relationship between recalled tourism experiences and emotion seems to exist. The first thing to note is that, with the exception of 'proud', all items are somewhere present with a significant sign. Looking more closely, an initial remarkable result is an almost total absence of significant differences when it comes to recalls of passive nature-based experiences as compared to active ones. The only relevant exception concerns passive moments, which leaves lasting strong impressions. Examples in the present data of such passive moments would be the sudden surprising sight of a wild animal or being for the first time on a high mountain peak, or again the first time seeing the Grand Canyon. Interestingly, 'impressed' is the single most important emotion across nature-based experiences and relevant for beach-based tourism as well, but with a negative sign. This contrasting result is regarded as a confirmation of the decision to not include beach vacation memories in the nature-based categories. Memories of active nature-based experiences return some significant signs for various emotional items, especially relating to active vacations. These evoke significantly higher excitement when recalled than all other kinds of nature-based memories. Finally, and probably slightly surprisingly, urban-based vacation recalls are the category with the largest number of significant emotions (including their average), and they all have a negative sign. This will be partly confirmed in the regression analysis below, where the *ceteris paribus* condition holds for the included categories. Overall, to the extent that these interdependent samples t-tests can be taken as an indication for the eventual existence of impacts, hypothesis 2 seems partly confirmed, especially with regard to a longer duration of the experience (nature-based vacation, beach-based vacation). Furthermore, hypothesis 3 seems to be confirmed, as significant differences in present-day emotions can be observed between the various kinds of recalls. However, note that independent samples t-tests cannot serve as evidence for a rejection or acceptance of a hypothesis, and that the tests are always against all other types, and not for differences among types. To do this, an ANOVA test was performed, returning a significant F-test only for the emotion 'impressed' where the post-hoc Tukey test was significant when comparing nature-based and beach-based.

Turning to SWL, Table 1.4 presents in an analogous way the statistically significant signs.

The discussion of these results can be kept short and will not refer to the research hypotheses, because the following regression will test in an appropriate way for causal impacts of recalled vacation experiences on SWL. Overall, active nature-based experiences (moments and vacations, the sum over all nature-based categories) and urban-based experiences demonstrate significant differences with respect to the whole sample differences and across the various items of the Diener *et al.* (1985) SWLS. A provisional conclusion would be that only intensive experiences among the nature-based ones, and urban recalls, make a relevant difference. A second observation concerns the SWLS items, where for the relevant categories one finds basically no difference in sign and strength among the items and their average. At least based on these preliminary tests, there is no clear evidence that the single items per se make a difference, which obviously is in line with the idea of the five SWL scale items forming a latent construct for life satisfaction. An ANOVA test was also performed for the five Diener *et al.* (1985) SWL scale items, resulting in an F-test significant on the 95% level, according to the post hoc Tukey test to a significant mean difference ($p = 0.02$) between nature-based and urban recalls, as well

Table 1.3. Significant differences in emotions by type of recalled experience.

	Nature-based							Other coded		
	Passive moment	Passive vacation	Passive total	Active moment	Active vacation	Active total	Nature-based	Beach-based	Urban-based	Locals-based
Relaxed	$+^{*}$									
Peaceful						$(+^{*})$				
Carefree									$-^{*}$	
Happy				$+^{**}$		$+^{**}$			$-^{*}$	
Fearless								$+^{*}$		$(+^{*})$
Satisfied					$(+^{*})$	$+^{*}$			$-^{**}$	$(+^{**})$
Cheerful					$(+^{**})$				$-^{**}$	
Excited					$+^{**}$					
Impressed	$+^{***}$		$+^{***}$		$(+^{**})$		$+^{**}$	$-^{**}$		
Proud										
Avg. emo.									$-^{*}$	

H_0 = no difference between a specific type and all others.

Table 1.4. Significant differences in SWL by type of recalled experience (H_0 = no difference between a specific type and all others).

	Nature-based							Other coded		
	Passive moment	Passive vacation	Passive	Active moment	Active vacation	Active	Nature-based	Beach-based	Urban-based	Locals-based
(1)					$+^{**}$	$+^{**}$	$+^{***}$	$+^{*}$	$-^{**}$	
(2)					$+^{**}$	$+^{***}$	$+^{***}$		$-^{***}$	
(3)				$+^{*}$	$+^{**}$	$+^{***}$	$+^{**}$	$+^{*}$		
(4)						$+^{**}$			$-^{**}$	
(5)			$(+^{**})$		$+^{**}$	$+^{**}$	$+^{**}$	$+^{*}$	$-^{**}$	
Avg. SWL				$+^{*}$	$+^{***}$	$+^{***}$	$+^{***}$		$-^{*}$	

Satisfaction with Life (SWL) scale by Diener *et al.* (1985):

(1) In most ways my life is close to my ideal.
(2) The conditions of my life are excellent.
(3) I am satisfied with my life.
(4) So far, I have gotten the important things I want in life.
(5) If I could live my life over, I would change almost nothing.

as a less significant mean difference ($p = 0.09$) between nature-based and beach-based vacation experiences.

Based on these tests, it is decided (a) to insert 'active' and the three control groups as covariates in the following regression, expecting a positive sign for active nature-based recalls and a negative one for urban-based experience recalls; and (b) to use only the average SWL scale, but not single items, as a dependent variable.

Regression analysis was undertaken using STATA 13 software. Regressions with robust standard errors were performed and tested separately for the eventual presence of multicollinearity and endogeneity of the 'emotions' variable. Both tests signalled the absence of the respective problems, in the case of multicollinearity with all variance inflation factors (VIF) being only slightly above 1, as well as for eventual endogeneity of the emotion variable with a Wu-Hausman F-statistic (1, 976) of 0.8492, $p = 0.3570$.

The socio-demographic and vacation descriptors used as controls and presented in Table 1.5 were chosen based on parsimony, being those with significant coefficients. In what follows, these will not be discussed in detail. A special mention is due to the Brief Inventory of Thriving (BIT) developed by Su *et al.* (2014), which was used as a covariate. As mentioned earlier, psychological wellbeing was measured as a comprehensive concept of human 'thriving' or 'flourishing'. Before participants were 'conditioned' on memories of vacation experiences, they filled out the BIT as the first survey question. This allows for controlling SWB independently of the subsequent survey questions and avoids specifying a long list of standard covariates for SWB in the final regression (note that eventual multicollinearity was controlled for). Inserting the BIT as a control variable implies that using the average SWL scale as a dependent variable, this study measured the impact of the covariates of interest on the end-of-survey SWL, given a general level of psychological wellbeing. Table 1.5 presents the results of the regression (descriptive statistics in Table A1 in the Appendix).

Overall, with a corrected R-squared of 0.464, the model performance is satisfactory. Regarding the controls, it is noted that the signs are according to expectations, probably with the exception of 'overseas'. Namely, domestic (Swiss) and European vacation recalls seem to create a stronger momentary satisfaction with life.

Table 1.5. Linear regression results.

Variable	Coefficient	Robust standard error
Socio-demographic controls		
Gender (female)	0.015 *	0.049
Income	0.069 **	0.018
Importance immaterial goods in consumption	0.005 **	0.017
Vacation descriptors		
Duration	0.005 ***	0.001
Year in past	– 0.010 *	0.005
Travel alone	– 0.240 ***	0.079
Overseas trip	– 0.175 ***	0.055
Other		
Emotions	0.074 **	0.034
A priori psychological wellbeing (BIT)	1.130 ***	0.088
Coded verbal memories		
Urban	– 0.122 *	0.072
Beach	– 0.077	0.066
Local community	0.140	0.135
Active nature-based experience	0.149 **	0.075
Constant	0.045	0.311
R-squared	0.464	

Dependent variable: end-of-survey Satisfaction with Life (five-item SWL scale by Diener *et al.*, 1985)

Looking at the variables of interest, a positive and significant sign for emotions is observed relating to SWL, which confirms the theoretical intuition. Furthermore, the regression yielded a significant negative impact of 'urban' and an equally significant but positive one of 'active' nature-based. While both signs confirm the expectations formed based on the descriptive analysis presented before, the positive impact of active nature-based experiences is taken as a relevant finding for research question 4, supporting the respective hypothesis. Not presented here, an even more significant positive sign (99%) was found for 'active vacation' for the 30 observations that represent intensive as well as long nature-based experiences. Moreover, and importantly, the results partly confirm H_5. Finding a positive and significant sign for 'active' and 'active vacations', but not for 'active moments' alone, and neither for 'passive total' (moments and vacations), confirms that the relationship between nature-based experiences and momentary SWL depends both on the duration and intensity of the recalled vacation experience.

Given that the coding used for characterizing recalled vacation trip experiences is text-based, the coding and therefore the results might be sensitive to the language used by the respondents (German or French). The model was therefore tested for both linguistic subsamples. Table 1.6 illustrates the results for the variables of interest for the discussion (the full estimation results are available from the author upon request). For completeness, the model was also tested by gender. The table contains the sign and level of significance only.

The results for the language and gender subsamples are rather striking. The main result of the regression analysis, namely the positive impact of recalled active nature-based experiences on SWL stems from the German-speaking subsample, and among these from the male subgroup. The results concerning urban recalls mirror this. Here, the negative effect seems again a result valid for the German-speaking subsample but not for the French-speaking one. Furthermore, positive emotions while recalling a vacation trip created momentary SWL only among the subsample of German-speaking participants, as well as the female respondents.

The mental involvement with individuals' personal vacation history and then with a specific past vacation trip while responding to the survey – the average response time was 13 minutes – influences their momentary SWL. It is folk wisdom that when being reminded of a past vacation trip, people will easily start talking about some episodes and thereby normally develop positive feelings and momentary happiness.

Table 1.7 reassumes the findings with respect to the research hypotheses.

Discussion and Conclusions

'Visiting natural environments could restore health and contribute to human sustainability', write Qiu *et al.* (2021, p. 1) in their recent review on visitors' recovery through NBT. While this statement finds easy acceptance among researchers and practitioners, the question is what forms of visits to what kind of natural environments can fulfil the expectations implied in it.

As a first relevant contribution to the literature, the present research therefore sets out to seek 'nature-based' as an ingredient in any form of tourism, and, as a second innovative contribution to the field, is seeking this evidence in vacation trip recall. Methodologically, the contribution to existing empirical research on

Table 1.6. Results for subsamples.

	Full sample (*N*=997)	German-speaking (*n*=757)	French-speaking (*n*=240)	Male	Female
Emotions	+**	+**			+***
Urban	–*	–*			–*
Beach			–*		
Local community					
Active nature-based experience	+**	+**		+*	

Dependent variable: end-of-survey Satisfaction with Life (five-item SWL scale by Diener *et al.*, 1985).

Table 1.7. Support for research hypotheses in the data.

Hypotheses		Supported by the data
H_1	Nature-based experiences are expected to represent a more frequently recalled vacation experience than, e.g. urban experiences.	Supported
H_2	The relationship between recalled nature-based vacation experiences and the intensity of emotions elicited is expected to be positive.	Partly supported: supported for an active and longer duration of the experience
H_3	The recall of passive experiences is expected to elicit different emotions as compared to recalling active ones.	Supported
H_4	Recalled vacation experiences that are nature-based and emotions elicited while recalling them are expected to have a positive influence on satisfaction with life.	Supported
H_5	Both the duration and intensity of a recalled nature-based vacation experience are expected to positively influence satisfaction with life.	Partly supported: the relationship between nature-based experiences and satisfaction with life depends on duration and intensity of the recalled experience.

NBT consists in researching the base in nature-based tourism in terms of intensity and duration of engagement with nature.

The present study finds that NBT experiences represent an important share of all vacation trip recalls, and that high-intensity experiences with nature, especially of a long duration, have a significant positive influence on SWL in the moment of recall. No other of the distinguished vacation experiences have a significant positive effect, while urban tourism experiences have a negative one, always compared to the reference group, namely all passive nature-based experiences and all non-coded experiences. Note that while these effects are momentaneous, they are here considered as relevant also in the long term, that is, over the lifespan. Namely, the effect is highly likely to be present every time you recall the same vacation trip experience. This is what in this research is termed long-lasting emotional memories.

Interestingly, the finding that a high-intensity and long duration experience with nature has a significant impact on SWB at the moment of recall does not find support in a non-tourism context. Recently, Oh *et al.* (2021) investigated the link between nature intensity, duration and frequency ('nature dose') and mental wellbeing (stress, depression and anxiety) as well as two physical health outcomes (high blood pressure and diabetes), finding no direct relationship. This may be due to (i) the research setting, namely a highly urbanized tropical city environment; (ii) the non-tourism context; and (iii) operationalization of the variables used. Concretely, nature intensity was measured by the item 'area of tree canopy within the most vegetated outdoor green space visited by each respondent and the proportion of that which is human-managed', while duration (of people's visits to green spaces) was measured by 'self-reported average number of hour(s) spent during each visit to public outdoor green spaces in the week prior to completing the survey' (Oh *et al.*, 2021, p. 5). This indicates that careful attention should be paid to the operationalization of duration and intensity, as well as to the specific environment (urban, beach, etc.) and context (tourism or not). Interestingly, Oh *et al.* (2021) found that people's connection to nature moderated the link between the time spent in green spaces and mental wellbeing outcomes. Therefore, connectedness to nature may be a relevant variable to include in further research on the association between nature-based experiences and SWB (Capaldi *et al.*, 2014, 2015) also in vacation settings, differentiating between duration, intensity and frequency of people's (recalled) nature-based tourism experience.

The current finding of a negative impact of a recall of urban vacation experiences on SWB, as compared to the reference group, is in line with the literature. Lehto (2013, p. 333) noticed that 'past research in environmental psychology appears to suggest that nature environments may be more effective in general for facilitating restoration than built environments'.

Given the decision to code beach tourism experiences separately, it is interesting to see whether they have similar effects on SWB as nature-based experiences. Firstly, the regression model does not find a significant impact of beach holiday experience recalls on SWL. This is in contrast to active nature-based experiences but in line with passive ones. Secondly, the independent samples t-test for beach recalls returns a significant negative sign for 'impressed', while it finds a significant positive one for active nature-based recalls – a striking difference. Thirdly, the independent samples t-test regarding the different items of the SWLS shows that beach shares the positive significant sign of three items with active vacations, only on a 90% level, but in contrast to the latter is not significant on average SWL. The ANOVA test returns a significant negative difference of recalled beach vacation experiences as compared to active nature-based ones. Not finding relevance of beach vacation recalls for SWL, the separate coding is clearly justified.

In an interesting study on the Maremma region in Tuscany, Bimonte and Faralla (2012), based on interviews on site, found a similar result when comparing the momentary happiness of visitors of a regional natural park and beach tourists. In a follow-up analysis of the data, the same authors confirmed this result, differentiating between visitors of the park only, beach tourists visiting also the park, and other beach tourists. Also remarkable is that applying an ordered regression, the level of happiness decreased 'continuously' from the first to the third category (Bimonte and Faralla, 2014).

When comparing the present results with the literature, it has to be considered that the measurement of subjective wellbeing here regards emotions and SWL at the moment of recalling a past trip experience – a 'memory'. Finding comparable results to those of on-site or end-of-vacation surveys (e.g. Bimonte and Faralla, 2012) implies that the present study contributes to the extant literature, demonstrating that, in contrast to previous research (De Bloom *et al.*, 2009; Chen and Petrick, 2013), there is a lasting impact of vacation trip experiences on subjective wellbeing via free recall.

A final consideration regards the way in which subjective wellbeing is modelled in the regression. Namely, using the five-item Diener *et al.* (1985) Satisfaction with Life Scale (SWLS) as a dependent variable, and Su *et al.*'s (2014) Brief Inventory of Thriving (BIT) as a covariate. Given the structure of the survey instrument with the BIT at the beginning and the SWLS at the end (before and after the interviewees responded to the vacation recall questions, respectively) permits the coefficients of the recall covariates to be interpreted as a 'pure' effect on SWL, given an exogenous level of psychological wellbeing. This can be considered a second relevant contribution to the literature on the effects of holidays on SWB, as it avoids asking respondents directly about their perceived impact of a vacation trip on their wellbeing (Buckley, 2020) or using simply the level of mental wellbeing and a long list of the usual socio-demographic and situational covariates determining happiness and wellbeing (Oh *et al.*, 2021).

While the research reported here produced interesting results, it has some obvious limitations. Firstly, the use of a Swiss standing panel will allow a conclusion beyond Switzerland only for a limited number of wealthy, mostly European, countries with a high travel intensity and an important public discourse on sustainability. Concerning the issue of being nature-based, the considerable under-representation of domestic tourism in vacation recalls might have negatively influenced the mention of such experiences as compared to, for instance, beach or urban tourism, both important in the present dataset. A similar consideration can be made for the interactions with local people, which can be assumed to be more frequent during domestic holidays, in a very small country like Switzerland. Future research should clearly implement a survey in parallel in different cultural and economic environments.

Secondly, the results are based on a survey not designed in view of nature-based tourism. This is in some sense a strong point because it permits the issue of nature-based to be analysed from an 'outsider' perspective. However, future surveys with the scope of providing representative evidence on nature-based tourism will need to increase the weight of qualitative and verbal information in the data regarding attitudes towards nature as well as the content of the vacation. Moreover, it would be important to compare the results of a free recall with those from instruments exploring experimental forms by combining surveys and experiments with selected groups; alternatively, by splitting the survey with different nudges towards nature-based behaviour. This way, the free recall would provide a sort of control group representing the status quo, while the results from nudges could provide interesting indications to stakeholders.

This contribution adopts a very broad definition of nature-based, thus finding that 26% of freely recalled trip experiences in the present sample are nature-based. Among these, memories of passive and active nature-based experiences appear of equal relevance, while nature-based moments are much more present than longer-duration vacation experiences. This might offer an opportunity for a debate, not taken up here, on whether one or more nature-based moments in an otherwise different vacation experience should be considered as being relevant in the discussion of nature-based tourism. Regarding the overall importance of NBT, this contribution confirms that a very narrow definition of NBT would represent a very small portion of recalled experiences of around 3%.

References

Arnegger, J., Woltering, M. and Job, H. (2010) Toward a product-based typology for nature-based tourism: a conceptual framework. *Journal of Sustainable Tourism* 18(7), 915–928. DOI: 10.1080/09669582.2010.485680.

Bimonte, S. and Faralla, V. (2012) Tourist types and happiness a comparative study in Maremma, Italy. *Annals of Tourism Research* 39(4), 1929–1950. DOI: 10.1016/j.annals.2012.05.026.

Bimonte, S. and Faralla, V. (2014) Happiness and nature-based vacations. *Annals of Tourism Research* 46, 176–178. DOI: 10.1016/j.annals.2014.02.002.

Buckley, R. (2020) Nature tourism and mental health: parks, happiness, and causation. *Journal of Sustainable Tourism* 28(9), 1409–1424. DOI: 10.1080/09669582.2020.1742725.

Capaldi, C.A., Dopko, R.L. and Zelenski, J.M. (2014) The relationship between nature connectedness and happiness: a meta-analysis. *Frontiers in Psychology* 5, 976. DOI: 10.3389/fpsyg.2014.00976.

Capaldi, C.A., Passmore, H.A., Nisbet, E.K., Zelenski, J.M. and Dopko, R.L. (2015) Flourishing in nature: a review of the benefits of connecting with nature and its application as a wellbeing intervention. *International Journal of Wellbeing* 5(4), 1–16. DOI: 10.5502/ijw.v5i4.449.

Chandralal, L. and Valenzuela, F.R. (2013) Exploring memorable tourism experiences: antecedents and behavioural outcomes. *Journal of Economics, Business and Management* 1(2), 177–181. DOI: 10.7763/JOEBM.2013.V1.38.

Chandralal, L. and Valenzuela, F.R. (2015) Memorable tourism experiences: scale development. *Contemporary Management Research* 11(3), 291–310. DOI: 10.7903/cmr.13822.

Chen, C.C. and Petrick, J.F. (2013) Health and wellness benefits of travel experiences. *Journal of Travel Research* 52(6), 709–719. DOI: 10.1177/0047287513496477.

Chen, C.C., Petrick, J.F. and Shahvali, M. (2016) Tourism experiences as a stress reliever: examining the effects of tourism recovery experiences on life satisfaction. *Journal of Travel Research* 55(2), 150–160. DOI: 10.1177/0047287514546223.

Chhetri, P., Arrowsmith, C. and Jackson, M. (2004) Determining hiking experiences in nature-based tourist destinations. *Tourism Management* 25(1), 31–43. DOI: 10.1016/S0261-5177(03)00057-8.

De Bloom, J., Kompier, M., Geurts, S., de Weerth, C., Taris, T. *et al.* (2009) Do we recover from vacation? Meta-analysis of vacation effects on health and well-being. *Journal of Occupational Health* 51(1), 13–25. DOI: 10.1539/joh.k8004.

Denkova, E.S., Dolcos, S. and Dolcos, F. (2013) The effect of retrieval focus and emotional valence on the medial temporal lobe activity during autobiographical recollection. *Frontiers in Behavioral Neuroscience* 7, 109. DOI: 10.3389/fnbeh.2013.00109.

Diener, E.D., Emmons, R.A., Larsen, R.J. and Griffin, S. (1985) The satisfaction with life scale. *Journal of Personality Assessment* 49(1), 71–75. DOI: 10.1207/s15327752jpa4901_13.

Elster, J. and Loewenstein, G. (1992) Utility from memory and anticipation. In: Loewenstein, G. and Elster, J. (eds) *Choice Over Time*. Russell Sage Foundation, New York, pp. 213–234.

Farber, M.E. and Hall, T.E. (2007) Emotion and environment: visitors' extraordinary experiences along the Dalton Highway in Alaska. *Journal of Leisure Research* 39(2), 248–270. DOI: 10.1080/00222216.2007.11950107.

Filep, S., Cao, D., Jiang, M. and DeLacy, T. (2013) Savouring tourist experiences after a holiday. *Leisure/Loisir* 37(3), 191–203. DOI: 10.1080/14927713.2013.842731.

Fossgard, K. and Fredman, P. (2019) Dimensions in the nature-based tourism experiencescape: an explorative analysis. *Journal of Outdoor Recreation and Tourism* 28, 100219. DOI: 10.1016/j.jort.2019.04.001.

Fredman, P. and Haukeland, J.V. (2021) Introduction. In: Fredman, P. (ed.) *Nordic Perspectives on Nature-based Tourism: From Place-based Resources to Value-added Experiences*. Edward Elgar Publishing, Northampton, Massachusetts, pp. xiii–xxv. DOI: 10.4337/9781789904031.

Fredman, P. and Tyrväinen, L. (2010) Frontiers in nature-based tourism. *Scandinavian Journal of Hospitality and Tourism* 10(3), 177–189. DOI: 10.1080/15022250.2010.502365.

Fredman, P., Wall Reinius, S. and Lundberg, C. (2009) *Turism i natur. Definitioner, omfattning, statistik*. Turismforskningsinstitutet ETOUR, Mittuniversitetet, Östersund. Rapport R2009:23.

Kim, H., Lee, S., Uysal, M., Kim, J. and Ahn, K. (2015) Nature-based tourism: motivation and subjective well-being. *Journal of Travel & Tourism Marketing* 32(sup1), S76–S96. DOI: 10.1080/10548408.2014.997958.

Kim, J. and Fesenmaier, D.R. (2017) Sharing tourism experiences: the posttrip experience. *Journal of Travel Research* 56(1), 2840. DOI: 10.1177/0047287515620491.

Knobloch, U., Robertson, K. and Aitken, R. (2014) (Mis)Understanding the nature of tourist experiences. *Tourism Analysis* 19(5), 599–608. DOI: 10.3727/108354214X14116690097891.

Knobloch, U., Robertson, K. and Aitken, R. (2017) Experience, emotion, and eudaimonia: a consideration of tourist experiences and well-being. *Journal of Travel Research* 56(5), 651–662. DOI: 10.1177/0047287516650937.

Larsen, S. (2007) Aspects of a psychology of the tourist experience. *Scandinavian Journal of Hospitality and Tourism* 7(1), 7–18. DOI: 10.1080/15022250701226014.

Lehto, X.Y. (2013) Assessing the perceived restorative qualities of vacation destinations. *Journal of Travel Research* 52(3), 325–339. DOI: 10.1177/0047287512461567.

Mannell, R.C. (1980) Social psychological strategies and techniques for studying leisure experiences. In: Iso-Ahola, S. (ed.) *Social Psychological Perspectives on Leisure and Recreation*. Charles C. Thomas, Springfield, Illinois, pp. 62–88.

Mannell, R.C. and Iso-Ahola, S.E. (1987) Psychological nature of leisure and tourism experience. *Annals of Tourism Research* 14(3), 314–331. DOI: 10.1016/0160-7383(87)90105-8.

Mather, M. and Carstensen, L.L. (2005) Aging and motivated cognition: the positivity effect in attention and memory. *Trends in Cognitive Sciences* 9(10), 496–502. DOI: 10.1016/j.tics.2005.08.005.

Newsome, D., Moore, S.A. and Dowling, R.K. (2012) *Natural Area Tourism: Ecology, Impacts and Management*, 2nd edn. Channel View Publications, Clevedon, UK. DOI: 10.21832/9781845413835.

Oh, R.R.Y., Fielding, K.S., Chang, C.-C., Nghiem, L.T.P., Tan, C.L.Y. *et al.* (2021) Health and wellbeing benefits from nature experiences in tropical settings depend on strength of connection to nature. *International Journal of Environmental Research and Public Health* 18(19), 10149. DOI: 10.3390/ijerph181910149.

Pearce, P.L. and Packer, J. (2013) Minds on the move: new links from psychology to tourism. *Annals of Tourism Research* 40, 386–411. DOI: 10.1016/j.annals.2012.10.002.

Qiu, M., Sha, J. and Scott, N. (2021) Restoration of visitors through nature-based tourism: a systematic review, conceptual framework, and future research directions. *International Journal of Environmental Research and Public Health* 18(5), 2299. DOI: 10.3390/ijerph18052299.

Scott, N. (2020) Cognitive psychology and tourism – surfing the 'cognitive wave': a perspective article. *Tourism Review* 75(1), 49–51. DOI: 10.1108/TR-06-2019-0217.

Skavronskaya, L., Moyle, B., Scott, N. and Kralj, A. (2020) The psychology of novelty in memorable tourism experiences. *Current Issues in Tourism* 23(21), 2683–2698. DOI: 10.1080/13683500.2019.1664422.

Speer, M.E., Bhanji, J.P. and Delgado, M.R. (2014) Savoring the past: positive memories evoke value representations in the striatum. *Neuron* 84(4), 847–856. DOI: 10.1016/j.neuron.2014.09.028.

Su, R., Tay, L. and Diener, E. (2014) The development and validation of the Comprehensive Inventory of Thriving (CIT) and the Brief Inventory of Thriving (BIT). *Applied Psychology. Health and Well-Being* 6(3), 251–279. DOI: 10.1111/aphw.12027.

Swiss Federal Statistical Office (2019) Anzahl Reisen mit Übernachtungen, Wohnbevölkerung ab 6 Jahren [Number of trips with overnight stays of the population aged 6 and over]. Available at: https://www.bfs.admin.ch/bfs/de/home/statistiken/kataloge-datenbanken/tabellen.assetdetail.10227090.html (accessed 8 July 2022).

Tung, V.W.S., Lin, P., Qiu Zhang, H. and Zhao, A. (2017) A framework of memory management and tourism experiences. *Journal of Travel & Tourism Marketing* 34(7), 853–866. DOI: 10.1080/10548408.2016.1260521.

UNWTO (2010) *International Recommendations for Tourism Statistics 2008*. United Nations, Department of Economic and Social Affairs Statistics Division, Studies in Methods Series M No. 83/Rev.1.

Van Boven, L. and Ashworth, L. (2007) Looking forward, looking back: anticipation is more evocative than retrospection. *Journal of Experimental Psychology. General* 136(2), 289–300. DOI: 10.1037/0096-3445.136.2.289.

Appendix

Table A1. Descriptive statistics of the variables in the model.

Variable	Mean (or %)	Std. dev.	Min.	Max.
Gender (female)	50.6%			
Monthly household income (in Swiss francs)				
< 3000	3.5%			
3000–4500	8%			
4501–6000	15.8%			
6001–9000	27.3%			
9001–12,000	23.4%			
> 12,000	22%			
Importance immaterial goods in consumption	5.418	1.263	1	7
Duration (in nights)	16.1	17.1	2	181
Year in the past (2018 = 0; 2017 = 1; etc.)	1.842	4.134	0	42
Travel alone	9.5%			
Overseas trip (outside of Europe)	36.4%			
Emotions (mean of 10 items on a 1–7 scale)	5.950	0.804	2.3	7
A priori psychological wellbeing (BIT) (mean of 10 items on a 1–5 scale)	4.069	0.535	1	5
SWL (mean of 5 items on a 1–7 scale)	5.389	0.993	1.2	7
Urban	16.15%			
Beach	13.93%			
Local community	4.03%			
Active vacation	13.62%			

2 Linking Tourism, Wellbeing and Ecosystems: Towards a Nature-based Solution for Improving Wellbeing through Social Tourism

Sara Duarte[1]*, Tim Taylor[2] and Carlos Ferreira[3,4]
[1]IGOT Institute of Geography and Spatial Planning, University of Lisbon, Portugal; [2]European Centre for Environment and Human Health, Department of Public Health and Sport Sciences, University of Exeter Medical School, Truro, UK; [3]University of Coimbra, CEGOT, Faculty of Arts and Humanities, Coimbra, Portugal; [4]CiTUR (Centre for Tourism Research, Development and Innovation), Estoril, Portugal

Abstract

The western world is facing increasing urbanization and an ageing population. Solutions are sought to respond to the need to ensure an active ageing process, while being attentive to the wellbeing and health of the elderly and strengthening connections of urban seniors to nature.

In this work, we attempt to clarify the potential contribution that nature can make to the wellbeing of urban seniors, through specific tourist and recreational offers in protected areas. Specifically, we set out to achieve the objective of contributing to the structuring and qualification of the tourist offer in natural areas for urban seniors, as a way of promoting their wellbeing and health. We present three axioms and a related theorem of the eclectic generosity of higher-quality natural spaces with the aim of presenting an approach that brings tourism, ecosystem services and wellbeing of urban seniors closer together.

Understanding Nature, Tourism and Wellbeing

When, in 2008, Richard Louv suggested that we live in a nature-deficient society, with direct consequences on our health (increasing rates of obesity, attention disorders and higher incidence of depression), he was referring to the decreased exposure that children have to nature, mainly due to the rhythm of modern life (Louv, 2008). In an increasingly urbanized society, this separation is almost inevitable (Hartig *et al.*, 2014), making it essential to act to make up for lost time and to identify strategies that facilitate life-long access to natural areas, particularly for the more vulnerable groups (Twohig-Bennett and Jones, 2018), notably children and seniors (Kabisch *et al.*, 2017).

There are numerous studies supporting the importance of nature for the health and wellbeing of society as a whole, which have been brought out by different meta-analyses focusing on a vast array of subjects and themes

*Corresponding author: saramduarte@campus.ul.pt

 Nature-based Tourism and Wellbeing: Impacts and Future Outlook (eds F. Niccolini *et al.*)
DOI: 10.1079/9781800621411.0002

(Martinez-Juarez *et al.*, 2015; McMahan and Estes, 2015; Husk *et al.*, 2016; Ohly *et al.*, 2016; Bosch and Sang, 2017; Gascon *et al.*, 2017; Britton *et al.*, 2018; Houlden *et al.*, 2018; Tillmann *et al.*, 2018; Twohig-Bennett and Jones, 2018; Roberts *et al.*, 2019; Pritchard *et al.*, 2020), and there are even a few dedicated to the impacts for older people (Orr *et al.*, 2016; Gagliardi and Piccinini, 2019).

It is therefore our understanding that we are no longer trying to understand whether nature is beneficial for human wellbeing in the western world – in particular, for subjective wellbeing (SWB), experienced and self-assessed by each individual (Deci and Ryan, 2006) – but rather on what are the mechanisms behind this relationship, in different populations and contexts.

This realization tends to deviate from the reality that is gradually imposing itself, on a planet where the population distribution is increasingly leaning towards urban spaces, with more than 50% of the world's population already living in an urban environment (UN, 2019b). At the same time, and for the first time in world history, people over 65 years old outnumber children under five, with increasingly ageing population projections for the world up to 2050 (UN, 2019).

The combination of ageing populations and urbanization requires effective action if we are to guarantee the future quality of life for this growing part of the population (WHO, 2020). The elderly themselves also perceive this need, as the number of adults at older ages who yearn to live in healthier environments is increasing (Carman, 2011). If it is generally established that nature promotes the health and wellbeing of populations, it is still very important to theorize about the importance that so-called blue and green spaces can assume as places where seniors can find opportunities to improve their physical health, psychological wellbeing and social relationships, aiming at increasing their quality of life (Finlay *et al.*, 2015). However, it is well known that there is still a limited number of studies fully dedicated to the experiences of seniors in their relationship with nature. In 2019, Gagliardi and Piccini found fewer than 50 research papers, with scattered themes as diverse as horticulture, gardening or visits to green spaces, of which only eight were related to what the authors called 'garden resting and views', which implies a tourism visit to green spaces.

It is therefore of great importance to study the role tourism plays as an activity that enhances the beneficial effects of nature on the lives of seniors, especially if we consider access to areas of greater ecological value, as is the case with protected areas. In what follows, we use a literature review to discuss the development of an overarching theoretical construct between tourism, ecosystems and the wellbeing of seniors.

Background

In protected areas, the perception of beauty, as a defining factor of a tourist site, seems to be the result of the combination of objectivist and subjectivist values, referred to by Frank *et al.* (2013), in a multi-dimensional logic of intrinsic values (such as richness in biodiversity, geological features and the presence, or perhaps more accurately the absence, of human intervention). This, combined with the individual and subjective perception that is made about that place, makes for a construct that, as stated by Brown and Reed (2012), is of great relevance for the planning and management of the offer.

Tourism attractiveness of protected areas is based on five generic components: tourism resources; tourism infrastructure; accessibility; local communities; and peripheral attractions, which are subdivided into several subcategories (Deng *et al.*, 2002). Tourism resources are subdivided into natural and cultural resources, from which physical factors and environmental factors emerge as an integral part of natural resources. Flora, fauna, water bodies and climatic phenomena are, thus, allied to the necessary sensations of safety and comfort, from which one should not dissociate management policies of the place and quality of the recreational experiences that are made available on site (Deng *et al.*, 2002; Bieling *et al.*, 2014).

It is in the higher-quality landscape settings, which are normally associated with protected areas (Beckena and Job, 2014), that the connection between natural landscapes and human wellbeing, in its true therapeutic sense, seems to be most apparent (Thompson, 2011).

By landscape, we mean the interaction between the various physical, biological and social components, which are part of the environment that surrounds us, and, of course, the perception we have of them (Ode *et al.*, 2008). Kaplan and Kaplan (1989) made important additions to what relates to the appropriation we make of a natural landscape and that are the foundations of attention restoration theory (ART): physical attributes; type of vegetation cover; informative variables; and perception variables. The ART approach to subjective wellbeing relates to the model of the '4 Es of the Experience Economy', by Pine and Gilmore (1999), whose dimensions – education, entertainment, esthetic and escapism – Hwang and Lyu (2014) argued, lie at the core of our perception of wellbeing, and that Jurowski (2009) had identified as a valid conceptualization in the optimization of tourism supply management.

Landscape management thus gains a new dimension, which is becoming more widely recognized for the therapeutic importance it can assume in our lives (Thompson, 2011). Thus, the link between subjective wellbeing and landscapes is rooted not only in the specific characteristics of the surrounding environment but also in the activities and experiences we develop while in contact with these environments (Bieling and Plieninger, 2013; Bieling *et al.*, 2014). Combined, the attributes that constitute them translate into aesthetic and recreational value, which will still have to be perceived by visitors or users (Zanten *et al.*, 2016). Subsequently, communicating landscapes, as excellent tourism resources which are expected to have an extended life cycle, must pass the test of authenticity and sustainability of the offer. The most obvious way seems to be through nature tourism activities, but we must go deeper, as the relationship with the restorative power of natural areas seems to increase the attractiveness of destinations. It brings together multiple concepts with attraction potential, such as 'slowing down' movements or outdoor recreational activities (Jarratt *et al.*, 2018).

Furthermore, the relationship and operationalization between the aesthetic value of landscapes and biodiversity are also relevant to the public. Science is still progressing in this matter but there is a need to invest in understanding the interconnection between this relationship and the classical descriptors of analysis and the descriptors of ecosystem ecology. This involves bringing together the way humans perceive ecological information and the way landscape components are interpreted (Tribot *et al.*, 2018).

We focus our research on the more passive than active experience of landscape gazing (Fig. 2.1), taking into account that the quality of the experiences we offer tourists is understood as critical for the sustainable management of the most relevant natural tourist areas (Moyle *et al.*, 2017).

Tourism should be understood as a fundamental vector of human interaction with 'green' and 'blue' spaces. The potential for novel interventions associated with the creation of tourism programmes that focus on promoting the simple acts of 'going' and 'enjoying' natural spaces must not be neglected, as it promotes reconnection with these spaces and thereby may enhance the health and wellbeing of visitors (Martinez-Juarez *et al.*, 2015).

The challenges to the tourism industry, particularly post-COVID-19, are shifting the consciousness of all stakeholders (Sheldon, 2020). We will have come full circle when tourism offers start to promote a deep human and environmental connection, based on processes of critical self-assessment and a real understanding of ourselves and our surroundings, ultimately contributing to real changes in human consciousness (Sheldon, 2020). As argued by Houlden *et al.* (2018), tourism allows visitors to develop an emotional relationship with nature, and this relationship will in turn consolidate a greater sense of environmental ethics (Ingólfsdóttir and Gunnarsdóttir, 2020).

Our connection level with nature seems to be the tipping point regarding its relationship with psychological wellbeing (Cervinka *et al.*, 2011; Capaldi *et al.*, 2014) and there are some important physiological processes that take part in this process.

Although we do not fully understand the mechanisms that lead our nervous system to generate subjective mental states, it seems that the set of neuronal components, known as the

Fig. 2.1. Senior visit to Arrabida Natural Park.

interoceptive nervous system (INS), is responsible for the integration of the large amount of information humans collect, concerning all the reconfigurations that systematically take place in the body, in search of a state of homeostasis (Carvalho and Damasio, 2021).

The changes we feel in our body when we face a particular stimulus are perceived by the INS and thus mapped by the central nervous system. This process, however, may either go unnoticed or be experienced in the conscious form of 'feelings'. And the interesting part is that these feelings seem to have been evolutionarily rewarded, because the psychological evaluation of a 'felt' experience facilitated the learning process regarding homeostatic imbalances and respective corrections, thus ensuring anticipation of future favourable or less favourable situations (Damasio and Carvalho, 2013).

We are, therefore, biophilic beings and this has implications. The biophilia hypothesis, set by Wilson (1984), revolves around the notion that human wellbeing – psychological, biological, cultural, symbolic and aesthetic – is fully interconnected with the ancestral need to interact with the natural environment in search of food, water or refuge and that the desire for a communion with nature has become innate throughout our evolutionary history (McMahan, 2018).

Recent studies have shown that there is a relationship between nature and the choices we make regarding the places where we prefer to relax and have fun, showcasing the importance of natural spaces as places to promote positive social contexts, good memories and general satisfaction with life, in support of this biophilia hypothesis (Chang *et al.*, 2020).

Putting all of this information together, around the need of facilitating access to nature for senior populations, we might also want to consider the framework of active ageing, in its actual reading given by the Political Declaration and Madrid International Plan of Action on Ageing, in which active ageing is defined as the process of optimizing opportunities that promote health, participation and security of older people, improving their quality of life (WHO, 2002).

This puts tourism as a central piece for social intervention, as a tool that promotes the welfare of all participants throughout life, but specifically the elderly, who are often kept apart from the offers that would benefit them the most.

The Context of Portugal

In Portugal, nature conservation is organized under the umbrella of the Fundamental Network for Nature Conservation, comprising the National System of Classified Areas (NSCA) and the continuity areas. The NSCA encompasses the National Network of Protected Areas (NNPA) and the Natura 2000 network. In mainland Portugal, the NNPA includes 47 protected areas, including 32 national protected areas (one national park, 13 natural parks, nine nature reserves, two protected landscapes and seven natural monuments), 14 regional/local protected areas (two nature reserves, 11 protected landscapes and one natural park) and one private protected area (ICNF, 2022).

The development of a National Programme for Nature Tourism (PNTN, 2015) aimed to contribute to the promotion and affirmation of the value and potential of protected areas, to provide for the creation of appropriate tourism products and services and to promote visitation.

Based on a review of the literature, we aim to contribute to the definition of strategies that might help to give nature and seniors a chance to reconnect. We realized that, associated with nature in its aesthetic features, natural and cultural richness and absence of man-made noise, there is a whole potential service, captured by a different human sense, which we term 'bioception', which refers to the ability for the apprehension of feelings of escape, enjoyment, knowledge, relaxation, creativity, spirituality, reflection and salutogenesis in a higher-quality natural environment, towards homeostasis.

The result of promoting tourism activities in natural spaces of superior quality, like protected areas, as a tool for accessing the countless services provided by these extraordinary spaces, translates, therefore, into the promotion of wellbeing in a powerful way with amplified benefits. The effects of nature services on wellbeing are boosted by the effects on wellbeing promoted by tourism activity itself (Balderas-Cejudo *et al.*, 2017), which translates into countless benefits, like rest – by reducing stress levels – attention and concentration restoration, intellectual development, increased creativity and, ultimately, prevention of several diseases that affect us in the course of our lives, with particular focus on the senior phase.

Bringing Tourism, Ecosystem Services and Wellbeing Together: Towards an Eclectic Generosity of Higher-quality Natural Spaces Theorem

Ecosystems are important in terms of the cultural services provided to people even in terms of their mere existence, by promoting feelings of escape, enjoyment, knowledge, relaxation, creativity, spirituality, reflection, salutogenesis, among others. Access to this service, however, is limited by the intervention of anthropogenic activities, such as tourism, other recreation and leisure activities, but also sports, education and research. These are the actual tools that promote access to the effective services that nature, culturally, provides us, such as the sense of evasion, fun, knowledge, relaxation, creativity, spirituality, reflection and salutogenesis, the experience of which leads to the positive effects on our wellbeing, through our sense of bioception. This interpretation is even more relevant if we realize that cultural ecosystem services (CES) are the only services provided by nature whose 'utility' is truly and only applied to us humans, for all living beings depend on provisioning services such as access to food, all living beings depend on nature's ability to regulate, and all see their existence depend on supporting services such as photosynthesis and the water and nutrient cycles, but not culture (Fig. 2.2).

Inspired by these myriad inter-relationships and aware of the pressing need to theorize about the incredible potential of what we have called the *good effect* of tourism, we have drawn on the methodology of the economics of ecosystems and biodiversity study (TEEB, 2010), to respond to its underlying invitation to adapt it and develop other knowledge, deepening approaches

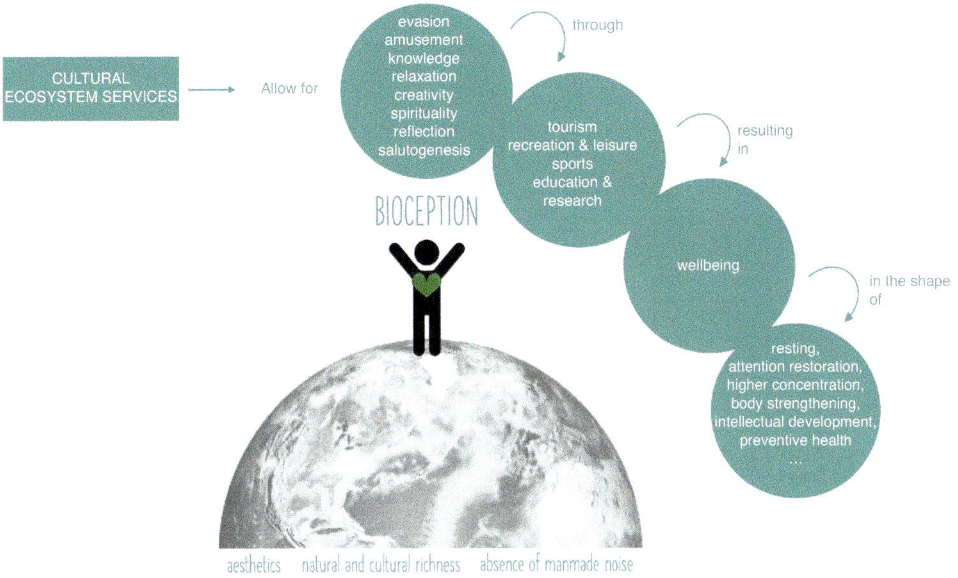

Fig. 2.2. CES and its relation with tourism, as one of the available tools that allow humans to have access to these services increasing SWB levels.

that translate into recommendations for more specific areas, such as tourism. This methodology has allowed us to present three axioms that will serve as the basis for the creation of a theorem that will support the theory of the importance of tourism for the wellbeing of society and seniors, which draws on the concepts of recognizing, demonstrating and retaining value. We describe these three axioms.

Axiom 1: It is necessary to recognize that humans are all part of nature

Recognition of the fact that humans are part of nature is at the basis of all processes of change.

Ecosystems, in their biotic and abiotic aspects, are at the basis of everything. This is true, whatever the view is on evolution. Some would argue, from a scientific perspective, proteins, lipids, carbohydrates and nucleic acids are the molecular bricks of nature's evolutionarily construction of beings that can be as simple as bacteria or as complex as humans. In an ancestral planet Earth – with characteristics similar to those imagined by Oparin (1924) and Haldane (1929) in their homonymous works (*The Origin of Life*), although written 5 years apart – the first forms of life would have appeared, in what became known as the primordial soup, a metaphor for the moment and the space in which the physical and chemical conditions for those beings to appear were gathered (BBC, 2020). From that somewhat fortuitous moment, countless organisms have evolved (Hug *et al.*, 2016). All of them, united by this common origin, evolved under the inexorable action of natural selection processes, but without ever losing the genetic memory of what unites them, in a primordial way. And human beings, tiny representatives of eukaryotic beings, belonging to the Opisthokonta branch, are no exception.

The species *Homo sapiens* is the result of its ancestral circumstances, which translated into extraordinary – but still not preponderant – characteristics over the equally particular characteristics that define the other eukaryotic species. In any case, by walking upright, improving the use of tools for hunting and mastering the use of fire, we adapted to the most varied environments, with physical consequences; for example, in the development of larger brains, promoted by the adaptive advantage conferred by the establishment of social networks – sharing food and the shared care of children – and by the development of language and symbols. This is how we changed the world (Smithsonian, 2020) without, however, dominating nature.

From other perspectives, in different religious worldviews, there are consistent suggestions of the intertwined nature of humanity with the surrounding world. From the Buddhist views of the need to respect the cycle and balance of nature to Christian views on the need for caring for the creation, it is clear that humans are part of nature.

Our condition as rational living beings offers us an apparent role as protagonists in the theatre of operations of biological processes. However, we are limited beings and whether one sees the hand of the Divine in making us aware of our limitations or the invisible hand of nature in revealing our lack of control over events, we need to realize that, ultimately, we are part of the natural world.

We are a threat to the overall balance if we do not understand that we are part of the solution, because we are also nature. This understanding, lost over decades of distance from the most basic knowledge regarding our survival on Earth – from our origin, to the support given to our very existence in the form of oxygen and water, to the regulation of trophic connections, of climatic phenomena and epidemics, to the fact that we see beauty in nature because it was in it that we evolved and it was in it that we learned to protect ourselves from threats – is basic and urgently needs to be recovered.

This acknowledgement is even more disturbing the further communities become from the natural environments that surround them, leading to an undeniable need to:

- promote the development of education models that allow children to have regular contact with nature, following an active purpose of nature literacy, with attribution of skills and knowledge that promote consciousness and active citizenship in this field;
- promote an active life process, which does not shy away from the long-term benefits that come from contact with natural spaces, with preventive health effects at the individual level, but also at the societal level, extending our biophilic conscience, through the continued training of our sense of bioception, with all the self and environmental advantages that may come out from it; and
- promote support systems around the elderly, defining strategies that do not confine them to their homes, promoting the so-called 'active ageing', opening, for example, their horizons to natural areas, thus guaranteeing them the benefits of contact with nature.

Axiom 2: There is a need to demonstrate that visiting high-quality natural tourism areas (HQNTAs) promotes the wellbeing of society and of senior citizens in particular

In a society intoxicated by fake news, there is no lack of science dedicated to finding the truth, based on the best knowledge and technologies available, including their advantages and limitations.

Demonstrating the importance of nature has several strands and can follow different understandings. Intrinsically, there is an ethical value of the right to existence that must be assumed (Pearson, 2016). Such recognition, from Axiom 1, of a moral value that goes beyond the utilitarian logic of human appropriation, places us on the same level of importance as all existing species, especially those that are facing extinction. At the other end of the spectrum comes the utilitarian value, organized in contributions that nature makes to human wellbeing, under the designation of 'ecosystem services', set by the Millennium Ecosystem Assessment report, which divides them into provisioning, supporting, regulating and cultural services (MEA, 2005). Putting nature's relevance in terms of its importance for human survival was a clear attempt to make them visible. However, more recently, a new typology of value was defined, called 'relational value'. It seeks to bring together the two concepts of intrinsic and utilitarian value, which is defined by the preferences, principles and virtues in the human–nature relationship, assuming a chain of several types of values, whose overlaps indicate that this value can be understood from different perspectives (Chan *et al.*, 2018; Himes and Muraca, 2018).

However, in this relational perspective of quality of life, the action-reaction process is implied, and using the strategy of caring for nature, directly and indirectly, has an underlying improvement in people's quality of life (Jax *et al.*, 2018).

Demonstrating the value of cultural ecosystem services (CES) is as much about common sense as it is about subjectivity. So much so that, despite all the emphasis given throughout this chapter to the value that nature assumes for human wellbeing, we should never ignore the fact that there is a considerable increase in phobias associated with natural spaces, which derive from the so-called 'nature deficit disorder', described by Louv (2008).

However, acknowledging this reality, we can work on its deconstruction by using quite simple tools. In fact, CES should be analysed from a semantic perspective of 'service', or rather the act of 'serving', to facilitate general understanding. Our interpretation is based on the idea that ecosystems provide services (cultural, not material) of promoting or inducing feelings of escape, enjoyment, knowledge, relaxation, creativity, spirituality, reflection and salutogenesis, all of which enhance levels of human wellbeing. Once again, the benefits inherent to these services are only attainable if duly intermediated by services provided by the human species itself, whether in the form of tourist activities, other recreational and leisure activities, the practice of sports or, of course, educational and research actions.

Tourism can act as a bridge that brings together concepts and needs of two different spheres which, although existing in interdependence, live conceptually apart – humans and nature – resulting in the promotion of the wellbeing of the former and the adequate conservation of the latter. Thus, the fascination, restorative effects and other motivations associated with visiting these spaces have not only a positive effect on SWB, but also on promoting pro-ecological behaviours (Hartig *et al.*, 2001; Mackay and Schmitt, 2019).

This is particularly true for the case of high-quality natural tourism areas in which levels of psychological restoration, inherent to a higher emotional attachment, are more intense and general health status improves compared to lower-value natural spaces (Wyles *et al.*, 2017; Martin *et al.*, 2020).

The HQNTAs should be understood as places of tourist interest located in areas of higher ecological and landscape value, normally protected, such as those that, in Portugal, are part of the National Network of Protected Areas or the Natura 2000 network. The value of these areas is automatically set by the national and European legislations, but there are others which can also be identified by a competent team of professionals that includes biologists (or others with technical skills in ecological and landscape quality assessment, as is the case with landscape architects), responsible for certifying the ecological value of the sites, and tourism professionals responsible for assessing the remaining aspects associated with a quality tourism offer, specifically for seniors.

Axiom 3: It is important to retain value by integrating tourism supply in high-quality natural tourism areas in the lives of seniors by preventing senescence

Retaining value implies defining concerted strategies.

Knowing the benefits associated with visiting HQNTAs is not enough if the means are not created so that the population recognizes the need for it, just as much as it recognizes the need to do physical exercise or eat healthy food.

We believe that tourism, in its global scope, represents a privileged tool in promoting wellbeing and preventing the state of disease, and should be considered as part of a strategy to enhance active ageing, with an effect on the reduction of senior dependency in the future, or at least on delaying human senescence (Carretero *et al.*, 2013).

We also believe that this type of qualified offer should be classified as a new branch of health and wellness tourism, along with thermal tourism, but with a wider territorial influence, promoting the restorative benefits of the HQNTAs in a preventive or therapeutic way, depending on the case (Shanahan *et al.*, 2019). It is intended to offer a prophylactic strategy to promote the wellbeing of seniors, with the expected outcomes of reducing the demand for National Health Service care and delaying the effects of senescence typical of this age.

The strategy to promote wellbeing of seniors by the HQNTA should be divided into three distinct intervention areas: development, exploration and promotion of specialized tourism products of wellbeing by nature.

In addition to support facilities and other ancillary services, the development of the specialized tourism product of wellbeing by nature should therefore consider the following components:

- adequate group size;
- identification of quality sites within the HQNTA; and
- identification of the most appropriate activities.

The exploitation of the specialized tourism product of wellbeing by nature is based on three fundamental factors:

- specialization of the offer;
- qualification of the guides; and
- operationalizing green prescriptions.

Lastly, promotion of the specialized tourism products of wellbeing by nature is paramount. First, as a sales promotion tool for these tourism products, and, second, as an informative tool to inform the population of the benefits associated with these new experiences. The strategic forms of communication are thus divided into:

- The private for-profit offer – the notion that individuals with higher income and education levels are more proactive in their demand for higher-quality natural areas (Duarte, 2022), associated with the growing trend of demand for qualified tourism supply that meets the specific needs of the senior market, leads us to believe that, in the post-COVID period, this trend will not only continue but will be reinforced, regarding the demand for nature-based visits. Therefore, it is up to the private sector to guide and set the pace in the process of innovation and improvement of the tourism offer to be developed in these tourism sites.
- The private offer provided by social economy organizations, known as the Third Sector. New trends, created by tour operators and tourist entertainment companies dedicated to health and wellbeing tourism, must be followed by social support institutions, or other non-profit institutions (e.g. University of the Third Age), which provide educational activities for seniors. It is essential that these types of tourist activities are no longer accessible only to those who can easily afford them, responding to the gaps identified in this and other studies, which found that, above all, older women, with low incomes and education, are kept apart from these opportunities to promote their wellbeing. Previous studies have shown that social tourism yields significant benefits (Ferreira, 2004).
- The public offer – in line with the private non-profit offer, there are public entities in the form of the social intervention bureaux of each municipality, whose responsibility increases as they have the responsibility to disseminate good salutogenic practices. Furthermore, in Portugal, we have around 22% of the national territory protected somehow, which means that there is always an HQNTA around or nearby.
- National Educational Campaign, 'Getting Closer to Nature' – the lack of information on nature conservancy and HQNTA must be addressed from childhood. School programmes should be adapted so that young citizens may have the tools to develop an effective emotional connection with these special places.
- Strengthening networks of influencing peers – in a world where the word 'influencer' has gained such relevance, it is quite interesting to think that the need to feel that we are part of a group comes from an intrinsic evolutionary need for safety and increased chance of survival. The leaders of the pack tended to be heard and followed and this is true to this day. So, if one can find the right peer influencer among the senior population it will be more likely that others will follow and HQNTAs will be psychologically easier to engage with.

The combination of these three axioms comes together in what we describe as the the eclectic generosity of higher-quality natural spaces theorem, which presents itself as an integrative theoretical instrument, with scientific bases, whose validation can be achieved if applied to future studies and which states the following (Duarte, 2022, p. 246):

> Natural spaces of high quality appeal to the most primal human instincts, namely because of the molecular and evolutionary proximity that interconnects us to all life on the planet, a fact that instigates our sense of bioception when exposed to their aesthetics – which we genetically recognize as safe and homeostatic – by means of accessible, concerted and qualified visitation plans, promoting natural services such as escape, enjoyment, relaxation, knowledge, creativity, spirituality, reflection and salutogenesis which, in turn, will translate into measurable effects on our wellbeing throughout life, with a special focus on enhancing general health and preventing short- and long-term senescence.

Conclusion

From the philosophical framework of the theoretical basis offered by this theorem of the eclectic generosity of higher-quality natural spaces, new models of action must arise, looking for an optimization of the process that interconnects nature services, with the welcome effects of our wellbeing, promoting health and preventing senescence, which can be achieved through a specialized tourist offer.

Thus, even if the prevention policies that are enacted for wellbeing and health start to take on a more prominent role and there is greater recognition that it is time to move beyond mere economic growth as a target for society, notably starting to recognize as imperative the incorporation of aspects of human wellbeing into the equation (Hartwell *et al.*, 2018), it is essential to define clear and achievable strategies.

The convergence between these top-down policies, with new bottom-up approaches which derive from a society more conscious and concerned with its holistic wellbeing, and the natural desire for personal fulfilment through tourism seems evident. However, it is essential to continue to promote research on this topic and to reinforce the conclusions on the relationship between tourism and health (Hartwell *et al.*, 2018), to which we add a third one, related to nature and its importance in this equation.

The theorem of eclectic generosity of higher-quality natural spaces aims to set the theoretical ground for this triad and lays the foundations for the development of solid, science-backed models of action. With this theorem we intend to contribute to restoring a sense of justice, by bringing nature back at the centre of our actions as citizens, with tourism serving as a facilitator, by educating and giving access to the wonders of protected areas.

References

Balderas-Cejudo, M.A., Leeso, G.W. and Urdaneta, E. (2017) Senior tourism: towards and active and healthy ageing. *Open Access Journal of Gerontology and Geriatric Medicine* 1(3), 555–563. DOI: 10.19080/OAJGGM.2017.01.555563.

BBC (2020) The secret of how life on earth began. BBC Earth. Available at: http://www.bbc.com/earth/story/20161026-the-secret-of-how-life-on-earth-began (accessed 4 December 2020).

Beckena, S. and Job, H. (2014) Protected areas in an era of global–local change. *Journal of Sustainable Tourism* 22(4), 507–527. DOI: 10.1080/09669582.2013.877913.

Bieling, C. and Plieninger, T. (2013) Recording manifestations of cultural ecosystem services in the landscape. *Landscape Research* 38(5), 649–667. DOI: 10.1080/01426397.2012.691469.

Bieling, C., Plieninger, T., Pirker, H. and Vogl, C.R. (2014) Linkages between landscapes and human well-being: an empirical exploration with short interviews. *Ecological Economics* 105, 19–30. DOI: 10.1016/j.ecolecon.2014.05.013.

Bosch, M. and Sang, A.O. (2017) Urban natural environments as nature-based solutions for improved public health: a systematic review of reviews. *Environmental Research* 158, 373–384. DOI: 10.1016/j.envres.2017.05.040.

Britton, E., Kindermann, G., Domegan, C. and Carlin, C. (2018) Blue care: a systematic review of blue space interventions for health and wellbeing. *Health Promotion International* 35(1), 50–69. DOI: 10.1093/heapro/day103.

Brown, G.G. and Reed, P. (2012) Social landscape metrics: measures for understanding place values from public participation geographic information systems (PPGIS). *Landscape Research* 37(1), 73–90. DOI: 10.1080/01426397.2011.591487.

Capaldi, C.A., Dopko, R.L. and Zelenski, J.M. (2014) The relationship between nature connectedness and happiness: a meta-analysis. *Frontiers in Psychology* 5(976), 1–15. DOI: 10.3389/fpsyg.2014.00976.

Carman, J. (2011) Elders' lifelong connection with the natural environment. *Educational Gerontology* 37(6), 488–498. DOI: 10.1080/03601277.2011.570203.

Carretero, F., Ferri, M. and Garcés, J. (2013) Accessible social tourism as a social policy strategy for healthy ageing: the relationship between tourism and functional health in older adults. In: Minnaert, L. and Miller, G. (eds) *Social Tourism: Perspectives and Potential*. Routledge, London.

Carvalho, G.B. and Damasio, A. (2021) Interoception and the origin of feelings: a new synthesis. *Bioessays* 43, 1–11. DOI: 10.1002/bies.202000261.

Cervinka, R., Röderer, K. and Hefler, E. (2011) Are nature lovers happy? On various indicators of well-being and connectedness with nature. *Journal of Health Psychology* 17(3), 379–388. DOI: 10.1177/1359105311416873.

Chan, K.M.A., Gould, R.K. and Pascual, U. (2018) Relational values: what are they, and what's the fuss about? *Current Opinion in Environmental Sustainability* 35, A1–A7. DOI: 10.1016/j.cosust.2018.11.003.

Chang, C., Cheng, G.J.Y., Nghiem, T.P.L., Song, X.P., Oh, R.R.Y. *et al.* (2020) Social media, nature, and life satisfaction: global evidence of the biophilia hypothesis. *Nature Scientific Reports* 10(1). DOI: 10.1038/s41598-020-60902-w.

Damasio, A. and Carvalho, G.B. (2013) The nature of feelings: evolutionary and neurobiological origins. *Nature Reviews – Neuroscience* 14(2), 143–152. DOI: 10.1038/nrn3403.

Deci, E.L. and Ryan, R.M. (2006) Hedonia, Eudaimonia, and Well-being: an introduction. *Journal of Happiness Studies* 9(1), 1–11. DOI: 10.1007/s10902-006-9018-1.

Deng, J., King, B. and Bauer, T. (2002) Evaluating natural attractions for tourism. *Annals of Tourism Research* 29(2), 422–438. DOI: 10.1016/S0160-7383(01)00068-8.

Duarte, S. (2022) Turismo, serviços culturais dos ecossistemas e saúde. PhD thesis, IGOT, Universidade de Lisboa, Portugal.

Ferreira, C.C.F. (2004) Portugal, destino turístico da população idosa Europeia – abordagem geográfica do turismo sénior internacional. PhD thseis, Faculdade de Letras da Universidade de Lisboa, Portugal.

Finlay, J., Franke, T., McKay, H. and Sims-Gould, J. (2015) Therapeutic landscapes and wellbeing in later life: impacts of blue and green spaces for older adults. *Health & Place* 34, 97–106. DOI: 10.1016/j.healthplace.2015.05.001.

Frank, S., Fürst, C., Koschke, L., Witt, A. and Makeschin, F. (2013) Assessment of landscape aesthetics – validation of a landscape metrics-based assessment by visual estimation of the scenic beauty. *Ecological Indicators* 32, 222–231. DOI: 10.1016/j.ecolind.2013.03.026.

Gagliardi, C. and Piccinini, F. (2019) The use of nature-based activities for the well-being of older people: an integrative literature review. *Archives of Gerontology and Geriatrics* 83, 315–327. DOI: 10.1016/j.archger.2019.05.012.

Gascon, M., Zijlema, W., Vert, C., White, M.P. and Nieuwenhuijsen, M.J. (2017) Outdoor blue spaces, human health and well-being: a systematic review of quantitative studies. *International Journal of Hygiene and Environmental Health* 220(8), 1207–1221. DOI: 10.1016/j.ijheh.2017.08.004.

Haldane, J.B.S. (1929) *The Origin of LIfe*. Rationalist annual. Available at: www.valencia.edu/~orilife (accessed 23 June 2023).

Hartig, T., Kaiser, F.G. and Bowler, P.A. (2001) Psychological restoration in nature as a positive motivation for ecological behavior. *Environmental Conservation* 33(4), 590–607. DOI: 10.1177/00139160121973142.

Hartig, T., Mitchell, R., de Vries, S. and Frumkin, H. (2014) Nature and health. *Annual Review of Public Health* 35, 207–228. DOI: 10.1146/annurev-publhealth-032013-182443.

Hartwell, H., Fyall, A., Willis, C., Page, S., Ladkin, A. *et al.* (2018) Progress in tourism and destination wellbeing research. *Current Issues in Tourism* 21(16), 1830–1892. DOI: 10.1080/13683500.2016.1223609.

Himes, A. and Muraca, B. (2018) Relational values: the key to pluralistic valuation of ecosystem services. *Current Opinion in Environmental Sustainability* 35, 1–7. DOI: 10.1016/j.cosust.2018.09.005.

Houlden, V., Weich, S., Porto de Albuquerque, J., Jarvis, S. and Rees, K. (2018) The relationship between greenspace and the mental wellbeing of adults: a systematic review. *PLoS One* 13(9), e0203000. DOI: 10.1371/journal.pone.0203000.

Hug, L.A., Baker, B.J., Anantharaman, K., Brown, C.T., Probst, A.J. *et al.* (2016) A new view of the tree of life. *Nature Microbiology* 1(5), 16048. DOI: 10.1038/nmicrobiol.2016.48.

Husk, K., Lovell, R., Cooper, C., Stahl-Timmins, W. and Garside, R. (2016) Participation in environmental enhancement and conservation activities for health and well-being in adults: a review of quantitative and qualitative evidence. *Cochrane Database of Systematic Reviews* (5), CD010351. DOI: 10.1002/14651858.CD010351.pub2.

Hwang, J. and Lyu, S.O. (2014) The antecedents and consequences of well-being perception: an application of the experience economy to golf tournament tourists. *Journal of Destination Marketing & Management* 4(4), 248–257. DOI: 10.1016/j.jdmm.2015.09.002.

ICNF (2022) *Rede Nacional de Áreas Protegidas*. ICNF. Consultado a 30 de setembro de 2022. Available at: https://icnf.pt/conservacao/parques/apambitonacionalregionalelocal (accessed 23 June 2023).

Ingólfsdóttir, A.H. and Gunnarsdóttir, G.û. (2020) Tourism as a tool for nature conservation? Conflicting interests between renewable energy projects and wilderness protection in Iceland. *Journal of Outdoor Recreation and Tourism* 29, 100276. DOI: 10.1016/j.jort.2019.100276.

Jarratt, D., Phelan, C., Wain, J. and Dale, S. (2018) Developing a sense of place toolkit: identifying destination uniqueness. *Tourism and Hospitality Research* 19(4), 408–421. DOI: 10.1177/1467358418768678.

Jax, K., Calestani, M., Chan, K.M., Eser, U., Keune, H. *et al.* (2018) Caring for nature matters: a relational approach for understanding nature's contributions to human well-being. *Current Opinion in Environmental Sustainability* 35, 22–29. DOI: 10.1016/j.cosust.2018.10.009.

Jurowski, C. (2009) An examination of the four realms of tourism experience theory. In: *International CHRIE conference*, University of Massachusetts. Available at: https://scholarworks.umass.edu/refereed/Sessions/Wednesday/23 (accessed 23 June 2023).

Kabisch, N., Bosch, M. and Lafortezza, R. (2017) The health benefits of nature-based solutions to urbanization challenges for children and the elderly: a systematic review. *Environmental Research* 159, 362–373. DOI: 10.1016/j.envres.2017.08.004.

Kaplan, R. and Kaplan, S. (1989) *The Experience of Nature: A Psychological Perspective*. Cambridge University Press, Cambridge, UK.

Louv, R. (2008) *Last Child in the Woods: Saving Our Children from Nature-Deficit Disorder*. Algonquin Books, Chapel Hill, North Carolina. Available at: http://richardlouv.com/books/last-child/ (accessed 23 June 2023).

Mackay, C.M.L. and Schmitt, M.T. (2019) Do people who feel connected to nature do more to protect it? A meta-analysis. *Journal of Environmental Psychology* 65, 101323. DOI: 10.1016/j.jenvp.2019.101323.

Martin, L., White, M.P., Hunt, A., Richardson, M., Pahl, S. *et al.* (2020) Nature contact, nature connectedness and associations with health, wellbeing and pro-environmental behaviours. *Journal of Environmental Psychology* 68, 101389. DOI: 10.1016/j.jenvp.2020.101389.

Martinez-Juarez, P., Chiabai, A., Taylor, T. and Gómez, S.Q. (2015) The impact of ecosystems on human health and well-being: a critical review. *Journal of Outdoor Recreation and Tourism* 10, 63–69. DOI: 10.1016/j.jort.2015.06.008.

McMahan, E.A. (2018) Happiness comes naturally: engagement with nature as a route to positive subjective well-being. In: Diener, E., Oishi, S. and Tay, L. (eds) *Handbook of Well-Being*. DEF Publishers. DOI: nobascholar.com.

McMahan, E.A. and Estes, D. (2015) The effect of contact with natural environments on positive and negative affect: a meta-analysis. *The Journal of Positive Psychology* 10(6), 507–519. DOI: 10.1080/17439760.2014.994224.

MEA (Millennium Ecosystem Assessment) (2005) Ecosystems and human well-being: synthesis. Available at: https://www.millenniumassessment.org/en/index.html (accessed 23 June 2023).

Moyle, B.D., Scherrer, P., Weiler, B., Wilson, E., Caldicott, R. *et al.* (2017) Assessing preferences of potential visitors for nature-based experiences in protected areas. *Tourism Management* 62, 29–41. DOI: 10.1016/j.tourman.2017.03.010.

Ode, Å., Tveit, M.S. and Fry, G. (2008) Capturing landscape visual character using indicators: touching base with landscape aesthetic theory. *Landscape Research* 33(1), 89–117. DOI: 10.1080/01426390701773854.

Ohly, H., White, M.P., Wheeler, B.W., Bethel, A., Ukoumunne, O.C. *et al.* (2016) Attention restoration theory: a systematic review of the attention restoration potential of exposure to natural environments. *Journal of Toxicology and Environmental Health, Part B* 19(7), 305–343. DOI: 10.1080/10937404.2016.1196155.

Oparin, A.I. (1924) *The Origin of Life*. Foreign Languages Publishing House. Available at: www.valencia.edu/~orilife (accessed 23 June 2023).

Orr, N., Wagstaffe, A., Briscoe, S. and Garside, R. (2016) How do older people describe their sensory experiences of the natural world? A systematic review of the qualitative evidence. *BMC Geriatrics* 16, 116. DOI: 10.1186/s12877-016-0288-0.

Pearson, R.G. (2016) Reasons to conserve nature. *Trends in Ecology & Evolution* 31(5), 366–371. DOI: 10.1016/j.tree.2016.02.005.

Pine, B.J. and Gilmore, J.H. (1999) *The Experience Economy: Work Is Theatre and Every Business a Stage*. Harvard Business School Press, Boston, Massachusetts.

PNTN (2015) Resolução do Conselho de Ministros n.o. 51/2015, de 21 de Julho. *Diário da República* 1.a série-N.o. 140.

Pritchard, A., Richardson, M., Sheffield, D. and McEwan, K. (2020) The relationship between nature connectedness and eudaimonic well-being: a meta-analysis. *Journal of Happiness Studies* 21(3), 1145–1167. DOI: 10.1007/s10902-019-00118-6.

Roberts, A., Hinds, J. and Camic, P.M. (2019) Nature activities and wellbeing in children and young people: a systematic literature review. *Journal of Adventure Education and Outdoor Learning* 20(4), 298–318. DOI: 10.1080/14729679.2019.1660195.

Shanahan, D.F., Astell-Burt, T., Barber, E.A., Brymer, E., Cox, D.T.C. *et al.* (2019) Nature-based interventions for improving health and wellbeing: the purpose, the people and the outcomes. *Sports* 7(6), 141. DOI: 10.3390/sports7060141.

Sheldon, P.J. (2020) Designing tourism experiences for inner transformation. *Annals of Tourism Research* 83, 102935. DOI: 10.1016/j.annals.2020.102935.

Smithsonian (2020) Human characteristics: what does it mean to be human. Available at: https://humanorigins.si.edu/human-characteristics (accessed 3 December 2020).

TEEB (2010) *The Economics of Ecosystems and Biodiversity: Mainstreaming the Economics of Nature: A Synthesis of the Approach, Conclusions and Recommendations of TEEB*. Available at: http://teebweb.org/wp-content/uploads/Study%20and%20Reports/Reports/Synthesis%20report/TEEB%20Synthesis%20Report%202010.pdf (accessed 23 June 2023).

Thompson, C.W. (2011) Linking landscape and health: the recurring theme. *Landscape and Urban Planning* 99(3–4), 187–195. DOI: 10.1016/j.landurbplan.2010.10.006.

Tillmann, S., Tobin, D., Avison, W. and Gilliland, J. (2018) Mental health benefits of interactions with nature in children and teenagers: a systematic review. *Journal of Epidemiology and Community Health* 72(10), 958–966. DOI: 10.1136/jech-2018-210436.

Tribot, A.S., Deter, J. and Mouquet, N. (2018) Integrating the aesthetic value of landscapes and biological diversity. *Proceedings of the Royal Society B: Biological Sciences* 285(1886), 20180971. DOI: 10.1098/rspb.2018.0971.

Twohig-Bennett, C. and Jones, A. (2018) The health benefits of the great outdoors: a systematic review and meta-analysis of greenspace exposure and health outcomes. *Environmental Research* 166, 628–637. DOI: 10.1016/j.envres.2018.06.030.

UN (2019b) World urbanization prospects – the 2018 revision. UN Department of Economic and Social Affairs – Population Division. Available at: https://www.un.org/development/desa/publications/2018-revision-of-world-urbanization-prospects.html (accessed 8 October 2023).

WHO (2002) Active ageing – a policy framework. Available at: https://apps.who.int/iris/handle/10665/67215 (accessed 23 June 2023).

WHO (2020) Ageing and health: a rapidly changing world. Available at: https://www.who.int/news-room/fact-sheets/detail/ageing-and-health (accessed 9 April 2020).

Wilson, E.O. (1984) *Biophilia*. Harvard University Press. Available at: https://www.hup.harvard.edu/catalog.php?isbn=9780674074422&content=reviews#:~:text=The%20Washington%20Post%20Book%20World,on%20life%20and%20lifelike%20process (accessed 8 October 2023).

Wyles, K.J., White, M.P., Hattam, C., Pahl, S., King, H. *et al.* (2017) Are some natural environments more psychologically beneficial than others? The importance of type and quality on connectedness to nature and psychological restoration. *Environment and Behavior* 51(2), 111–143. DOI: 10.1177/0013916517738312.

Zanten, B.T., Zasada, I., Koetse, M.J., Ungaro, F., Häfner, K. *et al.* (2016) A comparative approach to assess the contribution of landscape features to aesthetic and recreational values in agricultural landscapes. *Ecosystem Services* 17, 87–98. DOI: 10.1016/j.ecoser.2015.11.011.

3 Forests, Health and Tourism: Development of Sustainable Models for Nature-based Health Tourism in Austria and Bavaria

Sebastian Markov[1], Corinna Pippirs[1], Georg Christian Steckenbauer[1*], and Michael Bischof[2]
[1]Faculty European Campus Rottal-Inn, Deggendorf Institute of Technology, Germany; [2]Institute of Ecomedicine, Paracelsus Medical University Salzburg, Austria

Abstract
The increasing importance of forests as recreational places with scientifically proven positive effects on human health requires the creation of concepts for the sustainable use of this valuable and vulnerable natural resource in health tourism. Within the framework of a cross-border research and development project, network partners from Bavaria and Austria jointly developed pilot projects for sustainable tourism offers involving local forests. One essential part of the project is the implementation of representative population surveys in Austria and Germany in order to obtain the perception of forests, expectations and assessments of the health effects of visits to forests within the population as a basis for the development of tourism offers. The findings from the research project, the pilot studies and the stakeholder analyses are the basis for establishing a cross-border network, Forest Health and Tourism, which will promote further research and development in the future.

Introduction: 'Forest and Health'

For many people, a regular visit to a forest is one of the most important opportunities to experience nature; given its regenerative capacities, the forest allows people to distance themselves from their everyday life and to find serenity (Schraml, 2009; Ensinger *et al.*, 2013). Not least, the COVID-19 pandemic, with its restrictions, has led to forests gaining even greater significance, with people now seeking out forests at a far higher rate (Palm *et al.*, 2020).

The features and characteristics of a forest and the atmosphere dynamics associated with it (air, temperature, smell, haptics, etc.) can have positive psychological and physiological effects on people. Regularly visiting forests can therefore bring the following benefits: movement and activity in nature, social interaction, reduced noise and air pollution, lower stress levels and stimulated immune response. These are mutually reinforcing and can ultimately turn a forest visit into a health-promoting activity on many levels (Marušáková and

*Corresponding author: georg.steckenbauer@th-deg.de

DOI: 10.1079/9781800621411.0003

Sallmannshofer, 2019; Schuh and Immich, 2019; White *et al.*, 2019).

Increased awareness of the human body and its functions, consistently more differentiated personalized diagnostics, (digital) information and discussion opportunities, as well as new 'healing methods', are the reasons for today's far greater level of health consciousness (Zukunftsinstitut, 2022).

Considering this increasing orientation towards health, medical and scientific research into forests and their positive impact on people's health is taking on completely new dimensions – even though serious scientific studies in Europe are still in their infancy (Cervinka *et al.*, 2014; Schuh and Immich, 2019).

However, scientific knowledge and empirical knowledge are increasingly merging to form one integrated (medical) understanding (Zukunftsinstitut, 2022).

With these properties and developments, the forest is gathering serious attention and acquiring a new function as a 'public cure'. Spending time in the forest is increasingly becoming the priority, with health a driving motive – the forest is experiencing a change of meaning, of its essence (Kühne, 2019; Münderlein and Bruns, 2019), from a backdrop to a 'health space'. No longer is it just a source of subjective wellbeing, but, increasingly, it is a medically researched remedy. Medical and scientific research as well as this change are thus paving the way for the first evidence-based health tourism offers – while also making them necessary (Steckenbauer, 2017).

Case Study: Research Project Netzwerk Gesundheitstourismus Wald

Regarding the forest as a natural environment, initial trials are underway in Bavaria and Austria looking into using such an environment in a health tourism context. This potential – for residents and guests as well as forest owners, destinations or healthcare establishments – needs to undergo scientific investigation with a view to developing common solutions for their sustainable use.

Therefore, the project Netzwerk Gesundheitstourismus Wald (AB291) was established within the bounds of the INTERREG V-A programme 'Austria – Bavaria 2014–2020' in collaboration with four project partners, namely the Deggendorf Institute of Technology, the Medical University Salzburg, the Austrian Research Centre for Forests and the University of Applied Sciences Upper Austria. The aim of the project was to develop common approaches towards a sustainable use of local forest areas for health tourism purposes in five defined pilot regions in Bavaria and Austria (Fig. 3.1). The long-term goal is to develop a collaborative 'Forests, Health and Tourism' network and research field dedicated to the changing health and social demands placed on forest areas while also serving as the basis for further research, co-operation and follow-up projects.

The bundling of various scientific and practical skills in one cross-border co-operation, likewise, provides an innovative basis: the partners approach the project from a tourism, medical and forestry perspective taking into consideration scientific and practical aspects. The fundamental assumptions in all the measures are that these are sustainable and forest-preserving approaches, have a health-promoting effect on visitors to the forest and offer the stakeholders involved added value.

Methodology and Forest Perception

From rest and relaxation to social and sports activities, through to preventive measures or therapeutic interventions, the motives, expectations and assessments behind visiting forests are as broad as they are diverse (Marušáková and Sallmannshofer, 2019; Schuh and Immich, 2019). Such aspects are, likewise, not only crucial to understanding people's behaviour in the context of forests but also to being able to derive sustainable health tourism offers.

An online survey of the population was conducted in Germany and Austria as part of the project. The questionnaire survey is representative in terms of the age group (for people falling into the age category of 18–69), gender and background, with a sample of n = 3010

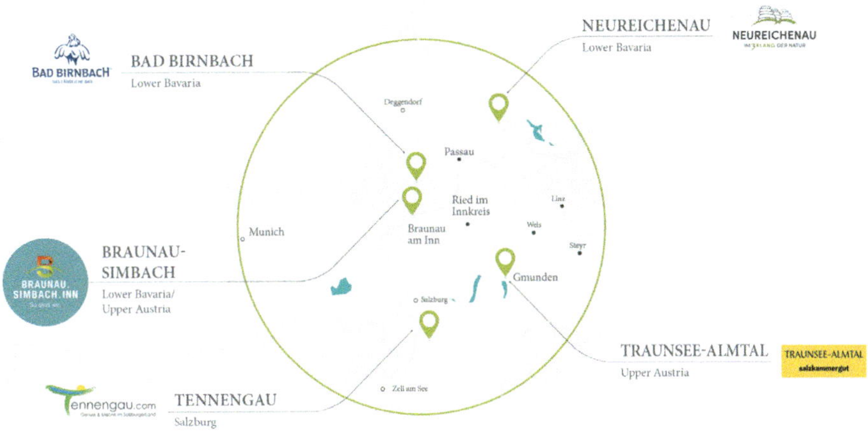

Fig. 3.1. Pilot regions of the project Netzwerk Gesundheitstourismus Wald (AB291).

participants in Germany and Austria. The questions, which covered, among others, the personal associations, motives and activities of those surveyed as well as their assessment of various criteria, aspects and attributes concerning forests and forest visits, sought to examine in greater detail the important interplay between forests, health and tourism.

Results of the survey

One of the core topics within the study was the question of what things are regarded as typical for a forest and what people spontaneously associate with the term forest. The most named associations are trees, nature and tranquillity. While the 'tree' association may appear banal, a more in-depth analysis of the findings reveals a very nuanced picture: a forest is more than a mere collection of trees and arises from a connection with other attributable features, most notably 'nature' and 'tranquillity'. The other key responses given are 'distinctive forest air', 'rest and relaxation', 'animals' and 'green'. Their broad spectrum, ranging from specific, objective things to personal feelings and social interpretations, gives a sense of the complexity of how forests are perceived (Markov *et al.*, 2023).

A similar image is revealed in terms of what people view as an ideal forest. The evaluation shows that the respondents, first and foremost, imagine 'a calm, quiet mixed forest where some form of water (stream, source, lake, etc.) can be found'. At the same time, the statements paint an, at times, excessively idyllic, almost fairy-tale-like image of forests. They simultaneously prove that the perception of a forest is nuanced and differentiated and includes many aspects and components.

The population's (increasing) affinity towards forests appears to be confirmed. Around three-quarters of all respondents, for example, stated that they have access to a forest or woodland in the vicinity of their living environment;

of those respondents with access to a forest, 30% visit the forest several times a week.

The activities undertaken in the forest appear to serve as a form of deceleration, as a conscious immersion and savouring of the moment – by way of example, the main motive cited for visiting a forest is 'to enjoy the tranquillity', followed by 'to enjoy the forest atmosphere and forest climate' and 'to experience and enjoy the (natural) beauty'. 'To boost recovery and regeneration' and 'to escape from the daily grind' are further motives named as playing a vital role.

Particularly in the interplay between forests and health, and thus the development of health tourism offers, it is interesting to note that respondents ascribe fundamentally positive health effects to forests: almost all statements in the survey on possible positive health effects of the forest were agreed with by the respondents – in some cases very clearly. Above all, the respondents confirmed that 'merely being in the forest is good for me' – they also attribute a personal 'relaxing effect' to forests and an 'inherent, health-boosting climate'. Despite the agreement regarding health effects of forests, the respondents still seem to be rather hesitant to make use of concrete offers; nevertheless, with about 45%, a potential of the population that should not be underestimated can imagine experiencing the forest with professional accompaniment (by trained forest educators, forest health trainers, hiking guides, etc.), and every fourth person would also be willing to pay for health tourism offers (e.g. accompanied forest health courses, relaxation exercises, preventive measures).

Based on the evaluations of the survey, various 'groups' can be identified as a result (Fig. 3.2a and b): among others, a group that is increasingly looking for 'staging'. By contrast, there is a group of 'health individualists', which only wants to move through the forest and draw their personal sense of wellbeing. For a group of 'spontaneous excursionists', brief breakouts from their everyday lives are important, while for 'connoisseurs of habits' the forest is part of their daily life.

While these individual groups may seem small in terms of their numerical quality considering the whole survey, they should not be underestimated when applied to the populations of Germany and Austria as a whole; groups that are ultimately differentiated convert to several million citizens and/or potential (forest) users (Markov *et al.*, 2022).

As varied as people's motives, interests and expectations are when it comes to forests and visiting forests, so are the groups that can be distinguished based on their thinking and behavioural patterns. It is vital that offers have such an attractive value that they meet the corresponding demand. At the same time, it should not be forgotten that residents, and not only (non-local) guests, patients or tourists, comprise a relevant target group and make use of local offers.

Forest-based Product Development

Forests are a 'polyvalent space' of very broad significance, ranging from a location where raw materials are extracted, to the provision of essential ecosystem services, through to their function for recreation (Berr and Jenal, 2022). Especially given the ongoing shift in significance of forests towards being a 'health space' (Claßen, 2016; Steckenbauer, 2017) and their increasing interpretation by society as a location for leisure activities, recreation and experiencing nature, forest-based health tourism activities and offers are now springing up in many places (Marušáková and Sallmannshofer, 2019; Suda *et al.*, 2022).

This shift with the accompanying provision of corresponding forest-related activities and offers ultimately means that different groups of people with very different motives and interests find themselves out and about in the forest at the same time, with potential differences and conflicts emerging as a result of this convergence of disparate interests (Jenal, 2019). It is therefore crucial to identify all relevant stakeholders (especially forest owners) and to ensure their involvement as well as their long-term commitment from the outset.

Offering health-promoting pursuits in the forest can, basically, work at any destination that has suitable forest areas. However, the relevant offers must be adapted to suit the conditions and prerequisites at the corresponding location and be attractively enhanced. In the

Fig. 3.2a, b. Target groups from the survey.

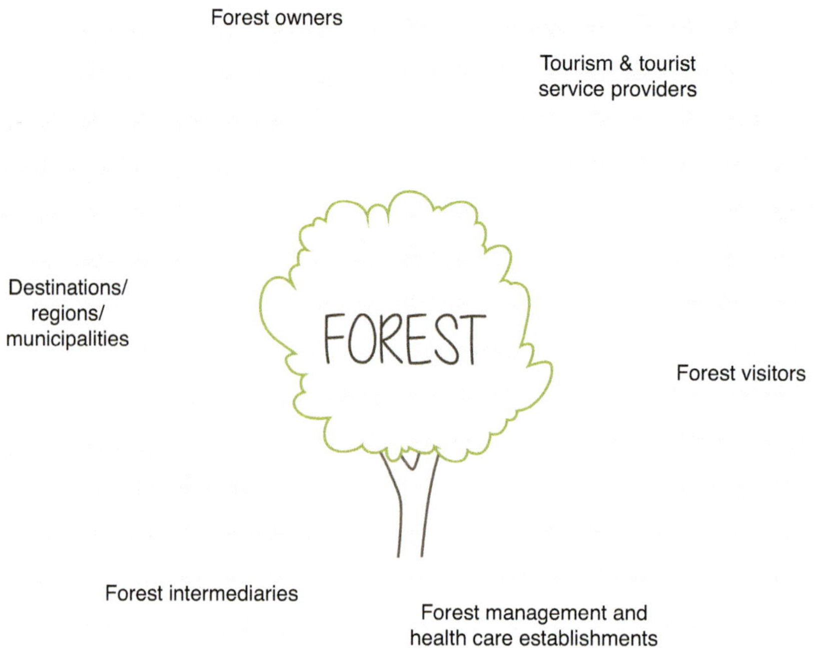

Fig. 3.3. Stakeholder groups involved in forest-based health tourism.

end it comes down to the marketability of the offer and to its demand. The medical evidence can support this and establish a unique selling point.

Exemplary Approach for a Project Region

Building on the current state of research, the findings and perspectives drawn from the project and corresponding surveys, approaches for a sustainable and health-promoting use of local forests were developed for the five pilot regions. These regions were selected for their heterogeneous nature.

The basis for the respective product development is a combination of movement and activity with specific applications, additional offerings or measures in the forest. Of elementary significance for all the implementation steps is the premise of sustainability, which secures value added for both the providers and the demanders and delivers ecological and socio-cultural benefits. Therefore, it is essential to ensure the ongoing and transparent involvement of all relevant local stakeholders (Fig. 3.3).

In accordance with these prerequisites, the pilot regions and their specific forest spaces were examined and tested for their usability and availability, local stakeholders were identified and involved, and overriding strategies were accounted for in the product development stage. A clinical study was conducted at two pilot regions and their findings incorporated into the developments. These findings on the use of different forest areas (forests without water vs forests with water, so-called 'green' and 'blue' space) can be viewed as part of the key findings of the project and are likewise of relevance from a general perspective.

Pilot Region: Bad Birnbach

The pilot region of Bad Birnbach in Lower Bavaria is situated south of the Danube and forms part of Lower Bavaria's so-called Spa Triangle, well known for its thermal baths.

Through the knowledge and findings of this project, the existing range of offers, comprising evidence-based and medical products, can be expanded. In addition to this, specific measures can be swiftly implemented in Bad Birnbach given the existing foundations and structures already in place.

Within the project, the Lugenz, a woodland area on the direct outskirts of Bad Birnbach, forms the basis for developing forest-related and health-promoting offers. A clinical study was conducted which examined the impact of a short-term intervention on people's health (a guided walk in the woodlands over a period of approximately 4 h). The aim of the clinical study was to analyse how a short-term intervention impacts people suffering from high stress levels. A standardized programme comprising exercise, mindfulness and nature-connection elements was developed, which was then conducted under the guidance of a fully trained forest therapist. Around this programme, various physiological and psychological measurements were taken from the test subjects (n = 30). The study was conducted in the exact same manner in the two pilot regions of Bad Birnbach and Golling (Salzburger Land, Austria) as this not only allowed the fundamental impacts on people's health to be analysed but also potential differences to be identified between the two different types of forests – without water (Bad Birnbach) and with ample water (Golling).

A core outcome of the study shows that the interventions have a positive effect on reducing stress and on the wellbeing of the participants. One example of the immediate impact of the intervention is the so-called Feeling Scale, which measures the current level of wellbeing on a scale of –5 (very bad) to 5 (very good). From a mean value for all participants of slightly higher than 1 prior to the intervention beginning, the participants' level of wellbeing rose to over 3 following a roughly 4 h intervention in Bad Birnbach (Fig. 3.4). This improvement in wellbeing is statistically significant.

At the same time, this positive effect not only manifested itself immediately following the intervention but also at least 2 weeks thereafter. For example, the levels of stress felt by study participants both at the time of signing up for the study and 2 weeks after the study was examined using a perceived stress questionnaire (PSQ). The PSQ was developed as a means of measuring the extent to which an individual perceives situations in their lives as being stressful. The scale of results ranges from 0 (no stress) to 100 (maximum stress level). The evaluations of the questionnaire reveal, when comparing the mean value of the two reference points (before, after), a significant reduction in levels of stress. The mean stress level across all participants stood at a good 53 points prior to the interventions and just 45 points 2 weeks after the interventions (Fig. 3.5). Overall, it was determined that stress levels therefore dropped by a good eight points. In turn, it was not possible to determine any effects in terms of the physiological measurement parameters.

Overall, the findings of the study reveal positive and statistically significant impacts in the context of psychological parameters. The swift implementability can be demonstrated using the following example: the intervention performed here as well as the proven positive impacts on health have since been incorporated into a 1week offering which can be booked in Bad Birnbach under the name 'Forests and bathing: experience the healing effects of nature'. In addition to applications at the local thermal bath, the offering also includes a visit to the forest under the guidance and supervision of a trained forest health trainer who was involved in the study.

Conclusion and Outlook

The project and the approaches resulting from the five pilot regions illustrate the multi-faceted

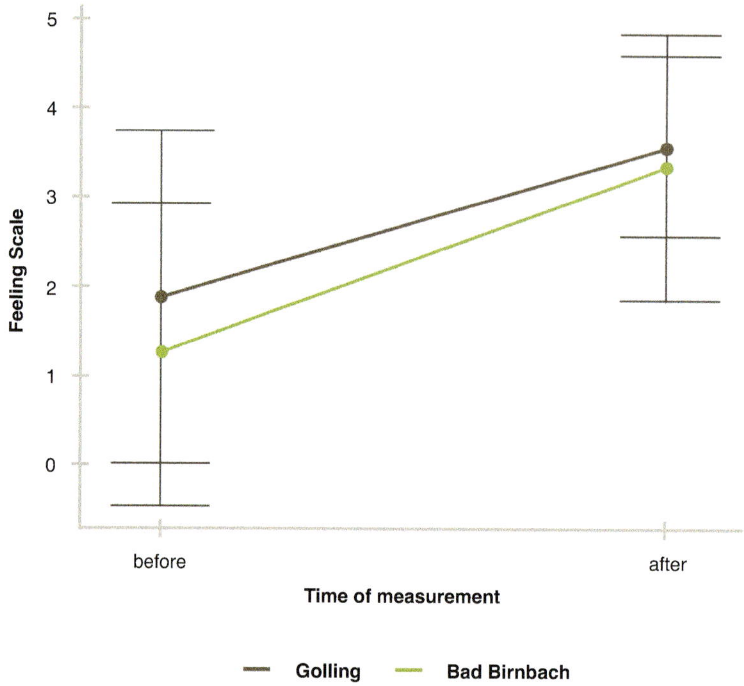

Fig. 3.4. Current level of wellbeing in study participants before and after the intervention (Feeling Scale).

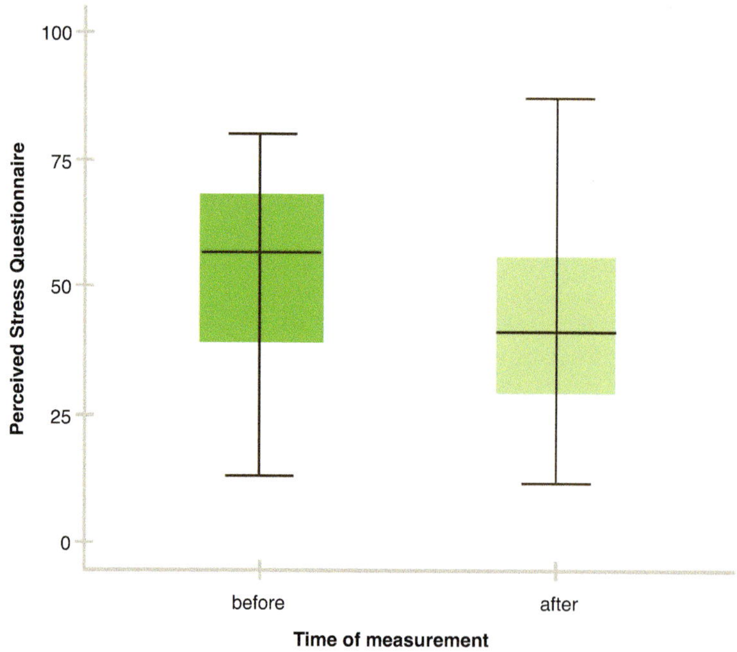

Fig. 3.5. Perceived levels of stress among study participants at the time of registration for the study (before) and 2 weeks after the study (after) (Perceived Stress Questionnaire).

nature of forest-related offers that can be developed based on the individually different parameters. Throughout the project, taking existing programmes and offers into account was (and fundamentally is) a prerequisite for developing health tourism products, as regional structures and offers need to be incorporated. Similarly, it is essential that existing and available competencies (e.g. forest owners, forest mediators, local therapists) are used.

The forest itself is increasingly evolving from a pure setting to a natural health space. As the population's health awareness grows, forests and forest visits will become more frequent and of increasing social importance.

Science is, equally, picking up on these trends and topics to an increasing degree. In particular, the connection between use of forests and people's health as well as the associated sustainable mechanisms of action are the central focus of the studies. The evidence-based nature lends weight and establishes credibility. In order to continue to preserve forests as a natural health space for the future, sustainability is the premise and guideline for development in all of its dimensions. This, equally, guarantees the creation of value for the participating stakeholders.

The aspects and challenges outlined here formed an essential part of the project's implementation, but were also collated and addressed in the concluding Practical Project Guide for practitioners (accessible at: www.netzwerk-gesundheitstourismus-wald.de [accessed 9 October 2023]). The guide does not strive or pretend to entirely solve these challenges but to serve as a point of reference and thought-provoking impulse for further research into the Forest, Health and Tourism topic (Markov *et al.*, 2022).

There is still a continued and great need for research to be conducted in the context of (European) forests before specific causal relations can better be understood and thus also be used, since serious scientific studies in Europe are still in their infancy. Furthermore, there is still high demand for sustainable business models in the field.

References

Berr, K. and Jenal, C. (2022) Wald in der Vielfalt möglicher Perspektiven – ein Überblick. In: Berr, K. and Jenal, C. (eds) *Wald in der Vielfalt möglicher Perspektiven*. RaumFragen: Stadt – Region – Landschaft, Springer VS, Wiesbaden, Germany, pp. 1–14. DOI: 10.1007/978-3-658-33705-6.

Cervinka, R., Höltge, J., Pirgie, L., Schwab, M., Sudkamp, J. *et al.* (2014) *Zur Gesundheitswirkung von Waldlandschaften*. Bundesforschungszentrum für Wald, Vienna.

Claßen, T. (2016) Empirische Befunde zum Zusammenhang von Landschaft und physischer Health. In: Gebhard, U. and Kistemann, T. (eds) *Landschaft, Identität und Gesundheit*. Springer VS, Wiesbaden, Germany, pp. 71–91. DOI: 10.1007/978-3-531-19723-4.

Ensinger, K., Wurster, M., Selter, A., Jenne, M., Bethmann, S. *et al.* (2013) Eintauchen in eine andere Welt – Untersuchungen über Erholungskonzepte und Erholungsprozesse im Wald. *Allgemeine Forst- und Jagdzeitung* 184(3/4), 70–83.

Jenal, C. (2019) *'Das Ist Kein Wald, Ihr Pappnasen!' – Zur sozialen Konstruktion von Wald: Perspektiven von Landschaftstheorie und Landschaftspraxis*. RaumFragen: Stadt – Region – Landschaft, Springer VS, Wiesbaden, Germany. DOI: 10.1007/978-3-658-25714-9.

Kühne, O. (2019) Zwischen Macht und Essenz, Konstrukt und Objekt? Wie Landschaftstheorien Deutungskonkurrenzen von natur Zeigen. *Stadt+Grün (2019)* 12, 24–27.

Markov, S., Bischof, M., Frank, V., Maier-Lehner, G., Gianordoli, I. *et al.* (2022) *Wissen schafft Wege – Praktischer Leitfaden für eine gesundheitstouristische Nutzung von Wäldern in Österreich und Bayern*. Netzwerk Gesundheitstourismus Wald, Deggendorf, Germany.

Markov, S., Steckenbauer, G.C., Herntrei, M. and Mouzhi, G. (2023) Zur sozialen Konstruktion von Wald und seinem Bedeutungswandel im Kontext von Gesundheit. In: Kühne, O., Freytag, T., Sedelmeier, T. and Jenal, C. (eds) *Landschaft und Tourismus*. RaumFragen: Stadt – Region Landschaft, Springer VS, Wiesbaden, Germany. DOI: 10.1007/978-3-658-39085-3.

Marušáková, L. and Sallmannshofer, M. (2019) *Human Health and Sustainable Forest Management*. Forest Europe, Bratislava, Slovakia.

Münderlein, D. and Bruns, D. (2019) Landschaft und Gesundheit. In: Kühne, O., Weber, F., Berr, K. and Jenal, C. (eds) *Handbuch Landschaft*. Springer VS, Wiesbaden, Germany, pp. 489–503. DOI: 10.1007/978-3-658-25746-0.

Palm, T., Weinbrenner, H. and Wirth, K. (2020) Die Bedeutung von Stadtwäldern während der Corona-Pandemie. *AZF – Der Wald* 22, 12–15.

Schraml, U. (2009) Erholung und Tourismus als Themen einer Zukunftsstrategie für die Waldnutzung in Deutschland. In: Seintsch, B. and Dieter, M. (eds), *Waldstrategie 2020*. Tagungsband des BMELV Symposiums, 10–11 December 2008, Johann Heinrich von Thünen-Institut, Berlin, Braunschweig, Germany, pp. 17–25.

Schuh, A. and Immich, G. (2019) Waldtherapie – das Potential des Waldes für Ihre Gesundheit. Springer, Berlin, Heidelberg, Germany. DOI: 10.1007/978-3-662-59026-3.

Steckenbauer, G.C. (2017) Gesundheitswälder – Touristische Inwertsetzung der Ressource Wald. *TourismusWissen Quarterly* 10, 259–264.

Suda, M., Gaggermeier, A., Koller, N. and Wagner, L. (2022) Waldnutzung zu Gesundheitszwecken – Beobachtung und analyse aktueller Entwicklungen für die strategische Positionierung forstlicher Akteure. TUM, Freising, Germany.

White, P.M., Alcock, I., Grellier, J., Wheeler, B.W., Hartig, T. *et al.* (2019) Spending at least 120 minutes a week in nature is associated with good health and wellbeing. *Scientific Reports* 9(1), 1–11. DOI: 10.1038/s41598-019-44097-3.

Zukunftsinstitut (2022) Healthness: Gesundheit wird ganzheitlich. Available at: https://www.zukunftsinstitut.de/artikel/healthness-gesundheit-wird-ganzheitlich/ (accessed 24 February 2022).

4 Nature-based Tourism in Resilient Destinations: Methodological Approaches and Pilot Applications for Product Development in Experimental Destinations

Anastasia Traskevich[1]* and Martin Fontanari[2]
[1]Belarus State Economic University, Minsk, Belarus; [2]International School of Management (ISM), Im Mediapark 5C, Cologne, Germany

Abstract
Nature-based and wellness tourism is visioned as the core tourism theme for development of resilient destinations. Crisis-driven perspectives are merged under the paradigm of long-term resilience with the focus on personality, wellbeing, community identity, responsible communication of people with nature, and ultimately on building on resilient and healthy livelihoods and meaningful society. The study considers the Code Red-Model of destination resilience to introduce the applied framework for resilient product development. The applied framework is verified by product development in the experimental nature-based destination Augustow Canal (Belarus). The portfolio of pilot resilient products elaborated includes combined canoeing tours 'Paddle and Pedal', outdoor adventuring and survival camps. As a methodological contribution, the present research concludes with the key principles of resilient product development for universal application in nature-based experimental destinations for the perspectives of sustainability, responsible regional prosperity and inclusive community-based entrepreneurship, innovative business models and alternative resilient value creation.

Introduction: Nature-based and Wellness Tourism as Core Tourism Themes for Development of Resilient Destinations

Destination resilience for transformation of the tourism industry

The more complex and interconnected the economic, technical and political elements of tourism systems in the destination are, the more challenging their management becomes and the less flexible the systems are in the case of unfavourable and destructive environment (Bürkner, 2010; Biggs *et al.*, 2012; Sharpley, 2012; Osztovics *et al.*, 2012; Backer and Ritchie, 2017; Sisneros-Kidd *et al.*, 2019). Therefore, proactive management efforts are required (Innerhofer *et al.*, 2018; Jiang *et al.*, 2018; Filimonau and De Coteau, 2020) to protect and develop the integrative qualities of sustainability and resilience for tourism regions (Cochrane, 2010; Davoudi *et al.*, 2012; Butler, 2017). This is with a view to making local stakeholders the key components of a holistic socio-economic system of tourism and destinations that build

*Corresponding author: anastasia.traskevich@akalem.de

DOI: 10.1079/9781800621411.0004

on the sustainable quality of resilience in the long term (Benli and Kuhzady, 2017; Hall *et al.*, 2018; Pechlaner *et al.*, 2022).

The current research interest is focused on the conceptualization of novel approaches for resilience-oriented product development. It aims to contribute to the sustainable and meaningful transformation of the tourism industry in line with the recent trend of tourism research (Farmaki and Pappas, 2021; Hall, 2022; Sharpley, 2022). In this context, this research approach emphasizes local identity and uniqueness to implement the required transformations (Calgaro *et al.*, 2014; Sheller, 2020) towards resilient product development. Therefore, the following general question arises: How can resilience orientation be understood expansively as an advanced approach to sustainability? (Espiner *et al.*, 2017; Kato, 2018). Sustainability, with its three pillars of ecological, economic and social orientation, represents any kind of action which is resource-conserving, environmentally and socially friendly, and which implies economic responsibility (Camilleri, 2016). Its aim is to live in line with a vision of long-term stability in terms of people's living conditions and their environment, with the intention that future generations can also claim the same qualities and conditions of physical, mental, cultural and spiritual vitality.

This holistic understanding is based on the maintenance of a status quo of sustainable economic activity. However, it does not take into account the destructive or critical forces of systemically relevant crises and their ability to cause the collapse of the entire system, to the point that it can no longer function (Prayag *et al.*, 2018). If a socio-economic system that has not acquired the quality of resilience is consistently destroyed, one can no longer speak about sustainable development. Yet, the concept of resilience explains precisely how to maintain a vital performance in one's supply functions and social systems (Traskevich and Fontanari, 2021), particularly in the case of both external and internal disruptive disturbances (Lew *et al.*, 2016). Therefore, the concept of resilience guarantees the minimum performance required for the lives of people in their environment, thus shaping the vital precondition for any further sustainable development goals (Powell *et al.*, 2018; Siakwah *et al.*, 2019; Sheller, 2020).

At the same time, resilient thinking and behaviour encompass the entire approach of 'advanced sustainability' (Holladay and Powell, 2013; Wall, 2018). Even the dependencies of economic and social systems can still be recognized as being quite sustainable solutions. However, they are not resilient at all and, therefore, they cannot guarantee long-term existence in any-case scenarios. Furthermore, the concept of resilience can be valued as a managerial concept for advanced sustainable regional development in the tourism industry (Traskevich and Fontanari, 2021) so long as it offers the following criteria: a deeper orientation of the region to local resources (Hartard, 2014) and competitive advantages; a functional autonomy (Mandal and Saravanan, 2019); the viability of regional natural (Ruiz-Ballesteros, 2011; Jamaliah and Powell, 2018; Jarratt and Davies, 2019), socio-cultural (Lew, 2014; Lew *et al.*, 2016) and economic systems (Jiang *et al.*, 2019); and, finally, the continuous efficient performance of businesses with all stakeholders compliant with different event scenarios in the region, the country and the world (Welter-Enderlin and Hildenbrand, 2006; Fontanari and Kredinger, 2017; Prayag *et al.*, 2018). These prerequisites shape the further conceptual orientation of this study.

Amid the 'pre-COVID-19 normality', tourism was mainly concerned with guest flows and their consumption contributions and, thus, the generation of regional impact to the gross domestic product, thereby creating tourism intensity indicators of the destination. However, for the post-COVID perspective, tourism can be much stronger strategically visioned as a cross-sectoral instrument of regional development, particularly in the seam of the formation of infrastructure, innovative projects, advanced recreation themes and local leisure products. In this context, authentic, native and unique tourism potentials of the destination help the region to become recreationally more attractive, which is advantageous for overall regional marketing and also enables derivation of unique selling propositions (USP) of the tourism product portfolio of the destination. At the same time, these well-recognized contributions of tourism, especially formation of infrastructure and emerging recreational services implemented as a result of innovative tourism development,

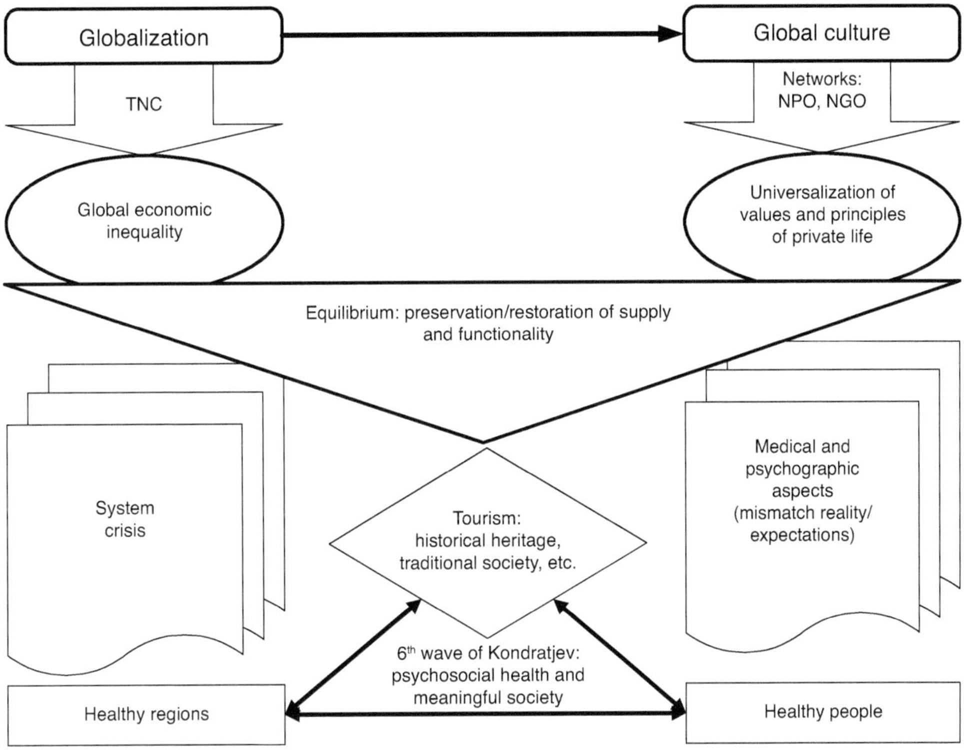

Fig. 4.1. Transformative framework conditions: perspectives for design of nature-based and wellness tourism products for destination resilience. TNC = transnational corporations, NPO = non-profit organizations, NGO = non-governmental organizations.

increase the locational quality of the municipality itself and, thus, the welfare of the local community (Bec *et al.*, 2015; Hughes and Scheyvens, 2018; Cheer *et al.*, 2019) and the entire touristic region.

Product development approaches in nature-based and wellness tourism for building on destination resilience

Nature-based and wellness tourism is visioned as the core tourism theme for development of prosperous, competitive and resilient destinations which operate within a promising long-term perspective in the framework conditions of the post-COVID-19 era (Kutzner, 2019; Fontanari *et al.*, 2021). These framework conditions should be analysed in the context of conceptual and empirical prerequisites which are to be elaborated on in tourism in the present crisis times for re-evaluation and transition of the global tourism business. Fig. 4.1 presents the transformative framework conditions and development prerequisites which provide the argumentation for the growing demand in resilience-oriented nature-based and wellness tourism products.

The transformative framework conditions for tourism development highlight that multinational, transnational and international companies that benefitted from the standardization of products (Benedikter, 2013) have brought critical economic impacts on culture and society: concentration and dependence on supply and sales processes; inequality in economic power and concentrated global control; the erosion of national economies and social, political and cultural institutions; emergence of a 'global culture'; and universalization of new personal values, with the keywords of 'career', 'success', 'welfare' and 'publicity' (Traskevich and Fontanari, 2018). These critical framework preconditions stand in the background of the present tremendous societal change and related disruptions of human health and longevity (such as loneliness, burnout, depression, unhappiness) generated by the COVID-19 pandemic and later crisis events (war in Europe, inflation, energy crisis, etc.). This is due to the fact that the framework conditions have had an enormous long-term impact on inherited, natural and personal aspects of life while the recent crises themselves have only worked as catalysts to reveal the collision of those social and cultural changes.

In this context, nature-based and wellness tourism products create the top tourism themes which work substantially against the aforementioned critical social and cultural developments. At the same time, nature-based and wellness tourism products are largely detached from global supply dependencies as they are effectively designed by relying on inner strengths and authentic potentials of the region and imply safer supply structures and healthier and more meaningful livelihoods (Chen *et al.*, 2019). This is how nature-based and wellness tourism themes can be envisaged as the core tourism themes for development of destination resilience. Furthermore, the new groundbreaking paradigm of 'healthy destinations' arises for transformative tourism development, which refers to the landscapes and recreational activities that allow both locals and guests to deal with present critical social, political and economic developments at different personal and social levels of understanding (Fontanari, 2016); this allows them to shape the new resilient awareness of the necessity of holistic health (Traskevich and Fontanari, 2018).

Tourism product development amid the paradigm of healthy and resilient destinations can attain the transformative meaning for the global tourism industry. This is on condition that the transformation is implemented within the sixth *Kondratjev wave* guided by the principle of *meaning* (Traskevich and Fontanari, 2018). The society that will overcome the present challenging framework conditions will further focus on meaningful life routines and life-fulfilling interactions as well as the preservation and reactivation of value-based social structures in a networking and sharing society. In this case, the impact of nature-based and wellness tourism themes will inevitably grow by highlighting the core transformation of touristic awareness towards destinations' historical heritage, traditional society, indigenous knowledge, and the identification of guests and locals; thus serving as the bridge for a new emotional and spiritual impetus for tourists and the local community (Powell *et al.*, 2018) bringing to the framework of a meaningful society.

Methodological Approach for Resilient Tourism Product Development: Code Red Model for an Applied Design Framework

In its theoretical background, this chapter considers the conceptual integrative models of destination governance targeted to long-term resilience and sustainability of tourism regions (Traskevich and Fontanari, 2021). Particularly, Code Red Model (Traskevich and Fontanari, 2018) is envisaged as a theoretical model for integrative holistic development of resilient destinations through nature-based and wellness tourism. The Code Red Model stands for a Convergent development model for the design of tourism products in resilient destinations. It presents the conceptual idea to elaborate on transitive qualities which provide movement from subjective personal characteristics of wellbeing to objective systematizing features of tourism products of nature-based and wellness tourism themes. At the same time, these products target the development of destination resilience. This way, the Code Red Model reveals the evolution of attributive characteristics for development of both destination resilience and USP of nature-based and wellness tourism products. The model establishes qualitative interconnections for applied design of tourism products as shown on Fig. 4.2.

Converging emotions and creativity and USP are designed in tourism products by addressing self-development, fulfilment and flourishing in tourism motivations. This convergence is implemented by the endeavour to exploit the abilities and skills of both tourists and local community with regard to exchanging and sharing their best human characteristics and values

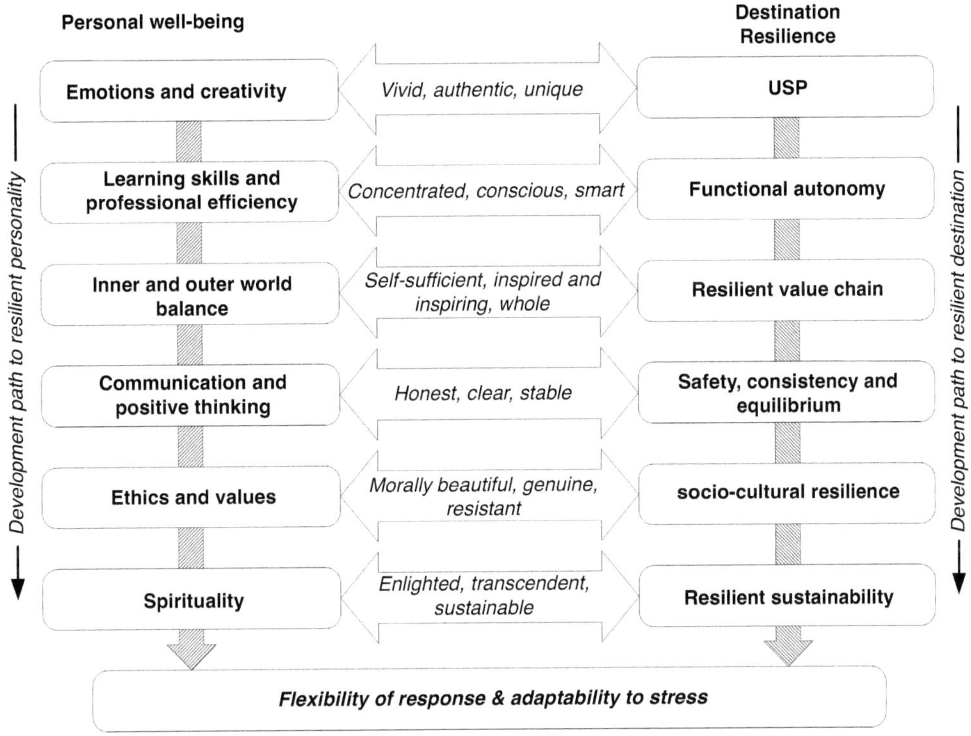

Fig. 4.2. Code Red Model: *c*onvergent *de*velopment model for the design of tourism products in *re*silient *d*estinations.

through recreational activities. Learning skills, professional efficiency and functional autonomy is implemented in the product design approach through consistent development of new nature-related skills and wellness experiences which are still targeted to lifestyle efficiency for the daily life of tourists, even when coming back home. Inner and outer world balance and resilient value chain is merged in the product design by means of creating and staging authentic livelihoods of the destination (Zheng *et al.*, 2020), thus expanding a tourist's spectrum of experiences and with it the idea of the holistic nature of things and states. Communication and positive thinking, converged with safety, consistency and equilibrium, are actualized in the product development approach through building on more stable and resilient relationships with other people.

Ethics and values and socio-cultural resilience are converged in the proposed product development approach by application of authentic indigenous knowledge and values in tourism product development (Amoamo *et al.*, 2018), which results in shaping for tourists and preserving for locals the alternative (authentic) lifestyle patterns. Spirituality converged towards resilient sustainability is visioned in the present design approach as the most intensive level of personal health and sustainable attitude to life (Cornelisse, 2019). Hence, the tourism products are designed with the overarching goal that residents and guests create a generic style of living and working together with higher goals than individual ones, offering thereby a new wellness experience of conscious lifestyle. The ambience of the nature-based and wellness tourism products implies the focused embedding of humanity in nature for guests to reflect themselves to be loving creatures of a creative force based on eternal love.

The core precondition of the applied implementation of the Code Red Model to the design of nature-based and wellness tourism products is that tourists are intensively involved at the product level, interacting positively with the local suppliers and a tourism community. The product development framework based on the Code Red Model is presented in Table 4.1.

It can be concluded that in terms of product development, personal resilience can act as both a clear pattern and the key generating factor for the development of destination resilience (Wang *et al.*, 2022; Barbhuiya and Chatterjee, 2022). At the same time, resilient characteristics of the destination inspire personal perceptions of wellbeing (Cohen, 2009) for the tourists. This way, the destination livelihood communicates to tourists the patterns and values of resilient lifestyle. Therefore, the transition and evolution of subjective qualities of personal resilience into more general and objective features of the destination resilience becomes essential. The reverse impact of a resilient destination on the wellbeing of the guest works inter-relatedly in product design. This way, the Code Red Model provides the methodological basis to elaborate on new products within the nature-based and wellness tourism themes. Moreover, these themes receive strategic thematic focus for destination resilience (Traskevich and Fontanari, 2018, pp. 200–202; Rauschen *et al.*, 2022).

Empirical Implications: Resilient Product Development in Experimental Nature-based Destinations in Belarus

The aforementioned design framework is applied for resilient product development in experimental nature-based destinations. Belarus is chosen as a useful case study for the application of the methodology because the tourism potentials as well as touristic profiles and images of the country remain unexplored and undefined for the outside world (Reshetnikov, 2011). At the same time, the local people still possess the inherited native knowledge on resilience and employ it in their daily lives. This creates potentials for free and creative experiments for resilient product development. And these inventive potentials are especially rich in the thematic orientation of nature-based and wellness tourism due to the diverse natural recreational resources of the emerging destinations in Belarus (Reshetnikov and Traskevich, 2011). In addition, rural destinations in Belarus have significant prerequisites for the implementation of functional autonomy and self-sufficiency of the value chain of tourism products. Unlike urban destinations, rural destinations of Belarus remain unattractive for foreign investors (international hotel chains, spa chains, etc.), which excludes excessive dependence on global suppliers and globalized structures formed in mass international tourism (Danskikh and Traskevich, 2018).

As a case-study illustration of the empirical possibilities to apply the presented design framework, the emerging nature-based destination Augustow Canal is chosen due to its high recreational value and resource potential. This includes both authentic cultural factors and natural assets which can generate the unique touristic identity and the potential diversity of genuine experiences.

The pilot products of resilient nature-based and wellness tourism were elaborated in the course of the international development project funded by the EU: 'Support to sustainable tourism development in Belarus, project No. ENPI/2015/365-203'. The main focus of the project was on rural areas of Belarus. The main project partner was the Ministry of Sports and Tourism of the Republic of Belarus. Other partners have been several relevant governmental institutions, regional and local authorities, and associations and NGOs linked to tourism.

Within the project consultation for the present case study, seven expert meetings were held in the period from 11 April to 25 April 2017, with service providers of Grodno region and the Augustow Canal. They were professionally affiliated with diverse locally based development projects and nature-based and wellness tourism products. Prevailingly, they came from the hotel industry and local tourism administration. At the same time, the special feature of the expert consultation process was the possibility to involve independent tourism enthusiasts from the local communities of the destination. The impact of this was higher motivation for effective locally based solutions, the overall friendly atmosphere for creativity, native co-operative commitment for product development, and precise targeted integration of the local knowledge within the development process. Two larger steering meetings followed in June and August,

Table 4.1. Design framework for resilient product development on the basis of the Code Red Model.

Areas of convergence		Tourism activities	Product development prerequisites
Emotions and creativity	USP	Cognitive thinking, emotional self-expression, creativity in revealing one's innate uniqueness	Establishment of self-identification of local communities with the core brand values of the destination and its USP portfolio so that their personal uniqueness corresponds with the USP of the destination
Learning skills, professional efficiency	Functional autonomy	Nature-based and wellness activities which trigger complete consciousness and engagement of tourists	Mental and emotional focus to be drawn away from ongoing lifetime challenges towards the real-life recreational activities
Inner and outer world balance	Resilient value chain	Recreational elements which trigger tourists' inner imaginary world for the mental and emotional balance and positive thinking in harmony with a tourist's views, experiences and values	Tourists and locals are required to align themselves self-sufficiently with their environment, thus contributing in a synergistic way to development of destination resilience
Communication and positive thinking	Safety, consistency and equilibrium	The recreational activities in which tourists learn to discover and express themselves	Elaboration on a stable society on the local level of the community: consolidation, stability, consistency, confidence and fruitful positive exchange; authentic community relationships
Ethics and values	Socio-cultural resilience	Targeted recreational activities evolve indigenous knowledge and beliefs: they are communicated within the tourism products and are encouraged to be further integrated into the daily routines of tourists	Intangible assets of the destination and its community are staged authentically but are not commoditized
Spirituality	Resilient sustainability	Nature-based and wellness recreational activities embrace the conscious exercise on perception of beauty, self-understanding and reflection on the surrounding world, setting up spiritual goals for life and acceptance of challenges for personal growth	Spiritual level of resilience development requires for the destination that the local residents themselves overcome 'individualism' and believe in the healing power of nature and the community of people who feel well and are happy to live their life in accordance with spiritual and natural laws

each attended by eight subject-matter experts from the region. This way, the product designs were evaluated and refined for feasibility.

As a result, the clear preconditions have been discovered to implement the concept of destination resilience. Particularly, the field studies identified the favourable opportunities for product design under the paradigm 'Back to nature and naturally healthy life'. New tourism products have been designed within the presented design framework on the basis of the Code Red Model (Table 4.1) to facilitate the practical implementation of the concept of resilience in the rural destination Augustow Canal. At the same time, the tourism products were targeted to increase the competitiveness and attractiveness of the product portfolio in this authentic natural region. However, the elaborated tourism products only remain conceptually embedded into the perspective tourism offer of the emerging destinations. This can be explained by the reluctant attitude and the lack of competence of the tourism policy-makers in Belarus (Slocum, 2021). Co-operative resilience commitment is further required for the real business implementation of these suggested products (Traskevich and Fontanari, 2018). Nevertheless, the portfolio of thematic nature-based and wellness tourism products introduced to the destination Augustow Canal are further described to illustrate the authors' methodological approach to tourism product design for formation of destination resilience.

Product 1: combined canoeing tours 'Paddle and Pedal'

At the destination Augustow Canal, the best combinations for water, bicycle and outdoor tourism are offered. Cycling paths parallel to rivers and canals, like the Augustow Canal, are ideal for the development of this nature-based and wellness product. Such a combination tour (e.g. in the context of a weekend package offer) could be as follows. From the hotel, the tourists travel to the starting point of the canoe tour with their own or borrowed bicycles and a special tour map. A transporter brings the bikes to the end point of the route during the canoeing so that the group can return to the hotel by bicycle. On the way, sufficient resting times and regional food are provided. The accommodation is in one or several simple to relatively comfortable hotels or rural farmsteads. This would meet the general demand for canoe travel to provide basic to more comfortable all-round service and package deals. At the same time, it can inspire the growth of entrepreneur activity among the local communities of the rural areas of the Augustow Canal.

Product 2: outdoor adventuring

Classic adventure tourism means exploring pathless terrain without orientation and is targeted at advanced nature-based tourists. The destination Augustow Canal offers special places for niche tourists who can, first and foremost, serve as tourism brand ambassadors for this experimental resilient destination. Nevertheless, this specific market niche does not deliver the expected economic value within the feasible range. Therefore, basic infrastructure and additional services are required to implement this product. Outdoor adventuring may attract groups rather than individuals. In this instance, a basic package relies on steady infrastructure and rental services.

One of the tourist's motives for these activities is the positive impact of co-operation experience for self-development. This idea can help to mitigate the typical gender- and age-related family problems by providing understanding that people cannot physically survive without each other. In this niche segment, the destination Augustow Canal could be specialized and differentiated from competing destinations by focusing on those interested in outdoor recreational activities who have little experience in this field. This target accepts a minimal standard of infrastructure located in a radius of 5 km, each situated on so-called 'base sites'. This is consistent with the current emphasis on environmentalism and meets the increasing significance of sustainability. Thus, the 'base sites' are designed for the present product with regard to the following core elements: basic routes and themes for 2 days' outdoor experience; weatherproof overnight accommodation with own sleeping bag and backpack; fireplaces; minimal sanitation; accessible water supply (drinking); high seats for night experiences (bird/animal watching); transportation service from homestead or hotel to the start of outdoor tour and pick-up; basic outfit/backpack for rental and related services. An environmentally friendly waste management concept is considered as outdoor tourists deal with waste responsibly. In the case of disorientation, an emergency transmitter should be available to rent from tourist information or local rangers. In the context of coaching or introduction to the outdoor standards, the rangers have to demonstrate the relationship between tourism and biodiversity, where duties and opportunities, but also dangers, lie.

Product 3: survival camps

A resilience-oriented segment that has so far been underestimated but is increasingly gaining demand is the tourism theme of survival. An increase in the outdoor-oriented form of the undisturbed and original 'naturally authentic' experience would be the settlement of survival camps. Here, the multi-day stay under dispensable conditions plays a central role. The core niche tourist motivation lies in experiencing one's own survival ability to adapt to the daily survival struggle in nature and also developing personal resilience skills applied to fire, drinking water, knotting and orienteering in the forest. Thus, an ultimate package for the tourism product is the outdoor survival camp (2–5 days) which includes training in theory and practice on various survival challenges. Particularly, the following product elements are considered: construction of emergency accommodation, orienteering in the forest and terrain, making fire under difficult conditions (wood drill, fire stone, etc.), instruction on ropes or knots, getting food from nature (searching the right classification of edible plants, flowers and roots), purifying water, production of tools and vessels for survival in the wild, rope-bridge construction, construction elements for rafting, riding in expedition vehicles, etc.; but also, survival psychology in the realm of psychological resilience (behaviour in emergency situations, motivation and survival strategies) as well as survival medicine, outdoor first-aid, medicinal plants and survival hygiene. These resilience skills are acquired by tourists in the form of edutainment. The range of variations and also innovations is conceivably large and can be put together individually according to group requirements (Fontanari, 2000). Other tourism products may be easily combined with the described one.

Discussion

Theoretical contribution: experimental and transformative destinations for tourism resilience

The first lessons learned from the empirical implementation of the presented methodological approach to design of resilient tourism products evidence that confident and intensive co-operation among stakeholders of nature-based and wellness destinations is essentially required. This should eventually shape a sufficient dimension of destination resilience – *socio-cultural resilience* (Diaferia *et al.*, 2022). While the previous research proves the multi-dimensional knowledge-based nature of destination resilience (Cochrane, 2010; Davoudi *et al.*, 2012; Butler, 2017; Traskevich and Fontanari, 2021), the present study on resilient product development proves that the focus of the destination portfolio on nature-based and wellness tourism products that are based on centuries-old knowledge and values of local people strengthens the resilience of the destination. The present empirical research discusses that such tourism products are resilient because they are timeless. They are not attractive in terms of the value system characteristic of a mass consumption culture, and are, therefore, not dependent on global production links, as well as fluctuations in consumer tastes and tourist demand. Tourists for whom such tourism products have attractiveness are characterized by high loyalty to destinations, returning even in spite of possible external shocks.

The previous fundamental research on resilience in tourism argues for integrative, positively oriented approaches to its long-term strategic implementation (Calgaro *et al.*, 2014; Dahles and Susilowati, 2015; Bec *et al.*, 2015;

Hall *et al.*, 2018; Innerhofer *et al.*, 2018; Fontanari *et al.*, 2021; Gössling *et al.*, 2021; Fontanari and Traskevich, 2023). In this context, the theoretical contribution of the present study is to bring relevant discussion on entrepreneurial resilience agility, which is to be developed through long-term planned and targeted managerial activities. These corporate resilience-oriented efforts include building on resilience-awareness, co-operative resilience commitment and resilience-competence advanced also among tourism stakeholders and local community, as well as local and national tourism administration. This sets up the core precondition for resilient product development which would still be realized in the strong realm of feasibility and competitiveness for the tourism product portfolio of the destination.

This way, the empirical findings highlight two new dimensions for integrative resilient product development which clarify the term of an experimental destination. The first dimension is comprised at the level of tourism products which go beyond the claims of a classic understanding of sustainability. A further dimension is constituted in integration of the locals in the service production. Such involvement is based on their native traditional, cultural and recreational competences, which are required to ensure feasibility of the implementation of resilient products. From this perspective, the present research *refines the definition of the experimental destination* as a tourism region which encounters a high potential of adaptation and advancement of the product concept due to the tourism experiences generated through authentic involvement of native service providers and locals.

Furthermore, the present study contributes to the academic understanding of the transformative approach to destination development (Cave and Dredge, 2018; Sigala, 2020; Farmaki and Pappas, 2021; Ateljevic and Sheldon, 2022; Dredge, 2022; Hall, 2022; Nandasena *et al.*, 2022; Sharpley, 2022) by highlighting the notion on transformative destinations which are characterized by the following iterative elements of resilient product development. According to the trial-and-error principle, when resilience-oriented products are designed and offered, the respective evaluation and acceptance on the part of service providers, tourists and guests is transferred into a synthesizing process of product development. This results in continuous optimization processes of the products themselves, which leads to a higher tourism acceptance on the part of the local community and a higher touristic demand on the part of the guests, thus bringing *the transformative destination* to growth and success. This also leads to a change or further development of the infrastructure and supply services within the destination. This changes the destination itself and transforms it into a more resilient state.

Industrial implications: practical replicability of the product development framework and wider design perspectives

The presented methodological approach is further analysed in terms of its business replicability and wider design perspectives for development of resilient tourism products of nature-based and wellness tourism. This is with the aim to introduce the methodology to targeted nature-based destinations which aim to follow the transformative perspectives of sustainability, responsible regional prosperity, inclusive community-based entrepreneurship, innovative business models and alternative resilient value creation for tourism products (Fontanari *et al.*, 2021; Thees *et al.*, 2022). These positive crisis-driven perspectives are merged under the paradigm of long-term resilience and resilient product development with the focus on personality, wellbeing, personal and community identity, responsible communication of people with nature; and, ultimately, on building of resilient and healthy livelihoods and meaningful society. Based on the Code Red Model introduced in the methodological part of this chapter and the pilot empirical findings, the following *key principles of resilient product development* can be identified for universal replication of the presented methodological approach:

- building of mutually beneficial partnerships between all the stakeholders in rural regions, implemented on the principles of trust, common vision and mission of tourism development in the destination;
- intensification of the use of local recreational resources and involvement of tangible and intangible objects of tourist

interest, which preserve the identity and uniqueness of the rural areas;
- functional autonomy of the rural destinations through optimization and 'localization' of the supply chain;
- modernization of infrastructure on the basis of the concept of resilience;
- creation and promotion of diversified portfolio of sustainable resilience-oriented tourism products capable of generating a complex of tourism experiences and meticulously oriented to the target markets of domestic and international tourism;
- independent locally based decision-making processes in the destination management;
- minimum influence of global suppliers on the value chain of the tourism products of nature-based destinations; and
- the ability of the destination to draw on local knowledge and expertise to create a diversified portfolio of USP products.

In this framework, product design can begin with the creation of strategic working groups involving the main stakeholders of the destination, active representatives of the local community, as well as interested tour operators. The primary task of such working groups is to identify the priority components of destination resilience to be addressed. Further, the objectives and strategy for implementation of the concept of destination resilience are defined in the close context of established destination management. In accordance with the medium- and long-term prospects for the implementation of the concept of destination resilience, the areas of investment attraction are developed. Tourism development projects are proposed and approaches to the design of resilient tourism products are defined with regard to the sustainability factor. This step requires the pooling of resources and funding, which is recommended through public–private partnerships. The final stage is the development of innovative processes of creating value chains of tourism products of a resilient destination. In this case, each link in the value chain is implemented through the active and effective involvement of all interested partners (Fontanari and Fontanari, 2001). This ensures high reliability and stability of the offer of a competitive portfolio of resilient tourism products in the destination.

The applied framework for resilient product development incorporates the toolkit of empirical methods which allow for iterations of the design of tourism products targeted to responsible tourism entrepreneurship in experimental nature-based destinations. The range of these empirical approaches is structurally covered by the following design blocks: inventory analysis and assessment; territorial planning; strategic analysis; merged tourism themes for product development (deductive and integrative approach); product design and implementation; marketing, promotion and branding; co-operation and co-ordination; quality management; sustainability and resilience. The applied framework for resilient product development points the practical way for real-business implementation of the imperatives of sustainability, responsible building on competitiveness, and, ultimately, long-term resilience.

To communicate the described ideas broadly and to ensure the replicability of the present methodological approach, workshops on strategic destination management and operational tourism product development should take place. It is necessary that local stakeholders understand the core idea of resilient tourism products and how they can contribute efficiently with their specific elements of the entire locally based product value chain. The innovative factor within the resilient product development is to apply more to the target's needs, especially the ones who are not experienced in the proposed field of activity but are heavily interested in it. The suggested portfolio of nature-based and wellness tourism products has the potential to mark uniqueness for the experimental destination and to help in destination branding in the direction of competitive differentiation. For this case, the destination should offer a variety of nature-based specialized tourism products targeted for resilience-oriented niche markets. The more the value chain adds on complementary elements for leisure, entertainment and wellness recreation, the more visitors will be attracted to try these specialized resilient tourism products. Both sides of the ongoing tourism development process (tourism product development and destination management) have to be developed carefully and with full participation of private stakeholders and official institutions. Stakeholders have to be sensitized for both development paths to be motivated and prepared for the co-operative action and finally to gain synergies.

Conclusion

Methodological contribution of this study suggests that the following functional dimensions of resilience should be synthesized to attribute to resilient product development in nature-based and wellness tourism: community and personality resilience, entrepreneurial resilience, technological, engineering and infrastructural resilience, socio-ecological and socio-cultural resilience. These functional dimensions should be integrated into the holistic design framework which is applied for resilient product development as the Code Red Model suggests. The present study envisages the empirical application of the aforementioned design approach for the experimental nature-based destinations. The research provides novel notions on the definition of experimental and transformative destinations for resilient tourism product development.

Pilot empirical findings and their further employment for the case study destination confirm that experimental rural destinations possess excellent starting positions to gain competitiveness in niche resilience orientation within the described fields of nature-based and wellness tourism. In this combination, the destination should profile itself without heavy investments and related risks. Using the natural conditions in a sustainable way by offering a minimum of infrastructure requires a co-operative way of participation by private stakeholders and tourism administration.

The research contributes to the academic discussion on the global mega-trend to bring people in their leisure time back to nature and authentically green environment by enhancing the transformative meaning of resilience in the long term for sustainable tourism development. This imperative inspires the resilient product design towards introducing the new forms of sports, adventure and wellness activities in nature. Within the concept of resilience, these trendy product solutions receive also additional meaning and value as follows: high social mobility; leisure as a source of identity; self-realization and personal growth as the comprehensive component of both physical and mental wellness; increasing shift away from everyday life during leisure time achieved through immersive tourist experiences; elimination of everyday risks; avoidance of standardized forms of global culture and mass (over-) consumption in tourism; growing need for authentic and unadjusted natural experience; individualization of resilient tourism products; innovative techniques and equipment to facilitate the product implementation.

This study argues that the empirical implementation of the suggested design framework allows for experimental and emerging destinations of nature-based and wellness tourism to benefit from the presented approaches to design of resilient tourism products. At the same time, mature and successful destinations can also apply for the presented methodological approach within the Code Red Model, in case they have very strong strategic motivation for transformation. This way, the suggested principles for resilient product development are applicable for both experimental and transformative destinations of nature-based and wellness tourism.

The presented empirical approach for resilient product development prioritizes the following core sub-themes of nature-based and wellness tourism: mental wellness, spiritual tourism, community-based cultural tourism, innovative approaches to ecotourism and alternative, responsible forms of nature-escape tourism. This aims at a portfolio of tourism products with the intangible added value based on the communication between tourists, nature and the community. The present methodological design approach identifies the highest priorities as being the preservation of healthy ambience and unique ecosystems, as well as personal physical, mental and spiritual wellbeing.

References

Amoamo, M., Ruckstuhl, K. and Ruwhiu, D. (2018) Balancing indigenous values through diverse economies: a case study of māori ecotourism. *Tourism Planning & Development* 15(5), 478–495. DOI: 10.1080/21568316.2018.1481452.

Ateljevic, I. and Sheldon, P.J. (2022) Guest editorial: transformation and the regenerative future of tourism. *Journal of Tourism Futures* 8(3), 266–268. DOI: 10.1108/JTF-09-2022-284.

Backer, E. and Ritchie, B.W. (2017) VFR travel: a viable market for tourism crisis and disaster recovery? *International Journal of Tourism Research* 19(4), 400–411. DOI: 10.1002/jtr.2102.

Barbhuiya, M.R. and Chatterjee, D. (2022) Just survive or thrive? Effect of psychological and organizational resilience on adoption of innovative strategies by hospitality sector post covid-19. *Tourism Planning & Development* 20(2), 188–211. DOI: 10.1080/21568316.2022.2121312.

Bec, A., McLennan, C. and Moyle, B.D. (2015) Community resilience to long-term tourism decline and rejuvenation: a literature review and conceptual model. *Current Issues in Tourism* 19(5), 431–457. DOI: 10.1080/13683500.2015.1083538.

Benedikter, R. (2013) Global systemic shift: a multidimensional approach to understand the present phase of globalization. *New Global Studies* 7(1), 1–15. DOI: 10.1515/ngs-2012-005.

Benli, S. and Kuhzady, S. (2017) Toward a new paradigm in sustainable tourism: trend and pattern analysis in tourism resilience researches. In: *Proceedings of the 1st International Sustainable Tourism Congress*, Kastamonu University, pp. 974–980.

Biggs, D., Hall, C.M. and Stoeckl, N. (2012) The resilience of formal and informal tourism enterprises to disasters: reef tourism in Phuket, Thailand. *Journal of Sustainable Tourism* 20(5), 645–665. DOI: 10.1080/09669582.2011.630080.

Bürkner, H.-J. (2010) Vulnerabilität und Resilienz – Forschungsstand und sozialwissenschaftliche Untersuchungsperspektiven [Vulnerability and resilience: state of research and social science perspectives] Working Paper 43.

Butler, R.W. (ed.) (2017) *Tourism and Resilience*. CAB International, Wallingford, UK.

Calgaro, E., Lloyd, K. and Dominey-Howes, D. (2014) From vulnerability to transformation: a framework for assessing the vulnerability and resilience of tourism destinations. *Journal of Sustainable Tourism* 22(3), 341–360. DOI: 10.1080/09669582.2013.826229.

Camilleri, M.A. (2016) Responsible tourism that creates shared value among stakeholders. *Tourism Planning & Development* 13(2), 219–235. DOI: 10.1080/21568316.2015.1074100.

Cave, J. and Dredge, D. (2018) Reworking tourism: diverse economies in a changing world. *Tourism Planning & Development* 15(5), 473–477. DOI: 10.1080/21568316.2018.1510659.

Cheer, J.M., Milano, C. and Novelli, M. (2019) Tourism and community resilience in the Anthropocene: accentuating temporal overtourism. *Journal of Sustainable Tourism* 27(4), 554–572. DOI: 10.1080/09669582.2019.1578363.

Chen, F., Xu, H. and Lew, A.A. (2019) Livelihood resilience in tourism communities: the role of human agency. *Journal of Sustainable Tourism* 28(4), 606–624. DOI: 10.1080/09669582.2019.1694029.

Cochrane, J. (2010) The sphere of tourism resilience. *Tourism Recreation Research* 35(2), 173–185. DOI: 10.1080/02508281.2010.11081632.

Cohen, M. (2009) Spas, wellness and human evolution. In: Cohen, M. and Bodeker, G. (eds) *Understanding the Global Spa Industry: Spa Management*. Butterworth Heinemann, Oxford, UK, pp. 4–25.

Cornelisse, M. (2019) Moral claims in sustainable tourism development. *Tourism Planning & Development* 17(4), 1–19. DOI: 10.1080/21568316.2019.1622591.

Dahles, H. and Susilowati, T.P. (2015) Business resilience in times of growth and crisis. *Annals of Tourism Research* 51, 34–50. DOI: 10.1016/j.annals.2015.01.002.

Danskikh, S. and Traskevich, A. (2018) Cooperative approaches between health- and agro-tourism stakeholders in increasing rural destination resilience in Belarus. In: Innerhofer, E., Fontanari, M. and Pechlaner, H. (eds) *Destination Resilience: Challenges and Opportunities for Destination Management and Governance*. Routledge, Oxford, UK, pp. 77–87. DOI: 10.4324/9780203701904.

Davoudi, S., Shaw, K., Haider, L.J., Quinlan, A.E., Peterson, G.D. *et al.* (2012) Resilience: a bridging concept or a dead end? 'Reframing' resilience: challenges for planning theory and practice interacting traps: resilience assessment of a pasture management system in northern afghanistan urban resilience: what does it mean in planning practice? resilience as a useful concept for climate change adaptation? the politics of resilience for planning: a cautionary note. *Planning Theory & Practice* 13(2), 299–333. DOI: 10.1080/14649357.2012.677124.

Diaferia, V., Traskevich, A. and Fontanari, M. (2022) Sociocultural dimension of destination resilience and its implication for innovative product design in experimental destinations of Germany, Italy and Ireland. In: Andriotis, K., Cardoso, C.P. and Stylidis, D. (eds) *Tourism Development and Planning in Western Europe*. CAB International, Wallingford, UK and Boston, Massachusetts, pp. 111–126. DOI: 10.1079/9781800620797.0000.

Dredge, D. (2022) Regenerative tourism: transforming mindsets, systems and practices. *Journal of Tourism Futures* 8(3), 269–281. DOI: 10.1108/JTF-01-2022-0015.

Espiner, S., Orchiston, C. and Higham, J. (2017) Resilience and sustainability: a complementary relationship? Towards a practical conceptual model for the sustainability–resilience nexus in tourism. *Journal of Sustainable Tourism* 25(10), 1385–1400. DOI: 10.1080/09669582.2017.1281929.

Farmaki, A. and Pappas, N. (eds) (2021) *Emerging Transformations in Tourism and Hospitality*. Routledge, London. DOI: 10.4324/9781003105930.

Filimonau, V. and De Coteau, D. (2020) Tourism resilience in the context of integrated destination and disaster management. *International Journal of Tourism Research* 22(2), 202–222. DOI: 10.1002/jtr.2329.

Fontanari, M. (2000) *Trends und Perspektiven für die Destinationsentwicklung und das Destinations management. Wettbewerb europäischer Destinationen [Trends and perspectives for destination development and destination management. Competition between European destinations]*. Gabler Verlag.

Fontanari, M. (2016) Zukunftsfaktor Resiliente Gesellschaft. [resilient society as a factor for the future]. *JA! – Magazin Für Positive Und Gesunde Lebenswerte* 10, 8–9–26–27.

Fontanari, M. and Fontanari, M. (2001) The role of tourism as an instrument for holistic regional development: regional integration, regional identity and economic regional development in Europe. *Tourism in the Europe of Regions* 4, 59–80.

Fontanari, M. and Kredinger, D. (2017) Risiko- und Resilienzbewusstsein. Empirische Analysen und erste konzeptionelle Ansätze zur Steigerung der Resilienzfähigkeit von Regionen [risk and resilience awareness. empirical analyses and first conceptual approaches to increase the resilience of regions]. *ISM Working Paper 9*.

Fontanari, M. and Traskevich, A. (2023) Smart-Solutions for handling overtourism and developing destination resilience for the post-Covid-19 era. *Tourism Planning & Development* 20(1), 86–107. DOI: 10.1080/21568316.2022.2056234.

Fontanari, M., Traskevich, A. and Kutsch, H. (2021) Corporate resilience within tourism enterprises. In: Valeri, M., Scuttari, A. and Pechlaner, H. (eds) Resilienza e sostenibilitá: Dinamiche globali e risposte locali [Resilience and sustainability: global dynamics and local responses]. G. Giappichelli Editore, Turin, Italy, pp. 77–94.

Gössling, S., Scott, D. and Hall, C.M. (2021) Pandemics, tourism and global change: a rapid assessment of COVID-19. *Journal of Sustainable Tourism* 29(1), 1–20. DOI: 10.1080/09669582.2020.1758708.

Hall, C.M. (2022) Tourism and the capitalocene: from green growth to ecocide. *Tourism Planning & Development* 19(1), 61–74. DOI: 10.1080/21568316.2021.2021474.

Hall, C.M., Prayag, G. and Amore, A. (eds) (2018) *Tourism and Resilience: Individual, Organisation and Destination Perspectives*. Channel View Publications, Clevedon, UK. DOI: 10.21832/9781845416317.

Hartard, S. (2014) Resilienz durch nachhaltige Ressourcenwirtschaft [Resilience through sustainable resource management]. In: Hartard, S., Schaffer, A. and Lang, E. (eds) *Systeme in Der Krise Im Fokus von Resilienz und Nachhaltigkeit [Systems in Crisis in the Focus of Resilience and Sustainability]*. Metropolis, pp. 201–223.

Holladay, P.J. and Powell, R.B. (2013) Resident perceptions of social–ecological resilience and the sustainability of community-based tourism development in the Commonwealth of Dominica. *Journal of Sustainable Tourism* 21(8), 1188–1211. DOI: 10.1080/09669582.2013.776059.

Hughes, E. and Scheyvens, R. (2018) Development alternatives in the pacific: how tourism corporates can work more effectively with local communities. *Tourism Planning & Development* 15(5), 516–534. DOI: 10.1080/21568316.2018.1478881.

Innerhofer, E., Fontanari, M. and Pechlaner, H. (eds) (2018) *Destination Resilience: Challenges and Opportunities for Destination Management and Governance*. Routledge, London. DOI: 10.4324/9780203701904.

Jamaliah, M.M. and Powell, R.B. (2018) Ecotourism resilience to climate change in Dana Biosphere Reserve, Jordan. *Journal of Sustainable Tourism* 26(4), 519–536. DOI: 10.1080/09669582.2017.1360893.

Jarratt, D. and Davies, N.J. (2019) Planning for climate change impacts: coastal tourism destination resilience policies. *Tourism Planning & Development* 17(4), 1–18. DOI: 10.1080/21568316.2019.1667861.

Jiang, Y., Scott, N. and Ritchie, B.W. (2018) Crisis management and strategic implications. In: Moutinho, L. and Vargas-Sanchez, A. (eds) *Strategic Management in Tourism.* CAB International. Wallingford, UK, pp. 264–279. DOI: 10.1079/9781786390240.0000.

Jiang, Y., Ritchie, B.W. and Verreynne, M.L. (2019) Building tourism organizational resilience to crises and disasters: a dynamic capabilities view. *International Journal of Tourism Research* 21(6), 882–900. DOI: 10.1002/jtr.2312.

Kato, K. (2018) Debating sustainability in tourism development: resilience, traditional knowledge and community: a post-disaster perspective. *Tourism Planning & Development* 15(1), 55–67. DOI: 10.1080/21568316.2017.1312508.

Kutzner, D. (2019) Environmental change, resilience, and adaptation in nature-based tourism: conceptualizing the social-ecological resilience of birdwatching tour operations. *Journal of Sustainable Tourism* 27(8), 1142–1166. DOI: 10.1080/09669582.2019.1601730.

Lew, A.A. (2014) Scale, change and resilience in community tourism planning. *Tourism Geographies* 16(1), 14–22. DOI: 10.1080/14616688.2013.864325.

Lew, A.A., Ng, P.T., Ni, C.-C) and Wu, T.-C) (2016) Community sustainability and resilience: similarities, differences and indicators. *Tourism Geographies* 18(1), 18–27. DOI: 10.1080/14616688.2015.1122664.

Mandal, S. and Saravanan, D. (2019) Exploring the influence of strategic orientations on tourism supply chain agility and resilience: an empirical investigation. *Tourism Planning & Development* 16(6), 612–636. DOI: 10.1080/21568316.2018.1561506.

Nandasena, R., Morrison, A.M. and Coca-Stefaniak, J.A. (2022) Transformational tourism – a systematic literature review and research agenda. *Journal of Tourism Futures* 8(3), 282–297. DOI: 10.1108/JTF-02-2022-0038.

Osztovics, W., Kovar, A. and Mayrbäurl, C. (2012) *Arena Analyse 2012 – Resilienz oder Katastrophe [Arena analyse 2012 – resilienz oder katastrophe]*. Kovar & Köppl.

Pechlaner, H., Zacher, D. and Störmann, E. (eds) (2022) *Resilienz als Strategie in Region, Destination und Unternehmen: Eine raumbezogene Perspektive [Resilience as a strategy in region, destination and company: a spatial perspective]*. Springer Gabler Wiesbaden. DOI: 10.1007/978-3-658-37296-5.

Powell, R.B., Green, T.F., Holladay, P.J., Krafte, K.E., Duda, M. *et al.* (2018) Examining community resilience to assist in sustainable tourism development planning in Dong Van Karst Plateau Geopark, Vietnam. *Tourism Planning & Development* 15(4), 436–457. DOI: 10.1080/21568316.2017.1338202.

Prayag, G., Chowdhury, M., Spector, S. and Orchiston, C. (2018) Organizational resilience and financial performance. *Annals of Tourism Research* 73, 193–196. DOI: 10.1016/j.annals.2018.06.006.

Rauschen, L., Traskevich, A. and Fontanari, M. (2022) Strategic considerations for sustainable tourism development of the micro-destination East Belgium. In: Andriotis, K., Cardoso, C.P. and Stylidis, D. (eds) *Tourism Development and Planning in the Western Europe*. CAB International, Wallingford, UK and Boston, Massachusetts, pp. 80–94. DOI: 10.1079/9781800620797.0000.

Reshetnikov, D. (2011) *География туризма Республики Беларусь [Tourism Geography in Belarus]*. Cheture Chetverti, Minsk, Russia. [In Russian].

Reshetnikov, D. and Traskevich, A. (2011) Modern trends in development and spatial Organisation of sanatorium and SpA facilities of Belarus. *Proceedings of the Faculty of International Relations* 2, 102–113. [In Russian]

Ruiz-Ballesteros, E. (2011) Social-ecological resilience and community-based tourism: an approach from Agua Blanca, Ecuador. *Tourism Management* 32, 655–666. DOI: 10.1016/j.tourman.2010.05.021.

Sharpley, R. (2012) Tourism and vulnerability: a case of pessimism? *Tourism Recreation Research* 37(3), 257–260. DOI: 10.1080/02508281.2012.11081714.

Sharpley, R. (2022) Tourism and development theory: which way now? *Tourism Planning & Development* 19(1), 1–12. DOI: 10.1080/21568316.2021.2021475.

Sheller, M. (2020) Reconstructing tourism in the Caribbean: connecting pandemic recovery, climate resilience and sustainable tourism through mobility justice. *Journal of Sustainable Tourism* 29(9), 1436–1449. DOI: 10.1080/09669582.2020.1791141.

Siakwah, P., Musavengane, R. and Leonard, L. (2019) Tourism governance and attainment of the sustainable development goals in Africa. *Tourism Planning & Development* 17(4), 355–383. DOI: 10.1080/21568316.2019.1600160.

Sigala, M. (2020) Tourism and COVID-19: impacts and implications for advancing and resetting industry and research. *Journal of Business Research* 117, 312–321. DOI: 10.1016/j.jbusres.2020.06.015.

Sisneros-Kidd, A.M., Monz, C., Hausner, V., Schmidt, J. and Clark, D. (2019) Nature-based tourism, resource dependence, and resilience of Arctic communities: framing complex issues in a changing environment. *Journal of Sustainable Tourism* 27(8), 1259–1276. DOI: 10.1080/09669582.2019.1612905.

Slocum, S.L. (2021) The influence of government controlled dmos on destination competitiveness: the case of Belarus. *Tourism Planning & Development* 20(5), 870–884. DOI: 10.1080/21568316.2021.1958915.

Thees, H., Störmann, E. and Pechlaner, H. (2022) Business modeling for resilient destination development: a multi-method approach for the case of destination franconia, Germany. *Tourism Planning & Development* 20(2), 212–235. DOI: 10.1080/21568316.2022.2121313.

Traskevich, A. and Fontanari, M. (2018) Mental wellness in resilient destinations. *International Journal of Spa and Wellness Industry* 1(3), 193–217. DOI: 10.1080/24721735.2019.1596656.

Traskevich, A and Fontanari, M. (2021) Tourism potentials in post-COVID19: the concept of destination resilience for advanced sustainable management in tourism. *Tourism Planning & Development* 20(1), 12–36. DOI: 10.1080/21568316.2021.1894599.

Wall, G. (2018) Beyond sustainable development. *Tourism Recreation Research* 43(3), 390–399. DOI: 10.1080/02508281.2018.1475880.

Wang, H., Liu, L. and Sha, H. (2022) Exploring how the psychological resilience of residents of tourism destinations affected brand ambassador behavior during the COVID-19 pandemic. *Behavioral Sciences*12(9), 337. DOI: 10.3390/bs12090337.

Welter-Enderlin, R. and Hildenbrand, B. (2006) *Resilienz – Gedeihen trotz widriger Umstände [Resilience – thriving despite adverse circumstances*. Carl Auer, Heidelberg, Germany.

Zheng, D., Ritchie, B.W. and Benckendorff, P.J. (2020) Beyond cost–benefit analysis: resident emotions, appraisals and support toward tourism performing arts developments. *Current Issues in Tourism* 24(5), 668–684. DOI: 10.1080/13683500.2020.1732881.

5 Reviving the Abandoned Village of Vamvakou in Greece through Community-based Tourism with the Support of Private Funding: A Difficult Balancing Act?

Maria Goutou[1]* and Nicholas Karachalis[2]
[1]Hellenic Open University, Patras, Greece; [2]University of the Aegean, Mytilene, Greece

Abstract
This chapter critically discusses the role of sustainable cultural and tourism strategies and funding schemes in the revival of villages in rural and mountainous areas. Drawing on literature and case studies that refer to sustainable tourism development, nature-based tourism and the role of the social economy, the chapter presents the strengths and weaknesses of a privately-funded, non-governmental, bottom-up approach towards reviving an abandoned village in Peloponnese in Greece. The case study is analysed on the basis of field research regarding the village of Vamvakou in Laconia. The research outcomes are based on an analysis of the effort for its revival by the social co-operative enterprise 'Vamvakou Revival' with guidance and financial support from Stavros Niarchos Foundation (SNF), whose role is discussed critically. Finally, the demand side is presented through the tourists' wellbeing, environmental awareness and sensibility.

Introduction

Despite the fact that the pandemic slowed down global tourism flows for a long period, tourism is bouncing back, and the industry is witnessing high levels of demand since 2021. Arguably, the circumstances and priorities have changed as tourism's significant environmental impact has opened up a discussion on regenerative tourism within a sustainable way of travelling. As Bellato *et al.* (2022) argue, the main goal of regenerative tourism for visitors is to have a positive impact on their holiday destination by actually contributing to its regeneration, a concept that goes beyond the 'not damaging' principle and leads to a positive impact on nature, local communities and economies. It supports the idea that tourism should leave a place better than it was before and that it could improve the liveability of a place. Tourism motivation can also follow a similar pattern. Nandasena *et al.* (2023) suggest the term 'transitionary tourism' to describe the demand for tourism experiences that have a healing or life-changing effect and go beyond the more traditional vacation for recreation, while Fossgard and Fredman (2019) connect this to 'slow tourism' and a general change in the patterns of demand for travel. Community-led tourism development and sustainability are two key elements which are central to this

*Corresponding author: mariagoutou@hotmail.com

 Nature-based Tourism and Wellbeing: Impacts and Future Outlook (eds F. Niccolini *et al.*)
DOI: 10.1079/9781800621411.0005

discussion, while the digital transition and new technologies often can play a role in the way tourists interact with the places they visit (Cavicchi *et al.*, 2018; Buhalis and Sinarta, 2019). Village revival, within this context, is often connected to the regenerative role of tourism. At the same time, it is also connected to the digital nomads trend and the post-pandemic reappreciation of life in a less stressful environment, which offers a better quality of life and the opportunities offered by traditional activities (Dologlou, 2021; Koumara-Tsitsou and Karachalis, 2021).

The case study of Vamvakou in the Peloponnese is presented and discussed for promoting a sustainable cultural and tourism strategy for its revival, which is implemented by the social co-operative enterprise (SCE) ‘Vamvakou Revival’ with the guidance and financial support of Stavros Niarchos Foundation (SNF), focusing on participatory processes, stakeholder management and funding.

The Peloponnese is a very popular tourism destination on the Greek mainland with many natural and heritage sites, such as Olympia, Epidaurus, Mycenae, the ancient theatre of Epidaurus and Sparta. Vamvakou is a small mountain village in Laconia, which was built entirely out of stone in the 15th century at an altitude of 903 metres, on the slopes of Mount Parnonas near Sparta, Laconia. The village is steeped in historical and cultural heritage and rich in natural resources. Vamvakou was once home to hundreds of inhabitants – in fact, it was known for its large families – and now hosts only a handful of permanent residents as its demographics were severely affected by World War II, the civil war, migration and urbanization. The first steps to revive the village and attract new inhabitants within a nature-based, sustainable tourism development strategy came only recently at a moment when the pandemic crisis forced changes in work and everyday life, thus making the whole effort even more interesting to follow and discuss.

The current research essay aims to contribute to analysing examples of similar cases related to sustainable cultural and tourism strategies of bottom-up village revival and the role of participatory processes. Revival strategies of villages such as Torri Superiore and Riace in Italy, as well as similar examples of revival in Greece, like the village of Kapetaniana and Milia in Crete or Koryschades in Evritania, have also been considered in the analysis. The main key findings of this project are presented, followed by a discussion on the bottom-up approach and the funding model of this project. Reflections on the operation and on the approach of the SNF and the way the local community is involved were the central point of the interviews with key informants in the period from the summer of 2021 until the end of 2022. A limitation of the current research is that, due to COVID-19 safety measures, the operation and programming of the initiative were strongly affected during its launch period, resulting in not being able to evaluate its full potential – both as an initiative overall and specifically Vamvakou as a destination. Additional research is needed to further evaluate the impact, preferably after a 3 year operation period of the initiative.

The Revival of Vamvakou

Vamvakou (Figs 5.1 and 5.2) aims to become a destination where nature-based tourist activities are envisioned, planned, regulated, guided and oriented in a sustainable direction and managed by the local community. Tourism, the agricultural sector and agritourism, innovation, culture and education are at the centre of the revival programme. A fundamental element of Vamvakou's revival is the improvement and support of a business climate, allowing for entrepreneurial ideas to develop, often in connection with the authentic history and traditional activities of the village.

The effort was planned as a bottom-up approach since the main stakeholder responsible for the revival of Vamvakou is the SCE ‘Vamvakou Revival’. The organization was founded in 2018 by a group of citizens with strong ties to Vamvakou. Vamvakou Revival was established with guidance and financial support from the largest private cultural fund in Greece, the SNF, which is also responsible, among other things, for the largest cultural complex in Athens (Vamvakou Revival Website, 2023). The village is of special importance to the Vamvakou Revival group as each member has a personal connection to the village, as does the SNF, since it is the birthplace of its founder, Stavros Niarchos.

Fig. 5.1. Vamvakou Village. (Photo: private collection)

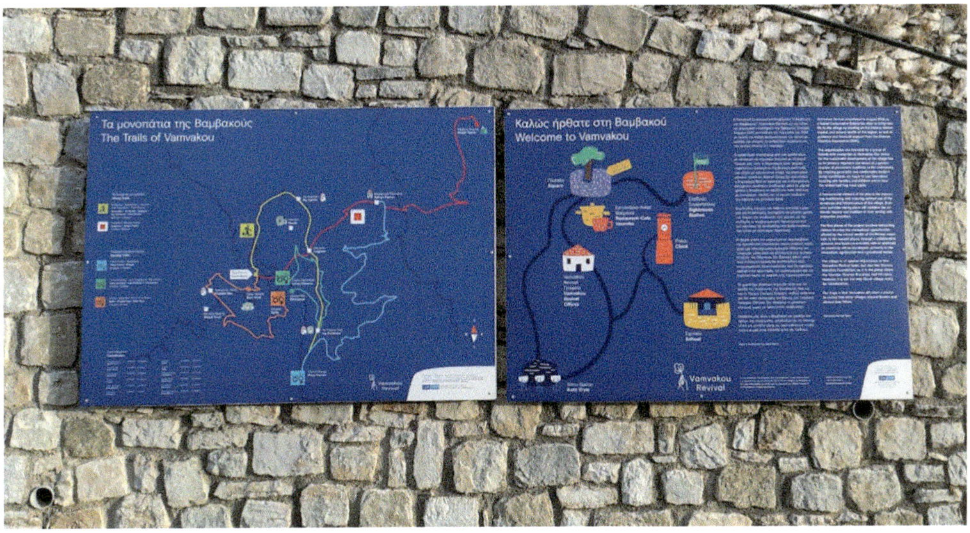

Fig. 5.2. Information Boards Vamvakou Village. (Photo: private collection)

The SNF 'adopted' the village and provided, through its network, support in educational and cultural programming. The idea was not to follow a top-down sponsorship programme but to support the local community in a bottom-up, participatory revival model. The vision for the sustainable development of Vamvakou included as a goal the gradual return of residents by creating jobs and living conditions based on wellbeing and environmental awareness. The goal is, in a few years, for Vamvakou to be inhabited by young people, families with children; therefore, a milestone is to reopen the village school, which closed down in 2008.

This chapter investigates the particular strengths and opportunities of the strategy of Vamvakou Revival. The content is based on a literature review of sustainable tourism practices, community-based tourism and its implications based on field research in Vamvakou conducted by the authors. The main research method leveraged information collected from initiators and participants through semi-structured interviews. It focuses on the elements that make this particular village's revival unique, in terms of management, and particularly interesting to examine further as a best practice. More particularly, 18 in-depth semi-structured interviews with selected key informants were conducted. The key informants principally represented the local community, the Vamvakou Revival and all the parties involved in this project, including residents, locals, local policy-makers, community leaders, experts in the field, professionals, visitors and artists. The main research question that was addressed is whether tourism can have a regenerative role in the revival of abandoned villages without opting for a hospitality-driven solution. Furthermore, two interviews with key informants from the Association for the Protection of Asterousia (APA) and experts from the region of Crete were conducted. The information was gathered from people who have relevant knowledge and can provide solid insights. At the end of each interview, a summary was prepared, and information was structured into key themes, issues and recommendations. As a result, several new points emerged: the regenerative role in the revival of abandoned villages with a bottom-up approach, the involving of local community regarding the village revival, and the role of a funding model.

Key Findings

Nature-based tourism as a solution

Comparing the case of Vamvakou with villages that have been revived elsewhere shows that it is quite a unique example since it follows a path focusing mostly on attracting inhabitants; it is not based on a tourism-centred approach. The benefits of connecting nature to wellbeing have been particularly clear from the beginning, as contact with nature and connectedness can offer self-development and transformative impacts (Clissold *et al.*, 2022). A key informant confirmed this, stating that 'back to village means back to nature and back to yourself'. Cases in Italy such as the ecovillage of Torri Superiore (Torri Superiore Website, 2023) follows a more experimental model in terms of everyday life with an emphasis on permaculture, while Riace follows a different path, mostly by attracting refugees as inhabitants (East, 2015; Driel and Verkuyten, 2020). Vamvakou's revival is based on a bottom-up approach and avoids turning the village into a place for tourism consumption only, while in other examples of revival in Greece, such as the village of Kapetaniana and Milia in Crete or Koryschades in Evritania, the hospitality function is stronger. Vamvakou sets a new paradigm where the visitors' economy is only used as one of the means of the revival, also agreed on by all the informants.

The visitors' economy plays a key role in Vamvakou's revival, without turning it into a hospitality business. Firstly, the revival of the village involves attracting visitors to enjoy the recreational opportunities offered by the natural wealth of the Parnonas mountain, while respecting the natural environment and improving the social and economic wellbeing of local people within a social economy logic. Importantly, Vamvakou is a case study where nature-based tourism can provide a well-structured format to facilitate opportunities for wellness. Vamvakou Revival organizes mountain activities, such as hiking throughout the trail network of Vamvakou, mountain cycling and tours by minibus or 4 × 4 vehicles (Fig. 5.2). For the last years, Vamvakou Revival has organized the 3 day Vamvakou Experience

Festival in the field of culture, sports, innovation and gastronomy with over 20 programmes and 1500 participants in 2022. The Vamvakou Experience Festival also includes the Vamvakou Mountain Run, a running mountain race with around 300 participants.

To accommodate visitors, Vamvakou now has two new guesthouses that operate as part of the Vamvakou Revival initiative, supported by the SNF. The average occupancy of the guesthouses reached approximately 35% for the first year of operation (Vamvakou Revival website), which means that its full potential has not been reached yet. Moreover, accommodation is offered by two other short-term rental apartments in the village as well as by other properties in the surrounding area.

Events, workshops and residencies

Vamvakou aims to be a place of inspiration and reflection, supporting a transformative experience for visitors, hosting events, workshops, residencies, etc. This has been organized partly due to the extensive network of the SNF, the Greek National Opera (GNO) and other organizations. The Tällberg Foundation hosted a 3 day workshop in Vamvakou in May 2022. The natural environment, technology, the war in Ukraine and the modern challenges of humanity were at the centre of the workshop discussions, which were hosted in the village in collaboration with Vamvakou Revival. The SNF partners, who are collaborating working from opposite sides of the European continent, made the village a place of open dialogue and organized a workshop which drew participants from 20 countries across a wide range of fields.

In September 2021, members of Special Olympics Hellas visited Vamvakou as part of its families programme. Special Olympics Hellas, which is a sport, educational, non-profit organization based on volunteering and offers training opportunities for people with intellectual disabilities, actively supports the effort of Vamvakou Revival to revive the village. In this context, they have started a collaboration with Vamvakou Revival, which will support the physical, spiritual and social development of the athletes through experiential educational and cultural activities implemented in the area as part of the revival initiative.

Infrastructure and programmes to support the revival

The village revival has been planned in a way to attract and sustain the local community. Overall, the school in Vamvakou has been renovated and transformed into a pioneering STEAM (Science, Technology, Engineering, Arts and Mathematics) centre named V.Lab (Vamvakou Technology Laboratory). In addition, the renovated building of the village school hosts creative activities for children and adults, such as educational programmes, workshops, art classes, board games, Lego bricks, an open lending library and cultural events. Moreover, the school premises are hosting speeches, lectures, movie screenings and theatrical plays, concerts and music events. It is a laboratory characterized by multi-functional spaces that can host a wide range of events and provide internet connection. This was part of a Rural Heritage Hub approach within a project called RURITAGE: it aims at operating as a living laboratory where local stakeholders and inhabitants co-operate to develop new heritage-led regeneration strategies for their territory (Conticelli *et al.*, 2021).

A residency programme for artists started in January 2020 with the contribution of SNF, Vamvakou Revival and the GNO, which joined forces to create a small 'hive' of arts and culture in Vamvakou. The Oros Ensemble, a group of young musicians who live and work in Greece and abroad, has taken up residence in the village of Vamvakou (Vamvakou Revival Website, 2023). The effort focuses on promoting active coexistence, fostering the artistic growth of the musicians, and reinforcing the cultural identity of the village and the wider area. The project is open to the public and enables creative interaction between artists and the community with several concerts and other events held in Vamvakou. Vamvakou Revival has also hosted students from the University of California, Los Angeles (UCLA) and the New York University (NYU). The V.Lab gives the opportunity to inhabitants and visitors to get acquainted with information technology.

Furthermore, an old abandoned building in the central square was renovated and is now a meeting place for the village, offering food and beverage services (the Voureiko Café-restaurant'). Moreover, Vamvakou Revival provides, as far as possible, health services for the residents by bringing a remote medical centre to Vamvakou, in collaboration with the Laconia General Hospital Molaoi Nursing Unit. The remote medicine programme includes conducting free preventive check-ups for residents of the village and the wider area. The Vamvakou Revival continues its efforts to achieve better health and wellbeing for the residents in the village and attempts to ensure the most modern living conditions are met, a fact that, according to most of the key informants, is connected to the new forms of remote working, the digital nomad phenomenon, urban density and housing shortages. Vamvakou constitutes an environment where remote workers will be able to avoid commuting and can therefore demonstrate greater productivity, effectiveness, flexibility and organization, without getting distracted by the difficulties of the traditional workplace. The provision of access for disabled people gives them equal opportunities to participate. Generally, the quality of life, the idea of 'working in a forest area in the mountains', has been mentioned as a key advantage by most key informants. This can be connected to the 'back to village' trend as rural areas have proven to be safe shelters, compared to urban agglomerations, during the pandemic, leading people to relocate (de Luca *et al.*, 2020).

Encouraging rural entrepreneurship

The Vamvakou Revival and the Mount Parnon Development Agency, through the 'New Agriculture for the New Generation' initiative, are collaborating in the wider region of Parnonas in an innovative project to strengthen agri-food activity, support local producers and farmers, but also regional development through targeted advisory and business support efforts. The two organizations, whose work is consistently supported by the Stavros Niarchos Foundation (SNF), join forces with the ultimate goal of creating employment opportunities through the activation of the agri-food sector, and a return to appreciating nature and local food production for residents, professionals and potential young entrepreneurs wishing to start a business in the area. More specifically, the effort includes advisory support for local producers with the aim of stimulating primary production, encouraging entrepreneurship, and creating a basket of high quality and distinctive local traditional products that will appeal to both Greek and international markets. According to Vamvakou Revival, the scientific and business support programme is implemented in collaboration with the Agricultural University of Athens, expert partners, and consultants in the development of agri-food entrepreneurship.

As a result, the Vamvakou's agricultural co-operative Parnon Poema was initiated, which consists of five groups of producers (farmers, beekeepers, nut and chestnut producers, producers of aromatic herbs and spices, and olive oil producers). At the same time, the women's co-operative Vamvakia was founded, which is going to process part of the raw materials of the region to create special local products and have a store for the manufacture and distribution of them. The production of the basket products of the area is carried out in certified partners that comply with all the required health standards, while the creation of a preparation store in the village is also planned.

In the context of encouraging entrepreneurship at local and regional levels, an incubator will be set up and operated in the village to host start-ups. This aims to act as a lever for attracting and developing new business activity in the area, while significantly contributing to the revitalization and quality upgrade of the local economy. The pillars of activation are the agri-food sector, tourism, technology, innovation and culture. The business incubator forms part of the Regional Operational Programme of the Peloponnese 2014–2020 and the Priority Axis 'Strengthening the competitiveness, attractiveness and extroversion of enterprises (particularly SMEs), transition to qualitative entrepreneurship spearheaded by the innovation and growth of Regional Added Value' (www.vamvakourevival.org [accessed 12 October 2023]). The project is co-financed by the European Regional Development Fund (ERDF). The creation and operation of this business incubator are supported by the SNF as part of the Vamvakou Revival initiative. The first cycle of the business support programme in Vamvakou Incubator includes five businesses to be established in Vamvakou and five businesses to be established in Peloponnese until the end of 2023. The programme offers to the beneficiaries free integrated business support, workspace for the period of support for the five businesses to be established in Vamvakou, and access to equipment (e.g. computers, high-speed internet connectivity, offices, printers and electronic equipment). It also offers training workshops, programmes, seminars and individual and group sessions of business coaching with specialized consultants, coaches and mentors, networking with pivotal bodies, organizations, and investors for the development of entrepreneurship and distinct business ideas. Moreover, the programme proposes to provide free consulting and monitoring by third parties (legal, accounting and managerial support, company establishment process, corporate identity branding design, webpage, social networks) and support to find accommodation in the area of Vamvakou and potential sources of financing. According to a key informant, a study is being carried out to plan the incubator's strategy and to map out the opportunities.

Discussions and Managerial Implications

Before having any discussion on repopulating rural areas based on a nature-based approach, it is necessary to consider the limited popularity of non-urban lifestyles for a large proportion of the population, a fact that changed significantly during the pandemic. González-Leonardo *et al.* (2022) refer to the 'urban exodus', but also point out the fact that this trend did not follow the same paths in all countries. Whether declining or stagnating rural areas will witness a further return is also in doubt. Connell and McManus (2011) refer to the role of place marketing in attracting attention, mentioning a national campaign called the Country Week in Australia. The discussion is connected to remote working and the opportunities that it brings to peri-urban, rural or even isolated areas and the role of a positive narrative could be crucial. Nevertheless, the discussion on rural revival after the pandemic rarely goes beyond broadband connectivity and the urban–rural digital divide, as if a fast internet connection will solve the problem automatically (Cowie *et al.*, 2020).

Cases such as Vamvakou follow this trend, promoting a narrative that fits into this new trend and attracting inhabitants and entrepreneurs and promoting an alternative, yet very fulfilling, lifestyle. Vamvakou can be seen as a case study that sets a new paradigm in terms of combining a bottom-up approach with support from a major institution but needs some time to be tested.

Overall, rural communities are increasingly interested in setting up inclusive and participatory regeneration processes, but participatory planning experiences in rural areas are still limited (de Luca *et al.*, 2021). However, Vamvakou offers a nature-based tourism model that tries to combine pleasure and hedonism (having fun, relaxation and rest), meaningful experiences (e.g. education, athletic activities and events, self-development or self-fulfilment), and activities connected to the empowerment of special population groups (e.g. training opportunities for people with intellectual disabilities by Special Olympics Hellas), local development and sustainability (i.e. being environmentally friendly and/or benefitting local communities). Concurrently, Vamvakou considers culture a transformative experience, encourages the artistic growth and reinforces the cultural identity of the village and the wider area (organizing concerts, hosting artists, collaborating with the GNO, etc.). Hence, the regenerative role of tourism and visitations is being promoted as a central element. From the interviews, it emerged that visitors have a positive impact on Vamvakou by contributing to its regeneration through tourism, a concept that leads to a positive impact on nature, local rural community and local economy.

Regarding the contribution of tourism, it is still too early to assess if the project is successful, although there are first indications (e.g. demand for overnights, city-break visits, group visits, digital nomads, artists). The research results from the interviews highlight the high expectations from all stakeholders from the revival project. Thus far, these seem to have been met: new residents have been attracted – in total 16 so far – the initial visitors' numbers are satisfactory and new funding opportunities have been secured for business development. This revival project is based on the residents' active role and leverages tourism as a vehicle for the visitors' economy, and the sustainability which plays a key role in the revival of Vamvakou. More particularly:

1. Opting for a regeneration model where nature-based tourism is central but not dominating the effort is agreed to be a valid choice, especially since the pandemic created a discussion on life outside the metropolitan areas, and with remote working the idea of attracting new residents became strengthened.
2. The Vamvakou Incubator aims for projects that make the most of Mount Parnon's natural wealth and the wider area of the Peloponnese to create agri-food products in accordance with the principles of sustainability. Furthermore, this project has a growth perspective for the wider area in nature-based tourism, technology, innovation and culture. The businesses help to cover the needs of the local community and/or address their problems, challenges and opportunities, and create job opportunities. Also, Vamvakou Incubator is looking for proposals, ideas and solutions regarding the management of resources, optimization of the existing production processes, improving access to basic services (e.g. civil protection, health, education, agricultural services, production optimization, circular economy). Entrepreneurship is often an important factor in the local development of rural areas and contributes to the sustainability of the local community.
3. The revival of Vamvakou is part of a wider effort and has become an integral part of the destination brand of the Peloponnese and strengthens the position of Laconia as an alternative sustainable tourism destination.
4. One should consider its particularities: its success is related to the key role of the SNF offering funding and solutions, but also threatening, to some extent, its bottom-up character. Although the 'adoption' of an abandoned village by a private fund or a sponsor could be a very efficient idea, it obviously cannot be applied everywhere and without issues of control. Therefore, despite its interest, Vamvakou can not be seen as a clearly bottom-up and community-driven initiative. Still, the model followed can be interesting as an approach for other cases as a way of dealing with insecurity. The creation of non-profit organizations and co-operatives from the local community can be an effective way of implementing the community-based initiative. The support of the SNF is seen as positive because of the strategy to allow stakeholder groups to take initiative, but also to foster responsibility. A dependency relationship is of course evident, but still the sharing of expertise and several advantages that have been analysed lead to the conclusion that private funding could follow a path that is sustainable.

The possibility of involving other agencies in the effort to revive the village of Vamvakou could be further investigated, so that the effort obtains a new dynamic and becomes financially sustainable in the future. The Vamvakou model aims to strengthen the local community and economy and small and medium-sized entrepreneurship, especially now that the ageing of the population in Greece has resulted in the abandonment of the regions and the villages.

Although the goal of this social experiment is to be sustainable and set an example for other areas in Greece and abroad, it is not certain that private funding and support will be available elsewhere to the extent that it has been in Vamvakou. Still, the choice to allow a social enterprise to take over, the choice not to only focus on the tourism/hospitality dimension, and the community-based approach that is finally followed, all offer valuable lessons. Further evaluation and impact assessment can be valuable to the discussion on sustainable rural development.

References

Bellato, L., Frantzeskaki, N. and Nygaard, C.A. (2022) Regenerative tourism: a conceptual framework leveraging theory and practice. *Tourism Geographies* 25(4), 1026–1046. DOI: 10.1080/14616688.2022.2044376.

Buhalis, D. and Sinarta, Y. (2019) Real-time co-creation and nowness service: lessons from tourism and hospitality. *Journal of Travel & Tourism Marketing* 36(5), 563–582. DOI: 10.1080/10548408.2019.1592059.

Cavicchi, A., Frontoni, E., Pierdicca, R., Rinaldi, C., Bertella, G. *et al.* (2018) Participatory location-based learning and ICT as tools to increase international reputation of a wellbeing destination in rural areas: a case study. In: Azara, I., Michopoulou, E., Niccolini, F., Taff, B.D. and Clarke, A. (eds) *Tourism, Health, Wellbeing and Protected Areas*. CAB International, Wallingford, UK. DOI: 10.1079/9781786391315.0000.

Clissold, R., Westoby, R., McNamara, K.E. and Fleming, C. (2022) Wellbeing outcomes of nature tourism: Mt Barney Lodge. *Annals of Tourism Research Empirical Insights* 3(2), 100077. DOI: 10.1016/j.annale.2022.100077.

Connell, J. and McManus, P. (2011) *Rural Revival? Place Marketing, Tree Change and Regional Migration in Australia*. Routledge, London. DOI: 10.4324/9781315607207.

Conticelli, E., de Luca, C., Santangelo, A., Tondelli, S., Perello, M. *et al.* (2021) Co-developing heritage-led regeneration plans in rural areas: the RURITAGE methodology for community-based heritage management and planning. In: *Le politiche regionali, la coesione, le aree interne e marginali.* Atti della XXIII Conferenza Nazionale SIU DOWNSCALING, RIGHTSIZING. *Contrazione demografica e riorganizzazione spaziale, Torino, 17–18 June*, Roma-Milano. Planum Publisher – Società Italiana degli Urbanisti, 03, pp. 179–185.

Cowie, P., Townsend, L. and Salemink, K. (2020) Smart rural futures: will rural areas be left behind in the 4th industrial revolution? *Journal of Rural Studies* 79, 169–176. DOI: 10.1016/j.jrurstud.2020.08.042.

de Luca, C., Tondelli, S. and Åberg, H. (2020) The Covid-19 pandemic effects in rural areas. *TeMA – Journal of Land Use, Mobility and Environment* 119–132. DOI: 10.6092/1970-9870/6844.

de Luca, C., López-Murcia, J., Conticelli, E., Santangelo, A., Perello, M. *et al.* (2021) Participatory process for regenerating rural areas through heritage-led plans: the RURITAGE community-based methodology. *Sustainability* 13(9), 5212. DOI: 10.3390/su13095212.

Dologlou, N. (2021) Could COVID-19 boost greece's village revival? *The Journal of Environmental and Tourism Analyses* 9(1), 5–21. DOI: 10.5719/JETA/9.1/1.

Driel, E. and Verkuyten, M. (2020) Local identity and the reception of refugees: the example of Riace. *Identities* 27(5), 614–632. DOI: 10.1080/1070289X.2019.1611075.

East, M. (2015) Community-led approaches and interventions for the regeneration of abandoned towns in southern Italy. *Ecocycles* 2(1), 18–25. DOI: 10.19040/ecocycles.v2i1.40.

Fossgard, K. and Fredman, P. (2019) Dimensions in the nature-based tourism experiencescape: an explorative analysis. *Journal of Outdoor Recreation and Tourism* 28, 100219. DOI: 10.1016/j.jort.2019.04.001.

González-Leonardo, M., Rowe, F. and Fresolone-Caparrós, A. (2022) Rural revival? The rise in internal migration to rural areas during the COVID-19 pandemic. Who moved and where? *Journal of Rural Studies* 96, 332–342. DOI: 10.1016/j.jrurstud.2022.11.006.

Koumara-Tsitsou, S. and Karachalis, N. (2021) Traditional products and crafts as main elements in the effort to establish a city brand linked to sustainable tourism: promoting silversmithing in Ioannina and silk production in Soufli, Greece. *Place Branding and Public Diplomacy* 17(3), 257–267. DOI: 10.1057/s41254-021-00200-y.

Nandasena, R., Morrison, A.M., Cai, W. and Coca-Stefaniak, J.A. (2023) Transformational tourism: a visionary approach to sustainable tourism? In: Morrison, A.M. and Buhalis, D. (eds) *Routledge Handbook on Trends and Issues in Global Tourism*. Taylor & Francis, London. DOI: 10.4324/9781003260790.

Websites

Milia Official Website (n.d.). Available at: https://milia.gr/ (accessed 24 April 2023).
Riace Citta Futura Official Website (n.d.). Available at: https://www.riacecittafutura.org/ (accessed 24 April 2023).
Ruritage Official Website (n.d.). Available at: https://www.ruritage.eu/ (accessed 24 April 2023).
Stavros Niarchos Foundation Official Website (n.d.). Available at: https://www.snf.org/ (accessed 24 April 2023).
Thalori Official Website (n.d.). Available at: https://www.thalori.com/ (accessed 24 April 2023).
Torri Superiore Official Website (n.d.). Available at: https://php7.torri-superiore.org/ (accessed 24 April 2023).
Vamvakou Revival Official Website (n.d.). Available at: www.vamvakourevival.org (accessed 24 April 2023).

6 How Shifts in Societal, Industry and Consumer Behaviour Shape New Strategic Realities for Nature-based Tourism Destinations

Simon Jones[1]* and Milena Nikolova[2]
[1]*NatureScapes, Raleigh, USA;* [2]*BehaviorSMART, Sofia, Bulgaria*

Abstract
This chapter discusses how shifts in society, as well as consumer behaviour and expectations, impact the nature-based tourism sector. Shifting trends and tourism market dynamics are investigated through consolidated third-party research findings and examples of tourism models that fit the new needs in the market. Three consumer personas are defined, and their unique characteristics explored to help tourism businesses and destinations understand the mindset of different traveller types and be able to adapt their tourism offer to the needs of each traveller.

Introduction

The COVID-19 pandemic forced people into their homes. The extra time on people's hands and the sense of global crisis triggered a shift in society and consumer thinking and new awareness on topics such as wellbeing, sustainability and climate (Hörisch, 2021). Many individuals emerged from the pandemic with a new sustainability consciousness and determination to modify everyday behaviour in ways that improve their impact on the future of the planet (Wei *et al.*, 2021). According to a study by McKinsey & Company, sustainability and climate change were one of four societal and business priority themes of 2021 (McKinsey & Company, 2021), while a Google survey the same year found that 82% of people say sustainability is more important and that there is a 70% increase in the number of people searching for sustainable travel options (Ireland, 2021). Academic studies also support findings that, as result of the pandemic, travellers' attitudes and demand patterns have shifted (González-Reverté *et al.*, 2022).

This affects daily lives as well as having a direct impact on nature-based tourism destinations where people recreate. Many feared that the sudden changes in visitation to natural and protected areas would put conservation and proper management at risk (Spenceley *et al.*, 2021). While the scale of the changes is significant and potentially challenging, they hold a considerable opportunity for nature-based destinations and businesses. Today's societal needs, the changing norms in tourism and recent enhancements in our knowledge about human

*Corresponding author: sjones@nature-tourism.com

DOI: 10.1079/9781800621411.0006

behaviour represent favourable conditions for investment in, and development of, new models in nature tourism.

These societal and traveller trends unfold along with a new readiness for change among tourism suppliers. For years, tourism industry thought-leaders talked about the radical changes that were needed to make progress on sustainability goals, community engagement and visitor use of natural landscapes (Higham *et al.*, 2021). In retrospect, pre-2020, it was unrealistic to believe that at a time of steady growth and profitability dramatic transformations in the tourism sector would actually take place (Torres-Delgado *et al.*, 2021). Profit and expansion fuelled a strong bias towards the status quo that kept sustainability progress limited (Nikolova, 2020).

Since 2020, the dynamic has dramatically changed (Azara and Foster, 2021). The profound disruption initiated by the COVID-19 pandemic forced changes at a scale and speed that pushed the boundaries of what used to be thought possible and acceptable. Changes that shape today's new realities for nature-based tourism destinations and businesses are occurring at two levels: (i) the broad socio-economic level; and (ii) the individual traveller level. As we will show in this chapter, these include a growth in popularity of nature and outdoor activities (Doubleday *et al.*, 2021); an increased focus and awareness on sustainability and climate issues (Adams *et al.*, 2017); an elevated importance of the benefits to local communities (Elkhwesky *et al.*, 2022) and the increasing importance of engaging communities (Di Franco *et al.*, 2020); new behaviours and expectations regarding health (safety) and wellbeing (happiness) (Purcell and Nikolova, 2021); emerging understanding that tourism service providers have influence over beneficial behaviour among travellers (Parapanos and Michopoulou, 2022); as well as other changes in how people travel and visit natural landscapes or seascapes.

We explore the major changes and trends that are likely to shape the future of the travel and tourism sector as it relates to sustainability and nature-based destinations. We then explore tourism 'personas' that define today's sustainable nature traveller and how these impact tourism destinations and suppliers.

Changes in Society

At the macro societal level, sustainability and climate change are no longer the focus of a small portion of the population, emerging as they did during the pandemic as priorities for societies around the world (Elkhwesky *et al.*, 2022). In recent years, the number of businesses, globally, that have taken significant steps to improve their sustainable footprint has also increased, evidenced in the tourism sector by the signatories to the Glasgow Declaration Climate Action in Tourism (2021) as well as many other efforts from individual tourism businesses.

The impetus for this shift comes from the markets that these businesses serve, that now expect companies to be committed to, and act on, sustainability issues. Many studies, such as the Booking.com *Sustainability Report* (2022) or the World Travel & Tourism Council consumer travel trends reports in 2021 and 2023 (WTTC, 2021; WTTC & Deloitte, 2023) have shown significant majorities of the travelling public prioritizing sustainability when searching for and booking travel, as well as expecting sustainability in the places they travel to and the companies they choose to travel with. A study by Expedia and Wakefield also showed that 90% of consumers looked for sustainable options when travelling (Expedia Group, 2022).

Large companies like Google are reacting to this trend, incorporating sustainability markers into information on accommodation and transportation options. These pre-screened certifications are likely to help build confidence within the travelling public in selecting more sustainable service providers and push more companies to become certified (Guix *et al.*, 2022).

In addition to the travelling public, host communities are also changing their perspective on tourism. Before the pandemic, a common mantra for destinations that host visitors was 'The customer (traveller) is always right'. Focused on how a destination, its businesses and its residents could support what visitors wanted, the metric for success was how many people visited the area. During the pandemic, especially in nature-based destinations where visitation was historically high, the approach switched from how to attract people to how to manage visitors better (Trunfio and Pasquinelli, 2021). This often put an emphasis on fewer people to the destination and greater focus on community wellbeing. This switch in narrative from how to attract and support visitors to how travellers support the community they are visiting has profound implications for tourism in nature-based destinations.

Communities that host visitors increasingly want to better define their relationship with tourism, to make it more equitable for residents, while still providing a great experience for guests (Di Franco *et al.*, 2020). In Hawaii, local residents' feelings towards tourism got worse during the pandemic, as people got used to the peace they enjoyed without large numbers of visitors. Two-thirds of residents in Hawaii agreed with the statement that the Islands are being 'run for tourists at the expense of local people' (State of Hawaii, 2022). This represented a historic low in sentiment towards tourism. Hawaii is now planning to limit visitor numbers to high-traffic sites, among other visitor-management measures.

Increased recognition of native communities and their important role in nature-based destinations has also occurred, especially in the world of conservation and land management. These communities are increasingly pushing for greater representation and control over the landscapes where they live as well as leadership in how tourism can benefit their communities.

Changes in Traveller Behaviour and Expectations

The last decade has seen a tremendous leap in our understanding of human behaviour, including how people act and make decisions as well as the factors that affect those decisions (Camerer, 2020). In this section, we explore shifts in traveller behaviour and how a realistic perspective on what they are can help to adapt to new visitor demands. These include an increased demand for nature, a focus on health and wellness and how/where people work.

Nature tourism, and more generally sustainable tourism, were on the rise globally before 2020, but this trend accelerated during the pandemic as people sought out open spaces during and after lockdowns (Doubleday *et al.*, 2021). Visitation to iconic national parks and other protected areas surged in countries like the USA, with Grand Teton and Zion National Parks seeing a 20% and 18% increase, respectively, in the summer of 2021, compared to 2019 numbers (Iacurci, 2021). Many protected areas have had to limit visitor numbers, institute reservation systems or employ other strategies to spread out visitation.

According to a 2020 Booking.com study, 69% of travellers surveyed look to appreciate more simple experiences, such as spending time outdoors, and 56% specifically look for rural, off-the-beaten-track destinations in the outdoors (Booking Holdings, 2021). Visitation to nature-based destinations continues to rise, as pandemic restrictions have eased. Those that discovered, or rediscovered, nature and stocked up on their outdoor recreation equipment now have a greater incentive to utilize this equipment. An increased desire for nature-based destinations is sustaining, but the impetus may now be less to escape the crowds and pandemic lockdowns and more focused inward to themselves and their families.

An increased focus on health and wellness, especially personal mental health, emerged during the pandemic years. Personal wellbeing (i.e. happiness) and the role travel plays in achieving this were brought into focus during the pandemic due to people's inability to travel and a realization of the mental impacts the pandemic had. Greater awareness of the importance of mental health and the link with travel is further driving demand for nature-based experiences, but with a refocused emphasis on wellbeing. In the Booking.com study, 62% of global travellers said they didn't realize how important travel was to their wellbeing until it was no longer an option (Booking Holdings, 2021). A further 79% affirm that travel helps their mental and emotional wellbeing more than other forms of rest and relaxation.

In southern Sweden, a study with a 72h cabin, a small glass structure positioned in a remote natural location and allowing visitors to totally disconnect from modern life for 3 days, has shown that a single stay in this unique cabin has a real impact on the health indicators of guests. Stress and anxiety levels decreased along with blood pressure and heart rate as well as a marked increase in creativity after the 72h period (Lublin and Klang, 2017).

Components that are core to travel experiences, such as the ability to disconnect from our daily lives and technologies (Moisa and Michopoulou, 2022); a more relaxed, quieter pace (Sthapit *et al.*, 2023); spending time with friends and family; learning or engaging in something new or achieving a life aspiration (climbing a mountain, seeing a bucket-list destination or charismatic animal) are all factors in an increased sense of happiness (Lõhmus *et al.*, 2021). In addition to frequently offering the above elements, nature-based destinations are also seen as a source of rejuvenation and wellbeing (Bočkus *et al.*, 2023). This provides an advantage for such destinations to adjust the tourism experiences offered to include greater elements of wellness for guests (Tiwari and Hashmi, 2022).

People's work-life balance was also an area of increased attention during the pandemic, with the scales shifting towards greater emphasis and importance given to non-work activities. Part of this shift was a focus on personal wellness, as previously discussed; another factor was a shift in many sectors to remote work and more flexibility over where employees could work. Many destinations received an influx of visitors, or part-time residents, looking for a place where they could work and recreate in nature.

As pandemic restrictions eased, some businesses recalled their workers to the office, or required a presence in the office for at least part of the time; but many maintained a fully remote or flexible system to give employees the option to decide where they work. Although greater flexibility in where people can work has created a growth in longer-stay visitors to nature-based destinations, not all travellers wish to blend work and holiday in this way. A study by Booking.com found that although 69% of travellers surveyed agreed that adding remote work before or after a holiday was a good way to extend their time in the destination, 73% wanted to re-establish the separation of work and holiday time, and 59% would prefer to spend less time on holiday if it meant they could totally switch off while there (Booking Holdings, 2021).

Although coming down from its peak, the number of people with more flexible work structures is likely to stay significantly higher than pre-pandemic times and some of these will seek out a mix of work and leisure in the places where they travel (Vyas, 2022). According to the Harris Poll survey, 20% of those who want to work remotely want the ability to travel and mix work and leisure (The Harris Poll, 2022). A Deloitte study on the subject also found that travellers mixing work and leisure booked twice as many trips in 2021 (2–4 trips) as those that kept work and leisure separate (1–2 trips) (Deloitte, 2022). For those mixing work and leisure, 38% added between 3 and 6 days, while 18% added more than a week to their trip.

Four Pillars of Sustainability

The wellbeing benefits of leisure experiences in natural environments have been known to researchers for some time (Azara *et al.*, 2018) but did not receive societal attention until the pandemic. A recently published report on sustainability and health by the World Travel and Tourism Council and Harvard University (Purcell and Nikolova, 2021), proposes a shift from the traditional thinking on sustainability as a 'triple bottom line' to a 'quadruple bottom line' that adds health and wellbeing to social, economic and environmental considerations in the application of sustainable principles in travel and tourism (Purcell and Nikolova, 2021). This additional pillar emphasizes the physical and mental wellbeing of ourselves (travellers) and the communities that host travellers as a critical component to the long-term sustainability of tourism destinations.

This adapted framework for sustainability fits with the trends described above, which can broadly be categorized into the four pillars of sustainability, especially as they relate to nature-based destinations. The first, an increase in interest in travel to nature-based destinations, supports the economic pillar, as travellers spend more time in the destination. The second, a desire to be more sustainable and climate-friendly in one's travel practices, supports the environmental pillar. The third, a focus on community wellbeing and communities having more control over the relationship with travellers, falls within the society pillar. An increased desire for personal wellbeing (happiness) and improved work-life balance fits within the new health and wellbeing pillar.

We propose that a focus on the quadruple bottom line for nature-based destinations will help these destinations and the tourism businesses within them to be more responsible stewards of their destinations as well as respond to societal and traveller shifts in demand. To do this, a better understanding of who the 'sustainable traveller' is will be helpful.

Sustainability Personas

For nature-based destinations the shift in market demand towards the four pillars of sustainability may be one of the more significant changes in the dynamic with travellers in recent times. These macro-level shifts have profound implications, but not every traveller is the same. Based on consumer trends and insights about sustainability awareness, we propose that there are roughly three segments of sustainability-aware consumers in post-COVID-19 societies.[1] We present them below as three personas developed on the basis of the numerous studies that analyse changes in sustainability attitudes. Then we look at what this means for destinations and businesses.

The KNOW-IT-ALL-er

Meet Sarah, the KNOW-IT-ALL-er. She is 29 and lives in a large city with her partner. She has no children yet and seeks new and diverse travel experiences on weekends and holidays near to where she lives as well as further afield. She was socially and environmentally aware before the pandemic but the additional stay at home time and opportunity to read more about the sustainability crisis and climate change sharpened her attitudes on the topic. She identifies as a responsible consumer and has made some deliberate changes in her everyday behaviour to align with what she knows about lowering negative impacts of modern everyday living, such as her carbon footprint.

Know-it-all Sarah in everyday life

Know-it-all Sarah identifies as a planet-conscious member of society with the responsibility to play a role in addressing the climate crisis and moving towards sustainable living. She has made an effort to eliminate single-use plastics from her life by incorporating reusable options by carrying reusable water bottles for use at the office and outside, reusable shopping bags to eliminate the need for single-use bags, carrying reusable containers for takeaway food, etc. Know-it-all Sarah is a label reader and always scans descriptions to make sure she is buying products that are in recycled or have recyclable packaging. She has found it easier to shop at smaller stores that carry products from responsible brands as she knows that she can choose from options that are already preselected by the store.

Know-it-all Sarah is aware that her diet choices come with different carbon footprints. She tries to eat mostly vegetarian, sometimes goes for local fish and tries to shop at local farmers' markets, opting for local seasonal products that are not loaded with additional carbon footprint from import transportation. She cooks at home at least two–three times a week.

Know-it-all Sarah has consciously modified her mobility preferences towards low-carbon options. She is more likely to walk to work and to take electric public transport when the weather or distance makes access by foot difficult.

An important aspect of Sarah's lifestyle choices is that she sees a direct relationship between the wellbeing of the planet and her own wellbeing. She recognizes that her sustainable eating choices have a lower footprint and are also healthier; mobility choices that involve more physical movement are better for physical wellbeing, etc.

As a young member of today's society, Sarah is savvy about technology and uses a set of mobile apps that help her navigate the complexities of determining the footprint of her choices and choose what is best for her and for the climate. She uses a mobility application that prioritizes green routes, a carbon footprint calculator for food products to guide her grocery shopping, as well as a mobile solution that allows her to offset the carbon footprint of everyday living on a monthly basis and offers tips for further lifestyle greening.

Know-it-all Sarah on travel

Know-it-all Sarah considers her impact when planning her leisure time and holidays. She has strong awareness of the carbon footprint of transportation modes so she would opt for trips involving flying only if there is absolutely no alternative and when she can fully offset her flights. While she used to enjoy travelling with friends and family to more urban destinations with rich culture and interesting heritage, the pandemic shifted her preference towards nature-based areas offering more opportunities for spending time outdoors.

Travel restrictions during the pandemic prevented her from pursuing her favourite international city breaks but led to a new appreciation for connecting with nature and green spaces. Now she often uses weekends to spend a few hours outdoors hiking or engaging in nature walks with her camera and close friends. She feels that spending time close to nature is important for physical and mental wellbeing, and powers her increased sensitivity towards the need to harmonize contemporary living with a healthier future for the planet. As a result of this post-COVID-19 preference shift, Sarah's holiday choices are now driven by the desire to engage in outdoor adventures. She has renewed appreciation for nature that sits close to cities but is eager to use her vacation days to discover nature-based destinations that are a bit further away and lesser known, especially if they are accessible by train or a rented electric vehicle.

Know-it-all Sarah is a self-planner as she knows she needs to make an extra effort to sort through options that balance footprint with convenience factors that are important for her. She uses some travel apps that help her search holiday options based on accessibility by train, availability of infrastructure and services supporting outdoor experiences, locations for local buying, etc. She always tries to book with accommodation providers and use food-service providers that are locally owned and that ensure that most of the money she spends stays in the local economy.

One of the main irritations for her is to see that service providers in nature-based destinations use single-use packaging, especially if they have not made the effort to at least introduce such options from natural and biodegradable materials. She also expects to find options for refilling her bottle with water (tap or filtered), or in case she opts for other refreshments, to see that packaging is being recycled.

The WANT-TO-BE-er

Lee is 48 years old and a WANT-TO-BE-er. She lives in a small rural town and is married with two kids. Before the pandemic Want-to-be Lee was familiar with sustainability and climate-change concerns but would rarely make a targeted effort to follow the topic. She would often dismiss opportunities to read or watch content on related topics with the argument that this is a complex global issue that should be resolved by governments and not by individual members of society such as herself. She has friends around her own age and younger who also think the same.

With more time on her hands during the pandemic, Lee came across some interesting readings and journalist reports on the carbon footprint of everyday lifestyle, the link between climate change and consumer choices, as well as other sustainability topics. These made a strong impression, making her rethink her philosophy that she has no role to play in addressing today's sustainability crisis. As a result, Want-to-be Lee has become more conscious about consumer choices and wants to identify as a responsible buyer, but may not have all the facts or know where to start.

Want-to-be Lee in everyday life

Lee desires to be perceived as a contemporary planet-conscious member of society that is aware of sustainable living principles. At the same time, she has not been very bold about making radical changes to her everyday life, often explaining to herself that her schedule or living situation does not allow her to make too many sacrifices in favour of lowering her footprint. She has made a point to have a reusable water bottle at the office; she would often justify choosing salads or vegetarian options for meals with the fact that they are greener.

On the other hand, Lee is less ready to make more significant shifts such as giving up driving when dropping off the children to school or going to work. Such shifts come with a time and inconvenience burden, which she feels cannot realistically be incorporated in her everyday lifestyle currently.

Want-to-be Lee is eager to make shifts towards more sustainable choices when shopping. She has never been a thorough label reader but has started giving priority to brands and products that carry symbols for recycled packaging or that are marked as responsible or circular. If a third party or the store has helped to identify local or sustainable products already, she is more likely to buy them. She has not changed her shopping habits significantly and continues to use mainstream supermarkets and convenience stores but with more attention to the sections with sustainable or greener products.

Lee is aware that different foods have different carbon footprints but her awareness is limited to some of the basics – she knows that imported foods have an extra carbon footprint due to the transportation and that vegetarian options are better for the environment than most meat-based alternatives. She has limited understanding beyond that and relies on labels or carbon footprint indicators if these are available on the product packaging. Some of the shifts in the product and food preference of Want-to-be Lee are related to the awareness that these are not only better for the climate and the environment but also for her and her family's health. She understands that consuming more seasonal vegetables is both healthier and climate-friendlier.

Want-to-be Lee is aware that there are some new applications and technology services that help people make more sustainable choices in everyday life. That said, she is unlikely to make a proactive effort to identify these and figure out how they can help her be more aligned with sustainable living principles. She has tried one or two that were recommended by younger colleagues at work but has not been exceptionally consistent with sticking to their daily use.

Want-to-be Lee on travel

Lee knows that travelling comes with a footprint and that some of the things we do during holidays violate sustainable living principles. That said, she sees holidays as the time when she needs to escape from the burdens of responsibility and feels that breaking a few sustainability principles is not going to make or break global climate-change efforts. The deal breaker of want-to-be Lee on holiday is for the sustainable options to be easy and to not require sacrifice of convenience or enjoyment, and not to cost extra.

She understands that transportation choices generate the biggest share of the carbon footprint of one's vacation but is unlikely to switch from plane to train unless the travel times are comparable and do not involve too many transfers. She appreciates the newly available carbon footprint information displayed by some search engines even though it has an effect on her choices only when she is considering options that are comparable in terms of price and convenience.

Want-to-be Lee is more eager to spend time outdoors since the pandemic. She and her family are more likely to travel to nature-based destinations and engage in some light outdoor physical activities such as nature walks. While they pursue more interactions with nature, these should not come at the expense of comfort, accessibility and connectivity.

Want-to-be Lee uses mainstream channels for travel planning such as large online travel agents and brands with which she has established profiles and membership cards. The factors that influence her planning are convenience and taste. Sustainability is not a consideration that influences her major decisions such as choice of destination, transportation mode and accommodation facility. She tends to notice sustainability communications on products and services during her travels but these translate to extra satisfaction derived from an accidental choice that turned out to be green without much intention or effort on her side.

The COULDN'T-CARE-LESS-er

Mary is 57, married and lives in a big city. When it comes to sustainability, she is a COULDN'T-CARE-LESS-er. Mary has one son, who is 31 and lives in a rural area about an hour's drive away from his parents, and holds similar attitudes towards sustainability as his mother.

Mary's awareness and approach to sustainability did not change much with the pandemic. She recognizes that it makes sense to care about the environment and the climate but is firmly convinced that this is an issue to be solved by governments and big organizations. She makes zero effort to adjust buyer decisions and lifestyle towards sustainability principles unless the specific situation requires it.

While Couldn't-care-less Mary is not proactive towards the environment, she is motivated by her own wellbeing. She might not give up steak over salad to lower her carbon footprint but she would go with a seasonal vegetarian meal because it is better for her physical wellbeing.

Couldn't-care-less Mary in everyday life

Mary does not link sustainability to her own identity in any way. She does not make conscious efforts to choose or live sustainably and does not care to be perceived as a responsible consumer. At the same time, she would go along with the stream when sustainability choices are expected or set as default. For example, while she would not make an effort to avoid using single-use plastic cups for coffee or tea, at her office she uses reusable cups because of her employer's efforts to keep the company plastics-free.

Mary does not consider making changes to her mobility behaviour. Switching to public transportation or another option that is greener than her car is out of the question. Changing shopping habits is also not on her radar screen. She has incorporated specific shopping spots in her weekly routine that remain stable and uninfluenced by any sustainability awareness. Her consumer response to an increasing concern about sustainability boil down to the fact that she acknowledges the fast growth of the responsible products section across shops and chains, and the fact that some of the products that she is used to buying now carry sustainability-related communication.

Couldn't-care-less Mary has limited awareness of the fact that different foods have different carbon footprints. She has come across some content on that but she has not retained any of the details as she has never really been ready to make changes to her diet. Mary is remotely aware that there are some new applications and technology services that help people make more sustainable choices in everyday life but has not considered looking into them.

Couldn't-care-less Mary on travel

Mary does not think about footprint during holiday planning or the holiday itself. If she comes across information or signs related to sustainability, it does not impact her decision-making. Her holiday is the time when she needs to escape from the burdens of responsibility, so worrying about the planet is not only not considered but openly ignored. The only scenario when Mary selects the sustainable option is when this is the default or easiest option or the only available option.

Couldn't-care-less Mary is aware that the transportation sector is criticized for not doing much on climate change. She assumes that airplanes are the worst contributors to carbon footprint but this has no effect on her consumer choices. She uses mainstream channels for travel planning such as large online travel agents and brands, which she is familiar with and with which she has history. The most important drivers of her planning choices are convenience and taste. Sustainability is not a consideration that influences any of her decision-making.

Mary makes a more proactive effort to be outdoors and engage with nature than before the COVID-19 pandemic. The main reason for that is her own wellbeing and the proven rejuvenation effects of green areas.

Persona Synthesis

	KNOW-IT-ALL-er	WANT-TO-BE-er	COULDN'T-CARE-LESS-er
Interest in nature-based travel	High	High	High
Sustainability priorities	High	Average	Low
Eager to make an effort to be more sustainable	Yes	No	No
Considers sustainability nice to have	Yes	Yes	No
Will select sustainable option when:	Always	If it is convenient and/or comparable to other options	If it is convenient (no other options)
Interest in environmental sustainability	Yes	Sometimes	No
Interest in social sustainability	Yes	Sometimes	No
Interest in health and wellbeing	Yes	Yes	Yes

What Does This Mean for Nature-based Destinations and Businesses?

As evidenced earlier, the trends in societal and traveller behaviours have changed to favour nature-based destinations, even among those who do not make much effort to align with sustainable living principles. How, then, do destinations and tourism businesses adapt to these changes to target the different personas described above and benefit from the observed shifts?

We propose that by focusing on what appeals to each persona and understanding the broader trends that influence that persona, destinations and businesses can make practical tweaks to their operations that will make them more competitive in the post-COVID-19 tourism marketplace. We divide these approaches based on the personas that they will target.

KNOW-IT-ALL Sarah

Sustainability, especially environmental issues, are the priority for this traveller. Sarah seeks out destinations and service providers that have strong sustainability programmes and will spend time doing her research on the right provider and place for a holiday. Based on 2022 trends, it appears that the Know-it-all-er and Want-to-be-er markets, combined, make up more than 70% of the international travel market; up by 10% from 2021.

To attract and engage these growing audiences, a business needs to be able to highlight what it is doing to minimize its footprint on the environment. This does not necessarily require large investment; it can start with smart targeted tweaks to operational models to take the first steps in improving sustainability and grow from there.

Initial sustainability steps for tourism businesses in nature-based destinations can start with a look at current operations. Businesses should ask themselves three questions:

1. What can we do to lower the consumption of energy coming from fossil fuels (or increase the share from renewable sources in our energy mix)?
2. What are we doing to minimize waste?
3. How can we (and our guests) consume more sustainable products?

Focusing on these three simple questions is a good place to start for any business looking to become more sustainable. The location of the business also makes a difference; some countries or destinations have strong and well-established sustainability programmes that support businesses, while others are more limited, and the responsibility lies more directly with individual businesses and their willingness to take action.

First steps to conserving energy within a business can be as simple as basic communications with guests. Signage asking guests to turn off lights when leaving the room, reuse towels to reduce the amount of washing required, and give up daily room cleaning are some ways to start. Accommodation can also avoid automatically having TVs or air conditioners running in the room while guests are not present (guests can turn them on as desired when they come into the room). Smart mobility optimizations are another area where easy improvements are possible: encouraging guests to take public transport, bike or walk as they explore the area all help to reduce carbon footprint. Beyond their guests, businesses can also make small policy changes that target the behaviour of their staff. These could involve encouraging staff to turn lights off when not in use, switching bulbs to LEDs and changing switches to motion-sensor lights. Businesses can also support programmes for staff car-pooling or use of public transportation, or even adjusting thermostats 1–2 degrees to minimize energy use in their buildings.

Local purchase of food and other supplies is also a way to reduce undesired footprint (and often packaging waste) and to support the local economy. Initial steps to minimize waste within a tourism business can include boosting recycling and composting efforts within the business and educating staff and guests on what is recyclable or compostable. This varies considerably by location, so steps need to be specific to the destination and locally available recycling programmes. Providing opportunities for guests and staff to rely on reusable containers, especially for water, rather than single-use plastic bottles also helps to reduce waste. The simplest step – replacing individual bottled water with filtered water receptacles for guests to refill their own reusable bottles – can make a significant difference.

While in most places moving to plastics-free operations is a business decision, there is a growing number of destinations that are accelerating sustainability processes by pushing the entire local travel ecosystem towards reusable alternatives. For example, California banned travel-sized plastic shampoo bottles from hotels in 2023. In Europe across the Nordics, most accommodation facilities have completely transitioned away from single-use packages for toiletries by installing refillable dispensers in every room. Many properties make cleaning products, soaps, etc. available that do not require single-use plastics, but rather provide concentrated tablets that dissolve in water to make the soap, bathroom cleaner or laundry detergent, and can be put in reusable bottles.

Encouraging travellers to transition to more sustainable consumption goes beyond the use of sustainable toiletries such as the examples provided above. Sustainable food choices are an area where consumer preferences can make a big difference. This can be achieved by adjusting menus to focus on locally farmed food (minimal transportation, maximum economic impact), and promoting more vegetarian meals. At its most basic level, reordering menu items to move more sustainable options to the top of the menu, or highlighting them through appealing descriptions, can help increase uptake of these items.

These simple steps can often make a big difference without requiring significant changes and investment. Further scale of impact can be achieved when multiple businesses take action within the destination and influence the preferences of visitors at different choice points along their journey. A hotel can provide guests with a reusable bottle to use during their trip, an attraction can offer savings to visitors bringing their reusable bottle and a restaurant can offer free refills to clients with bottles. By reinforcing the desirability of sustainable options at different points along their journey, different companies can reinforce responsible behaviour.

Nature-based tourism destination governments and tourism bodies are well-placed to support the scaling up of sustainability efforts as well. They can encourage sustainability as a norm through smart marketing communications as well as support businesses in employing smart behaviour changes that support new trends in the post-COVID-19 realities.

Destination-wide initiatives can include bringing businesses within the destination together to learn about and take first steps in sustainability as a group. This collective approach can lead to a greater sense of ownership of the issues among all businesses (people don't feel like they are doing it alone), promote cross-learning among businesses and also generate a friendly sense of competition on who is doing best with their sustainability efforts.

Although individual businesses should promote their own sustainability efforts to their guests, a destination can often add weight to that message. Destination authorities can help highlight individual businesses' efforts to a broader audience as well as tell the story of the whole destination and its sustainability initiatives. Collectively, these communication channels help to strengthen the messaging that will attract Know-it-all Sarah to the destination.

WANT-TO-BE Lee

Lee wants to be sustainable, but only if it is convenient to do so and doesn't impact her overall holiday experience. The same sustainability actions can be taken by businesses and destinations as described for Know-it-all Sarah; however, with this market the focus is more on how to make it easy for them to make the sustainable choice.

As with Know-it-all Sarah, good communications highlighting what the destination and individual businesses are doing regarding sustainability is critical. However, for Want-to-be Lee, this information is one of the factors in impacting decision-making, balanced alongside other considerations, such as cost and convenience. Businesses can help Lee make the sustainable choice by organizing information materials to feature sustainable options at the top, structuring pricing information below the description so that choice is based on information about the product or service rather than solely on price, and by providing comparison with other products to show the value of each. This may take several forms: it could highlight (if applicable) that the sustainable option is the same price as other options, which would make it more likely for Want-to-be Lee to select this option. Another price-based tactic is to include sustainable options in the main price rather than as an add-on. For example, lunch at a local farm can be included in the main price of a half-day hike rather than provided as an option along with other alternatives.

This market may also respond to information that highlights their potential impact in a tangible way and applies peer pressure encouraging sustainable choices. Examples of social pressure include signage that may highlight that 80% of guests bring their own water bottle on a tour, or the majority of patrons to a restaurant opt for the 'sustainable' market menu. These tactics can help nudge Want-to-be Lee to make a suitable choice.

COULDN'T-CARE-LESS Mary

The sustainability actions and information described for Sarah and Lee will have little impact on Couldn't-care-less Mary, but there are still ways in which to engage her. One way is simply by setting the sustainable option as default. Another alternative is to eliminate unsustainable alternatives where applicable, which will spare the guest a decision.

A focus on tourism experiences, products or services that support physical and mental wellness, especially personal happiness, will likely have greater appeal for Couldn't-care-less Mary. Destinations and businesses can appeal to more personal aspects of sustainability, such as personal health and wellness, to make the connection with this market. Promoting healthier eating, slower travel or taking a bike for exercise can achieve the same result, without a heavy focus on environmental sustainability messages.

Concluding Remarks

With over 70% of the global tourism market interested in more sustainable travel options and a growing demand for nature and outdoor experiences (Booking Holdings, 2021), the opportunity for nature-based destinations is significant. Focusing on the four pillars of sustainability described earlier in this chapter that align with traveller behaviour trends will help destinations and tourism businesses adapt to the post-COVID-19 realities while achieving broader sustainability and community development goals.

Travellers vary in their prioritization of sustainability issues, from the KNOW-IT-ALL-er who holds sustainability action as a top priority, through the WANT-TO-BE-er who pays attention to sustainable choices but is unwilling to give up comfort and accessible price, and the COULDN'T-CARE-LESS-er who ignores sustainability in their decision-making. By understanding distinct behavioural patterns in these new market realities and finding smart ways to address environmental, social, economic and wellbeing issues, nature-based destinations and businesses operating within them can achieve market goals and sustainability targets at the same time.

The conclusion of this review of market dynamics regarding sustainability implies that by integrating sustainability action into tourism operations, businesses and destinations can make themselves more appealing to at least the majority of the market, and in doing so strengthen their viability in the long term. Insights about how different traveller groups make decisions and form attitudes towards sustainability can make a big difference in helping nature-based destinations and businesses make the best out of the current realities. They have a chance to recover after the COVID-19 crisis by benefitting from new market opportunities and aligning with the needs of the sustainability transformation that is required in the global tourism economy.

Note

[1] The sustainability personas presented in this chapter are the original work of the authors. The personas are developed based on review of a series of research reports and trends analyses published immediately before or during the COVID-19 crisis (OECD, 2018, 2021; Langford *et al.*, 2019; Oliver Wyman, 2019; Euromonitor International, 2020, 2021; World Economic Forum, 2020; Booking Holdings, 2021; ETC, 2021; European Commission, 2021; Innovation Norway, 2021; Remes *et al.*, 2021; Shridhar, 2021).

References

Adams, S.A., Font, X. and Stanford, D. (2017) All aboard the corporate socially and environmentally responsible cruise ship: a conjoint analysis of consumer choices. *Worldwide Hospitality and Tourism Themes* 9(1), 31–43. DOI: 10.1108/WHATT-11-2016-0061.

Azara, I. and Foster, C. (2021) Editorial: Re-imaging a life after Covid-19. *International Journal of Spa and Wellness* 4(2–3), 114–117. DOI: 10.1080/24721735.2021.2015142.

Azara, I., Michopoulou, E., Niccolini, F., Taff, B.D. and Clarke, A. (eds) (2018) *Tourism, Health, Wellbeing and Protected Areas*. CAB International, Wallingford, UK. DOI: 10.1079/9781786391315.0000.

Bočkus, D., Tammi, T., Vento, E. and Komppula, R. (2023) Wellness tourism service preferences and their linkages to motivational factors: a multiple case study. *International Journal of Spa and Wellness* 6(1), 78–108. DOI: 10.1080/24721735.2022.2139899.

Booking Holdings (2021) Sustainability Report 130. Available at: chrome-extension://oemmndcbldboiebfnladdacbdfmadadm/https://www.bookingholdings.com/wp-content/uploads/2022/03/BKNG-Sustainability-Report-2021.pdf (accessed 26 October 2023).

Booking Holdings (2022) Sustainability Travel Report 2022. Available at: https://www.bookingholdings.com/wp-content/uploads/2023/04/BKNG-Sustainability-Report-2022.pdf (accessed 11 October 2023).

Camerer, C. (2020) Challenges for behavioral economics: cumulating and distilling what we know. In: Samson, A. (ed.) *The Behavioral Economics Guide 2020*. pp. vii–xv. Available at: https://www.behavioraleconomics.com (accessed 11 October 2023).

Deloitte (2022) Deloitte travel outlook: the winding path to recovery. Available at: https://www2.deloitte.com/content/dam/Deloitte/us/Documents/consumer-business/us-2022-travel-outlook.pdf (accessed 11 October 2023).

Di Franco, A., Hogg, K.E., Calò, A., Bennett, N.J., Sévin-Allouet, M.-A. *et al.* (2020) Improving marine protected area governance through collaboration and co-production. *Journal of Environmental Management* 269, 110757. DOI: 10.1016/j.jenvman.2020.110757.

Doubleday, A., Choe, Y., Busch Isaksen, T., Miles, S. and Errett, N.A. (2021) How did outdoor biking and walking change during COVID-19?: A case study of three U.S. cities. *PLoS One* 16(1), 1–13. DOI: 10.1371/journal.pone.0245514.

Elkhwesky, Z., Salem, I.E., Varmus, M. and Ramkissoon, H. (2022) Sustainable practices in hospitality pre and amid COVID-19 pandemic: looking back for moving forward post-COVID-19. *Sustainable Development* 30(5), 1426–1448. DOI: 10.1002/sd.2304.

ETC (European Travel Commission) (2021) Europeans' attitudes towards responsible travel choices. Available at: chrome-extension://oemmndcbldboiebfnladdacbdfmadadm/https://etc-corporate.org/uploads/2021/06/ETC_Europeans-Attitudes-towards-Responsible-Travel-Choices_vs-final.pdf (accessed 11 October 2023).

Euromonitor International (2020) Accelerating travel innovation after coronavirus. Available at: https://go.euromonitor.com/white-paper-travel-201110-travel-innovation-after-coronavirus.html#download-link (accessed 11 October 2023).

Euromonitor International (2021) Understanding the traveller journey: critical for a sustainable recovery. Available at: https://go.euromonitor.com/white-paper-travel-210817-traveller-journey.html (accessed 11 October 2023).

European Commission (2021) Scenarios towards co-creation of transition pathway for tourism for a more resilient, innovative and sustainable ecosystem. Commission Staff Working Document (SWD(2021) 164). Available at: https://ec.europa.eu/docsroom/documents/45977 (accessed 11 October 2023).

Expedia Group (2022) Sustainable travel study April 2022. Available at: https://go2.advertising.expedia.com/sustainability-study-2022 (accessed 11 October 2023).

González-Reverté, F., Gomis-López, J.M. and Díaz-Luque, P. (2022) Reset or temporary break? Attitudinal change, risk perception and future travel intention in tourists experiencing the COVID-19 pandemic. *Journal of Tourism Futures* 1–21. DOI: 10.1108/JTF-03-2021-0079.

Guix, M., Ollé, C. and Font, X. (2022) Trustworthy or misleading communication of voluntary carbon offsets in the aviation industry. *Tourism Management* 88, 104430. DOI: 10.1016/j.tourman.2021.104430.

Higham, J., Font, X. and Wu, J. (2021) Code red for sustainable tourism. *Journal of Sustainable Tourism* 30(1), 1–13. DOI: 10.1080/09669582.2022.2008128.

Hörisch, J. (2021) The relation of COVID-19 to the UN sustainable development goals: implications for sustainability accounting, management and policy research. *Sustainability Accounting, Management and Policy Journal* 12(5), 877–888. DOI: 10.1108/SAMPJ-08-2020-0277.

Iacurci, G. (2021) National parks are booming: that may ruin your next trip. CNBC. Available at: https://www.cnbc.com/2021/08/22/national-parks-are-booming-that-may-ruin-your-next-trip.html (accessed 11 October 2023).

Innovation Norway (2021) National tourism strategy 2030. Available at: chrome-extension://oemmndcbldboiebfnladdacbdfmadadm/https://assets.simpleviewcms.com/simpleview/image/upload/v1/clients/norway/Nasjonal_Reiselivsstrategi_engelsk_red_c59e62a4-6fd0-4a2e-aea1-5e1c6586ebc6.pdf (accessed 11 October 2023).

Ireland, B. (2021) Google data shows growing consumer demand for sustainable travel. *Travel Weekly*, 30 November.

Langford, G., Weissenberg, A. and Gasdia, M. (2019) US travel and hospitality outlook. Deloitte. Available at: https://www2.deloitte.com/content/dam/Deloitte/us/Documents/consumer-business/us-consumer-2019-us-travel-and-hospitality-outlook.pdf (accessed 11 October 2023).

Lõhmus, M., Stenfors, C.U.D., Lind, T., Lauber, A. and Georgelis, A. (2021) Mental health, greenness, and nature related behaviors in the adult population of Stockholm county during Covid-19-related restrictions. *International Journal of Environmental Research and Public Health* 18(6), 3303. DOI: 10.3390/ijerph18063303.

Lublin, P. and Klang, M. (2017) The 72 hour cabin in Sweden documentary. Visit Sweden. Available at: https://visitsweden.com/where-to-go/southern-sweden/vastsverige/72hcabin/(accessed 11 October 2023).

McKinsey & Company (2021) McKinsey's Global Surveys 2021: a year in review. Available at: chrome-extension://oemmndcbldboiebfnladdacbdfmadadm/https://www.mckinsey.com/~/media/mckinsey/featured%20insights/mckinsey%20global%20surveys/mckinsey-global-surveys-2021-a-year-in-review.pdf (accessed 11 October 2023).

Moisa, D. and Michopoulou, E. (2022) IT and wellbeing in travel and tourism. In: Xiang, Z., Fucha, M., Gretzel, U. and Höpken, W. (eds) *Handbook of e-Tourism*. Springer, New York. DOI: 10.1007/978-3-030-48652-5.

Nikolova, M.S. (2020) *Behavioral Economics for Tourism: Perspectives on Business and Policy in the Travel Industry*. Academic Press, Cambridge, Massachusetts.

OECD (2018) Analysing Megatrends to better shape the future of tourism. *OCDE Tourism Papers* No. 2018/02. DOI: 10.1787/d465eb68-en.

OECD (2021) G20 Rome guidelines for the future of tourism. OECD report to G20 Tourism Working Group. OECD Tourism Papers. Available at: https://www.oecd.org/cfe/g20-rome-guidelines-for-the-future-of-tourism-d11080db-en.htm (accessed 11 October 2023).

Oliver Wyman (2019) The experience revolution: what matters for the future of tourism. Available at: https://www.oliverwyman.com/our-expertise/insights/2019/feb/the-experience-revolution.html (accessed 11 October 2023).

Parapanos, D. and Michopoulou, E. (2022) Innovative mobile technology in hotels and the use of gamification. *Tourism Planning & Development* 20(2), 1–26. DOI: 10.1080/21568316.2022.2107563.

Purcell, W.M. and Nikolova, M.S. (2021) Behavioral economics in travel and tourism: getting service providers and travellers to choose sustainable options. WTTC x Harvard Learning Insights. Available at: chrome-extension://oemmndcbldboiebfnladdacbdfmadadm/https://wttc.org/Portals/0/Documents/Reports/2021/WTTC-Harvard-LearningInsight-BehaviouralEconomics.pdf?ver=2021-06-17-110543-890 (accessed 26 October 2023).

Remes, J., Manyika, J., Smit, S., Kohli, S., Fabius, V. *et al.* (2021) The consumer demand recovery and lasting effects of COVID-19. McKinsey & Company. Available at: www.mckinsey.com/mgi.%0Ahttps://www.mckinsey.com/industries/consumer-packaged-goods/our-insights/the-consumer-demand-recovery-and-lasting-effects-of-covid-19

Shridhar, A. (2021) Understanding the path to purchase: 2021 global consumer types. Available at: https://go.euromonitor.com/white-paper-EC-2021-Top-10-Global-Consumer-Trends.html (accessed 26 October 2023).

Spenceley, A., McCool, S., Newsome, D., Báez, A., Barborak, J.R. *et al.* (2021) Tourism in protected and conserved areas amid the COVID-19 pandemic. *Parks* 27, 103–118. DOI: 10.2305/IUCN.CH.2021.PARKS-27-SIAS.en.

State of Hawaii (2022) Resident Sentiment Survey Spring 2022 Highlights. Available at: chrome-extension://oemmndcbldboiebfnladdacbdfmadadm/https://www.hawaiitourismauthority.org/media/9744/dbedt-resident-sentiment-spring-2022-hta-board-presentation-accessible.pdf (accessed 11 October 2023).

Sthapit, E., Björk, P. and Coudounaris, D.N. (2023) Towards a better understanding of memorable wellness tourism experience. *International Journal of Spa and Wellness* 6(1), 1–27. DOI: 10.1080/24721735.2022.2136120.

The Glasgow Declaration: A Commitment to A Decade of Tourism Climate Action (2021). Available at: https://www.oneplanetnetwork.org/programmes/sustainable-tourism/glasgow-declaration (accessed 11 October 2023).

The Harris Poll (2022) Harris Poll COVID-19 Tracker Wave 126. Available at: https://theharrispoll.com/briefs/covid-19-tracker-wave-126/ (accessed 11 October 2023).

Tiwari, R. and Hashmi, H. (2022) Integrating concepts of destination image, travel motivations, expectation, and future behavior to create a model of wellness travel intentions. *International Journal of Spa and Wellness* 5(2), 185–206. DOI: 10.1080/24721735.2022.2057714.

Torres-Delgado, A., López Palomeque, F., Sanz, B.E. and Urgell, X.F. (2021) Monitoring sustainable management in local tourist destinations: performance, drivers and barriers. *Journal of Sustainable Tourism* 31(7), 1–21. DOI: 10.1080/09669582.2021.1937190.

Trunfio, M. and Pasquinelli, C. (2021) Smart technologies in the Covid-19 crisis: managing tourism flows and shaping visitors' behaviour. *European Journal of Tourism Research* 29, 1–20. DOI: 10.54055/ejtr.v29i.2437.

Vyas, L. (2022) 'New normal' at work in a post-COVID world: work–life balance and labor markets. *Policy and Society* 41(1), 155–167. DOI: 10.1093/polsoc/puab011.

Wei, X., Li, L. and Zhang, F. (2021) The impact of the COVID-19 pandemic on socio-economic and sustainability. *Environmental Science and Pollution Research International* 28(48), 68251–68260. DOI: 10.1007/s11356-021-14986-0.

World Economic Forum (2020) Markets of tomorrow: pathways to a new economy. Available at: https://www.weforum.org/reports/markets-of-tomorrow-pathways-to-a-new-economy/ (accessed 11 October 2023).

WTTC (2021) Trending in travel: emerging consumer trends in travel tourism in 2021. World Travel Tourism Council. Available at: chrome-extension://oemmndcbldboiebfnladdacbdfmadadm/https://wttc.org/Portals/0/Documents/Reports/2021/Trending_In_Travel-Emerging_Consumer_Trends-231121.pdf (accessed 11 October 2023).

WTTC & Deloitte (2023) A world in motion: shifting consumer travel trends in 2022 and beyond. Available at: https://www2.deloitte.com/content/dam/Deloitte/fr/Documents/consumer-business/Publications/wttc-consumer-trends-report-2023.pdf (accessed 11 October 2023).

7 From Consumption to Esteem: Birdwatching Tourism for Overcoming the Effects of the Exploitation of Natural Resources in Nagaland, India

Ben Wielenga*, Akke Folmer and Stefan Hartman
European Tourism Futures Institute, NHL Stenden University of Applied Sciences, Leeuwarden, The Netherlands

Abstract
Birdwatching tourism is one of the fastest-growing subsectors within nature-based tourism worldwide. If well managed, the phenomenon may benefit both nature and society. Communities in Nagaland, situated in northeastern India, adopted birdwatching/ecotourism to overcome negative effects to nature and wildlife. While tourism is in its infancy in Nagaland, it was found that birdwatching tourism has specific advantages and disadvantages for local communities, such as the generation of more income, mostly next to traditional professions; cultural exchange and improved biodiversity; yet also a lack of valid guidelines, ill-managed ecotourism and financial leakages. This case study adds to the existing body of knowledge of birdwatching impacts on the wellbeing of local communities and provides useful examples to nature (conservation) and community organizations and (tourism) policy-makers on what birdwatching tourism can mean to local communities.

Introduction

Birdwatching tourism is one of the sectors that is covered by the nature-based tourism umbrella. It includes the viewing and observation of, listening to and/or photographing birds in their natural habitats (Janeczko *et al.*, 2021). The subsector is gaining increasing popularity, and before the COVID-19 pandemic, the global birdwatching tourism market grew yearly (Janeczko *et al.*, 2021). Important source markets, such as the USA and UK, have been seeing an increase in birdwatchers. In the USA, the number of birdwatchers grew from 11.1 million in 2006 to 15.2 million in 2020 (Statista, 2022). In the UK, birdwatching became the number one hobby in recent years. The country has approximately 6 million birdwatchers (White, 2019). Many other countries have also witnessed increases, which are, for example, reflected in the growing number of subscribers to observation portals (such as the Dutch waarneming.nl and the international eBird (Geelen, 2020; EBird, 2021). It is expected that birdwatching tourism will continue to grow in the years to come. In combination with travellers' demand for new, exciting, authentic, educative and off-the-beaten-track travel experiences, birdwatching is still far from reaching saturation. Remote and unknown tourism destinations see visitor numbers gradually increasing, and with that

*Corresponding author: ben.wielenga@nhlstenden.com

 Nature-based Tourism and Wellbeing: Impacts and Future Outlook (eds F. Niccolini *et al.*)
DOI: 10.1079/9781800621411.0007

the offering of tourism products, too, among which is birdwatching tourism.

Given the expected growth in birdwatching tourism in remote and underdeveloped regions, the concept of responsible development ties in, as the wellbeing of those at the destination (locals, hosts, activity providers) must be accounted for. Wellbeing, defined as 'the combination of social, economic, environmental, cultural and political conditions identified by individuals and/or their communities as essential for them to flourish and fulfil their potential' (Wiseman and Brasher, 2008, p. 358), is important, yet it is under pressure, as birdwatching tourism often takes place in fragile nature areas, in lower- and middle-income countries and within small communities' territories. Birdwatching tourism, when consciously established, managed and monitored, functions to work on community development, improving livelihoods, wellbeing and biodiversity conservation (Biggs *et al.*, 2011).

Effects of birdwatching tourism on local communities

The future of birdwatching (tourism) looks promising, and the sector aims to contribute to a different way of travelling given its approach and underlying principles of conscious travel. By some scholars, birdwatching is considered one of the most sustainable practices within nature-based tourism (Li *et al.*, 2013; Kronenberg, 2014; Janeczko *et al.*, 2021) as it may lead to several benefits outlined in, among others Sekercioglu (2002) and Basnet *et al.* (2021). Birdwatching can also contribute to a stronger affective bond with birds' natural habitat, leading to more public support for nature protection (Folmer *et al.*, 2016). But it is not all positive; birdwatching tourism may also bring about negative effects to nature areas and communities' wellbeing. In Sekercioglu (2002), Collins-Kreiner *et al.* (2013), Kronenberg (2014) and Düzgün and Kurt (2021), a variety of those impacts are described. A complete understanding of the advantages and disadvantages is crucial for future implementation (Callaghan *et al.*, 2017). Therefore, this chapter adds to the existing body of knowledge of impacts of birdwatching on the wellbeing of local communities by looking at two specific cases in Nagaland, India.

Aim and structure of the chapter

The aim of this chapter is to explore the impacts of birdwatching tourism on wellbeing in two communities in Nagaland, India, by the central question, 'In what way does birdwatching tourism affect the wellbeing of local communities in Wokha and Khonoma?'. The lessons learned are meant as a showcase to other bird-rich regions that wish to develop birdwatching tourism to improve community wellbeing and preserve or strengthen natural values.

The chapter first briefly outlines the study area. Second, the methodology is presented. Third, the results are outlined. The conclusion gives an answer to our central question and provides an outlook on how to organize birdwatching tourism so that it contributes to satisfying and rewarding experiences for all stakeholders involved and, as a result, works towards resilient livelihoods and sustainable economies of local communities.

Study Area

Nagaland

Nagaland is situated in India's far north-eastern corner at the border with Myanmar (Fig. 7.1). The area primarily consists of hills, mountains, rivers, valleys and waterfalls (Kinny and Lanusosang, 2016) that attract significant bird species. While official records show over 500 different bird species (Nagaland State Biodiversity Board, 2022), the latest unofficial records from multiple sources report up to 731 different bird species. Such numbers are high compared to other bird-rich destinations. Moreover, there is a wide variety of near endemic and very rare birds in Nagaland that cannot be seen in any other place, which makes the state highly interesting for birdwatchers.

Land and resources in Nagaland are mostly possessed by tribal communities. They

Fig. 7.1. Location of Nagaland. (From *Encyclopaedia Britannica*, 2022.)

have profound connections with nature, and cultures and customs play an important role in understanding ecosystems services and biodiversity management. Nevertheless, forests are sometimes degraded and habitats lost. Although several measures have been taken on different levels to prevent negative impacts on nature, wildlife hunting (e.g. for food, rituals) is still an ongoing issue. While some of Nagaland's communities increasingly pursue birdwatching activities, others are still involved in bird hunting, which has been a way of life for a long time (Edake, 2016); it has become part of traditions and livelihoods and is, thus, difficult to banish.

Tourism in Nagaland

Tourism in the state is in its infancy; however, visitor numbers are rising steadily every year, from approximately 22,000 in 2010 to 126,000 in 2019 (Nagaland Tourism, 2022). Visitors are mostly involved in ecotourism or cultural or agricultural tourism and activities (Ezung, 2011). Birdwatching is an upcoming activity in Nagaland. For instance, the (Indian) Hornbill Festival, which is held annually in Nagaland, sees increasing visitor numbers, from approximately 5,200 visitors in 2010 to 58,600 in 2019 (Nagaland Tourism, 2022). Nevertheless, political tensions and insurgencies are regularly taking place in Nagaland, negatively affecting the tourism industry from time to time.

Where, in the past, mostly researchers would come to study birds and go birding, now small numbers of birdwatchers (primarily domestic, but the international share is growing), 'ecotourists', hardcore travellers and groups that specifically want to see the amur falcon migration are starting to visit Nagaland. The amur falcon migration is seen as one of the largest congregations of birds of prey (Sinha, 2017, 2018). Besides, Nagaland received significant attention as a result of its successful conservation of amur falcons, which were almost extinct due to hunting. Due to this globally known conservation success and its rich biodiversity (Nagaland State Biodiversity Board, 2022), National Geographic listed Nagaland among the world's finest destinations for birdwatching in 2018 (Strycker, 2018).

Nagaland's cultural and natural qualities, remoteness and unexplored character, in combination with global trends in tourism, make birdwatching a promising endeavour. As tourism is nascent in Nagaland, it is important to gain insight into the impacts birdwatching has on local communities' wellbeing and how the future may, or should, look. Two communities stand out in their approach to birdwatching tourism: Wokha and Khonoma. In both communities, the development of birdwatching improved the wellbeing of locals to a certain degree.

Methodology

To unravel the effects of birdwatching tourism on wellbeing, seven semi-structured interviews were carried out (with a wildlife conservationist, scientist, community member, birdwatching tour guide, tour company, forest ranger and policy adviser) in spring and summer 2022 using phone and online communication tools. We asked participants a variety of questions relating to characteristics of the birdwatching tourism market in Nagaland, important birdwatching sites, impacts of birdwatching tourism on nature conservation and wellbeing of locals, and the degree of involvement of local communities in birdwatching tourism.

Results

Transformation in attitudes and behaviour towards birds

In Wokha, where most of the amur falcon population rests and forages for a specific period in a year (generally October–December), the conservation programme resulted in an attitude and behaviour change among locals towards the bird. The programme was led by conservation parties and community representatives and contained several (educational) initiatives and activities that were set up to draw communities' attention to the bird, nature conservation and community wellbeing. Community conservation to increase awareness of and support for nature, to enhance local livelihoods and wellbeing, and maintain and develop healthy environments was messaged in combination with organizing a range of activities, among which was small-scale birdwatching. Awareness gradually increased and eventually resulted in changing attitudes, behaviour and respect as communities learned about the bird's uniqueness and how it could contribute to their livelihoods and wellbeing by combining conservation with tourism. Locals started to embrace the conservation of the bird, thereby counteracting the exploitation of natural resources. The falcon programme eventually became a flagship project and stimulated other communities to work on conservation.

The village of Khonoma sets an example regarding community conservation merged with tourism. Birdwatching and ecotourism helped the community to gradually transform from exploiters of birds to a community that developed esteem for birds and nature. Witnessing small successes such as changing community views on conservation, growth in visitor numbers and (tourism) facilities, community involvement in tourism and improvements in waste management (Khonoma was the first green village in India) ultimately led to encouragement and willingness to contribute to nature conservation and tourism among the wider community. The effects that conservation and tourism had on communal mindsets and wellbeing caused a significant impact on nature in the surroundings of the community, too:

> 'Last year, a survey was carried out, and more birds were observed than earlier, more butterflies. Same is true for Khonoma; it has a lot of birds. You see the direct impact. ... In places where there is nature tourism, we see that the number of birds is higher.'
>
> (nature conservationist)

These two cases show that birdwatching tourism, albeit small-scale, can be seen as a tool that contributes to environmental wellbeing of communities in a threefold way: (i) mindsets towards nature and birds gradually change leading to; (ii) richer biodiversity and better living environment for local communities; and (iii) improved community wellbeing. Communities become more eager to maintain and/or enhance environmental qualities themselves when they see the benefits of birdwatching tourism for their community. Giving communities authority over the environment they live in stimulates awareness of and support for nature and birds, enhances local pride and improves living standards. Especially in remote areas where villagers depend on each other, where social ties and cultures are strong and where collaborations are needed for the survival of the whole community, birdwatching tourism may bring a much-needed balance in gaining an income while simultaneously protecting nature, celebrating the local culture and creating a sense of pride based on the rich local biodiversity.

Awareness of best practices facilitated other communities in seeing what conservation and birdwatching/ecotourism could mean to their wellbeing. As such, catalyst effects are now seen in a few other villages, too. This has led to less bird hunting and an increase in conservation, protection and nature tourism activities.

Economic socio-cultural and environmental effects of birdwatching tourism on communities

Birdwatching tourism in Nagaland has resulted in the setting up of homestays and guesthouses by local communities. In addition, commercial resorts have been starting up. For families that run homestays, this has generally become an important (extra) source of income, albeit in some cases highly seasonal. During the amur falcon season, at least 1000 birdwatching/ecotourists visit the small communities in and around the area where the falcon settles down. Outside this season, a few hundred tourists visit the communities, mainly for butterflies and seeing the place. Most visitors need to have high spending power, as it takes considerable financial effort to reach the state and travel around.

Birdwatching tourism generates more and easier income than that which is obtained from traditional professions (such as agriculture and fishing) or by hunting, but is still mostly supplementary income. Birdwatching trips are provided for at least 500 Indian rupees (around 5 euros); this is a considerable amount given the provision of wages in agriculture and fishing in Nagaland. With the income that families earn by providing homestays and activities, they are enabled to finance basic needs (such as education for their children), and certain facilities (such as proper toilet facilities) that not only meet the needs of travellers but also improve the living standards and wellbeing of locals. In addition, part of the money is invested in conservation, training, trail maintenance, facilities and village management, as well as infrastructure, which is inefficient in some regions and places are poorly connected. Moreover, locals are now involved in guiding, while they were hunters in the past, and the yields that are earned with activities are invested in improving existing infrastructure to establish low-impact homestay ecotourism with a low carbon footprint.

Next to financial benefits, being involved in birdwatching tourism has given local communities the opportunity to meet outsiders, helping them become visible to the outside world, being recognized, appreciated and respected, creating a sense of honour and building on self-esteem through host–guest relationships. Visibility has led to attracting government schemes (e.g. in agriculture, ecotourism) aimed at improving community welfare. Furthermore, cultural exchange has been established between hosts and guests: local communities have become aware of cultural differences in perspectives on birds. This has created awareness and helped communities to dissociate themselves steadily from hunting.

The two communities have been using yields from birdwatching tourism for funding conservation as well as for the documentation of biodiversity in unexplored natural areas. In Nagaland, a great part of the forest is still unexplored and various types of birds that are present in nearby Myanmar and China, and possibly in Nagaland, too, have not been discovered yet. Birdwatching enables the discovery of the unknown part of Nagaland.

Negative effects of birdwatching tourism in Nagaland

Apart from the positive impacts birdwatching tourism has on communities in Nagaland, the negative impacts should not be ignored as they can potentially take off when tourism increases. There is a lack of valid strategic guidelines for ecotourism and unsystematic, poorly managed examples have been observed. This may result in tourism without respecting the 'eco' part of ecotourism when tourism grows further in Nagaland. Moreover, misusing the principles of ecotourism is seen. Locals are sometimes denied the little revenue they deserved for their effort in conservation, and external tour operators and tour guides ignore employing the locals, thus generating more revenue for themselves. Traffic intensity that comes with increasing tourism is another issue. Especially during festivals, cars sometimes enter roosting areas and drive deep into the forest. Tourism also brings garbage and attracts birds that chase off amur falcons. Other negative impacts are the destruction of plants by trampling and the use of call playback to attract birds. Birds might get confused, distracted and disoriented and change their behaviour.

Conclusions

Effects of birdwatching on community wellbeing

The examples of Wokha and Khonoma show that birdwatching/ecotourism has been a successful counter-movement for overcoming negative effects to nature, and in particular birds. It may act as a catalyst for sustainable interactions with nature's resources through community awareness on nature values and transformed community mindsets, and, most importantly, creating value for birds as living creatures rather than them being captured for other purposes. The case shows that birdwatching/ecotourism functions as a vehicle to resilient livelihoods by providing communities a sustainable (supplementary) source of income, helping diversification by limiting dependability on traditional (exploitative) sectors and contributing to Wiseman and Brasher's (2008) dimensions of community wellbeing. It stimulates wellbeing in a variety of ways among individuals and communities: social (developed self-esteem and sense of pride); economic (higher income, job creation, finance opportunities, improvement of living standards); environmental (nature conservation, documenting biodiversity); and cultural (exchange of cultures in host–guest encounters). Birdwatching/ecotourism has led to community capacity building by partly overcoming collective problems and negative wildlife stories, transforming the issue of traditional hunting into bird esteem among the wider community. This has resulted in a sense of ownership and empowerment, giving communities greater control over their own future development.

The future of birdwatching tourism in Nagaland

Bird hunting remains an issue in Nagaland, also in communities where tourism is growing but where it is still more a seasonal activity than a structural one. Some communities are sensitive to relapsing into traditional behaviours and have picked up hunting again in times outside the tourist season and/or during major events such as COVID-19 and regional tensions and insurgencies, when visitor numbers severely dropped. As a result, pathways should be explored to make birdwatching/ecotourism a means to generate structural income and work on awareness and support raising, as birdwatching tourism has proved effective. Seeing what birdwatching tourism can do in local communities, albeit sometimes small-scale, helps in raising awareness and support for the activity, increasing enthusiasm, starting up new initiatives in tourism and/or conservation, and encouraging other community members to become involved, too. However, breaking with traditions and transforming communities completely from consuming birds to establishing bird esteem takes time and needs conscious attention.

Establishing birdwatching activities may thus be a profitable activity, both for enhancing community wellbeing in different ways and conservation of nature and landscape qualities, when it is organized and managed consciously and with the involvement of local communities. However, the state government, together with local parties, should be extremely careful in balancing the effects of birdwatching tourism and local wellbeing, given the absence of structural monitoring. Incentivizing local tourism, improving infrastructure and setting carrying capacity (also for festivals) should be a priority. The following factors are at all times key to successful birdwatching tourism in remote areas: tourism and birdwatching infrastructure; ensuring employment opportunities for community members; using community knowledge on birds and nature; setting up and combining birdwatching activities with other natural and cultural attributes of a region; preventing financial leakages; (over)commercialization; overexploitation; cultural and natural degradation; monitoring human visits to nature areas and birdwatching sites; and the effects on communities structurally. Connecting to other important bird areas in states nearby to join forces can strengthen tourism products, create stable markets, strengthen marketing campaigns and help in destination branding.

Limitations and implications of the study

Quantitative data on the effects of birdwatching tourism on communities in Nagaland would have deepened our study. Nevertheless, this case study provides useful examples for nature (conservation) organizations, community organizations and (tourism) policy-makers in regions where they deal with similar issues such as wildlife/bird hunting, as they outline how negative wildlife stories may be overcome by tourism in which communities are educated and trained in becoming nature ambassadors, given control and involved in decision-making processes.

References

Basnet, D., Jianmei, Y., Dorji, T., Qianli, X., Lama, A.K. *et al.* (2021) Bird photography tourism, sustainable livelihoods, and biodiversity conservation: a case study from China. *Mountain Research and Development* 41(2), D1–D9. DOI: 10.1659/MRD-JOURNAL-D-19-00054.1.

Biggs, D., Turpie, J., Fabricius, C. and Spenceley, A. (2011) The value of avitourism for conservation and job creation-an analysis from South Africa. *Conservation and Society* 9(1), 80–90. DOI: 10.4103/0972-4923.79198.

Britannica (2022) Nagaland. Available at: https://www.britannica.com/place/Nagaland (accessed 13 July 2022).

Callaghan, C.T., Slater, M., Major, R.E., Morrison, M., Martin, J.M. *et al.* (2017) Travelling birds generate eco-travellers: the economic potential of vagrant birdwatching. *Human Dimensions of Wildlife* 23(1), 71–82. DOI: 10.1080/10871209.2017.1392654.

Collins-Kreiner, N., Malkinson, D., Labinger, Z. and Shtainvarz, R. (2013) Are birders good for birds? Bird conservation through tourism management in the Hula Valley, Israel. *Tourism Management* 38, 31–42. DOI: 10.1016/j.tourman.2013.01.009.

Düzgün, E. and Kurt, A. (2021) An assessment of bird watching in Turkey as a recreational activity. *Journal of Tourism and Gastronomy Studies* 9(1), 67–83. DOI: 10.21325/jotags.2021.777.

EBird (2021) Year in review: eBird, Merlin, Macaulay Library, and Birds of the World. Available at: https://ebird.org/news/2021-year-in-review#:~:text=In%20May%202021%2C%20eBird%20passed,observations%20submitted%20this%20year%20alone (accessed 12 July 2022).

Edake, S.S. (2016) Community conserved areas of Nagaland: a case study. Available at: https://www.teriin.org/opinion/community-conserved-areas-nagaland-case-study#:~:text=The%2016%20major%20featured%20tribes,Liangmai%20(Zeliang%20and%20Pochury) (accessed 20 July 2022).

Ezung, T.Z. (2011) Rural tourism in Nagaland, India: exploring the potential. *International Journal of Rural Management* 7(1–2), 133–147. DOI: 10.1177/0973005212459833.

Folmer, A., Haartsen, T., Buijs, A. and Huigen, P.P.P. (2016) Wildlife and flora and the perceived attractiveness of green places: a comparison between local and national green places. *Journal of Outdoor Recreation and Tourism* 16, 16–23. DOI: 10.1016/j.jort.2016.09.001.

Geelen, J.P. (2020) Tien miljoen dieren gespot in een jaar: Wij blinken uit in slechte foto's. Available at: https://www.volkskrant.nl/nieuws-achtergrond/tien-miljoen-dieren-gespot-in-een-jaar-wij-blinken-uit-in-slechte-foto-s~be861db6/ (accessed 8 June 2022).

Janeczko, E., Łukowski, A., Bielinis, E., Woźnicka, M., Janeczko, K. *et al.* (2021) 'Not just a hobby, but a lifestyle': characteristics, preferences and self-perception of individuals with different levels of involvement in birdwatching. *PLoS One* 16(7), e0255359. DOI: 10.1371/journal.pone.0255359.

Kinny, A. and Lanusosang, T. (2016) Exploring the potentials for Ecotourism and sustainable tourism in Nagaland, India. *International Journal of Applied Research* 2(9), 156–160.

Kronenberg, J. (2014) Environmental impacts of the use of ecosystem services: case study of birdwatching. *Environmental Management* 54(3), 617–630. DOI: 10.1007/s00267-014-0317-8.

Li, F., Zhu, Q. and Yang, Z. (2013) Birding tourism development in Sichuan, China. *Tourism Economics* 19(2), 257–273. DOI: 10.5367/te.2013.0201.

Nagaland State Biodiversity Board (2022) *Biodiversity of Nagaland*. Available at: https://nsbb.nagaland.gov.in/ (accessed 17 June 2022).

Nagaland Tourism (2022) *Tourist inflow statistics in respect of department of tourism, government of Nagaland*. Available at: https://tourism.nagaland.gov.in/index.php/tourist-inflow-statistics-in-respect-of-depatment-of-tourism-government-of-nagaland/ (accessed 15 June 2022).

Sekercioglu, C.H. (2002) Impacts of birdwatching on human and avian communities. *Environmental Conservation* 29(3), 282–289. DOI: 10.1017/S0376892902000206.

Sinha, N. (2017) The amur falcon goes to school. Available at: https://www.thehindu.com/sci-tech/energy-and-environment/the-amur-falcon-goes-to-school/article22239227.ece (accessed 15 June 2022).

Sinha, N. (2018) *Recommendations for amur falcon conservation with specific reference to Amur Falcon Festival and eco-tourism. Doyang Reservoir Important Bird Area and Pangti Forest, November 2012*. Bombay Natural History Society. Available at: chrome-extension://oemmndcbldboiebfnladdacbdfmadadm/https://www.bnhs.org/public//pdf_documents/1551264707.pdf (accessed 12 October 2023).

Statista (2022) Number of participants in birdwatching in the United States from 2006 to 2020. Available at: https://www.statista.com/statistics/191207/participants-in-birdwatching-in-the-us-since-2006/ (accessed 12 October 2023).

Strycker, N. (2018) The world's best destinations for bird-watching. Available at: https://www.nationalgeographic.com/travel/article/best-places-to-see-birds-world-birding-destinations (accessed 16 June 2022).

White, J. (2019) How popular is birdwatching? Available at: https://birda.org/how-popular-is-birdwatching/ (accessed 12 October 2023).

Wiseman, J. and Brasher, K. (2008) Community wellbeing in an unwell world: trends, challenges, and possibilities. *Journal of Public Health Policy* 29(3), 353–366. DOI: 10.1057/jphp.2008.16.

8 Sustainable Management of Invasive Alien Species in Protected Areas? The Future Design Dilemma of 'Rabbit Island', Japan

Thomas Jones[1] and Rie Usui[1]*

[1]Ritsumeikan Asia Pacific University (APU), Beppu, Japan

Abstract

This chapter focuses on feral rabbits on Ōkunoshima, Japan's 'Rabbit Island', which is located inside Seto Naikai National Park. The humble rabbit is often overlooked by researchers and remains a polarizing subject in a nature-based tourism (NBT) debate that often seeks 'win-win' solutions to maximize economic impact while simultaneously maintaining conservation goals. Rabbits elicit intense emotions due to their cute appearance while visitors misconstrue feeding as a contribution to their wellbeing without overt consideration of their status as an invasive alien species. Ultimately, the challenge for NBT on Ōkunoshima is how to implement sustainability guidelines by limiting the number of visitors, rabbits or food. In 2022, Japan's Ministry of the Environment announced the Ōkunoshima Island Future Design Supporter system, which calls for volunteers to participate in rabbit-related stewardship activities. Future research will investigate this supporter system to see if Rabbit Island afficionados can co-create a long-term plan for sustainable NBT.

Introduction

This chapter focuses on Rabbit Island, which presents a conundrum for nature-based tourism (NBT). Due to their cute and cuddly image, rabbits represent cultural icons that are used to promote NBT despite simultaneously being listed as an invasive alien species (IAS). This dilemma echoes many familiar trade-offs between economic development and nature conservation that continue to pose management questions for protected areas (PAs) (Azara *et al.*, 2018). The designation criteria of most PAs such as national parks and nature reserves assume static distribution of biodiversity. But climate change, globalization and the proliferation of IAS are forcing a rethink in management priorities (Scott *et al.*, 2002). To date, most IAS research has focused on the shifts in species distribution, or the invasion process and subsequent impacts on the native flora and fauna. Few studies have sought to address the socio-economic dimensions of such trade-offs, such as the demand for NBT. Yet these anthropological preferences often underpin the direction of IAS policy in reality. For example, a survey by Ohsawa and Jones (2017) investigated six IAS examples in a Japanese national park context and asked stakeholders if

*Corresponding author: rieusui@apu.ac.jp

DOI: 10.1079/9781800621411.0008

they approved of systematic eradication or not. Results indicated significant differences in attitudes of park rangers and conservation experts that favour the arrival in Hokkaido of certain non-native species (e.g. beech trees[1]) while resisting others (e.g. cockroaches). Likewise, public support for the removal of non-native species tended to be weaker in other cases where the IAS in question was deemed to be iconic or aesthetically valuable (Fraser, 2006; Bremner and Park, 2007). Beyond such socio-economic imperatives linked to market demand for NBT, anthropological associations can lead to contradictions whereby the same IAS deemed 'pests' in one PA are dubbed 'cute' drawcards in another park, inviting controversial questions over the positioning of IAS in PA management plans.

This chapter will investigate one such phenomenon using the case study of Ōkunoshima Island in Japan's Hiroshima prefecture, where hundreds of free-ranging feral rabbits form the main attraction for 200,000–400,000 annual tourists (Usui *et al.*, 2018; Tomikawa, 2022). Visitors to Rabbit Island can interact freely with the introduced European rabbits (*Oryctolagus cuniculus*) that roam the island. Most partake in some form of petting and hand feeding from bags of pellets or vegetables brought purposely from the mainland (Usui and Funck, 2020). In fact, the pre-pandemic tourism boom to Rabbit Island was driven primarily by a self-perpetuating social media cycle fuelled by images and video footage of international and domestic visitors covered from head to toe with bunnies (Usui *et al.*, 2018). Our chapter draws on a bilingual literature review combined with primary data drawn from meeting minutes, PDF slides and other records of the workshops, triangulated with key-person interviews and ethnographic insights into sustainable management for Rabbit Island. The first section reviews the literature and outlines the research framework before investigating the Japanese Ministry of the Environment's (MOE) stance in managing IAS inside different categories of PA. The ramifications for PA management across Japan's national parks are raised and some of the possible future design solutions for Ōkunoshima are discussed, including a limit on the number of tourists and/or rabbits on the island.

Research Parameters and Literature Review

Management goals in diverse protected area categories

This section begins by summarizing the conceptual definitions and providing an overview of biodiversity (BD), invasive alien species (IAS) and nature-based tourism (NBT) in protected areas (PA). To navigate these complex and overlapping conundrums, we will start with a short review of management priorities courtesy of the International Union for Conservation of Nature (IUCN), the foremost authority and global watchdog for PAs. The IUCN's (2004) system was intended to allow international standardization of different national networks by establishing common criteria to enable comparisons to be drawn between different PAs. According to the aforementioned system, the top three management priorities of Category II national parks are (i) preservation of species and genetic diversity; (ii) maintenance of environmental services; and (iii) tourism and recreation. However, the IUCN's framework has no legal jurisdiction but instead serves as a guideline for the environmental agencies of respective sovereign nations (Pellizzaro *et al.*, 2015). Each country is thus left to implement the guidelines autonomously under the national or local legislation that governs the respective PAs. In reality, this leads to considerable discrepancies at the site-management level, not least because many so-called national parks do not correspond to the IUCN's definition of a Category II but instead represent a Category V Protected Area. PAs such as Seto Naikai National Park, where Rabbit Island is located, have a broader separate set of objectives including 'protection of natural and cultural characteristics' and 'maintenance of cultural attributes' alongside 'tourism and recreation'.

Protected areas as bastions of biodiversity

PAs such as national parks and nature reserves host national conservation strategies and international treaties including the Convention on Biological Diversity (CBD). As biodiversity comes increasingly under threat in an era of the '6th mass extinction', PAs are vital bastions to safeguard biodiversity along with ecosystem services such as providing clean air and water (Stolton *et al.*, 2015). But the addition of biodiversity to PA management plans is surprisingly recent. The first global trailblazers and early PA templates emerged in large mountainous national parks in North America such as Yellowstone (USA) and Banff (Canada), set up from 1872 onwards (Jones *et al.*, 2021). At the time, designation drivers focused on NBT at the expense of biodiversity as certain 'undesirable' species such as the wolf were hunted, removed or deliberately eradicated from the parks (Emel, 1995). As PAs were subsequently designated all around the world, six categories were developed by the IUCN to accommodate PA management goals that have evolved and expanded over time (Dudley, 2008). At one extreme of the IUCN spectrum, a few sites such as the Swiss National Park are labelled category 'Ia', with limited tourist access and a management agenda that prioritizes scientific research. IUCN category 'II' corresponds most closely to the original Yellowstone model of large, mountain national parks, while category 'V' PAs encompass traditional, inhabited landscapes and seascapes that were made or modified by human influences such as farming, forestry and fishing. The IUCN system thus reflects the chronological evolution of PA management objectives that shifted away from the original Yellowstone model, whereby large areas were set aside with few permanent populations of people but significant tourism resources including spectacular geological heritage (e.g. waterfalls, hot springs and geysers) and iconic flora and fauna. This shift from sightseeing to biodiversity was recognized in the set-up of the CBD that followed the 1992 Earth Summit in Brazil (Stolton, 2010). In 2012, the Fourth World Congress on National Parks & Protected Areas echoed the CBD's aim to designate at least 12% of the Earth's terrestrial surface as PAs by 2000 (Mose and Weixlbaumer, 2007). The CBD framework included Signature Programme No. 2 to 'unlock the potential of protected areas, including indigenous and community conserved areas, to conserve biodiversity while contributing towards sustainable development'. This, in turn, paved the way for the Aichi Biodiversity Targets, wherein Target 11 sought to expand PAs to 17% of terrestrial and inland water areas and 10% of coastal and marine areas by 2020 (CBD Secretariat, 2015).

Biodiversity and invasive alien species' position within protected area management goals

Effective management of PAs is challenged by invasive species, regarded as those posing a particular threat to PAs' 'native' BD compared to other types of 'non-native' species that were introduced by humans (Sandlund *et al.*, 1999 cited in Koichi *et al.*, 2012). Although public perceptions of IAS have been explored in the field of biology conservation (Schüttler *et al.*, 2011; Sharp *et al.*, 2011), not much prior research has addressed visitor perceptions of IAS or investigated how IAS-related NBT can affect or informally guide the direction of national park policy.

Nature-based tourism in protected areas

NBT is an umbrella term to describe 'any type of tourism that relies on attractions directly related to the natural environment' (Weaver, 2001, p. 16). A conceptual definition is provided elsewhere in this book. Briefly, some see it as a *doppelganger* for environmental or ecotourism that is 'primarily concerned with the direct enjoyment of some relatively undisturbed phenomenon of nature' (Valentine, 1992, p. 108). The precise positioning of NBT on the spectrum of casual to committed ecotourists is controversial for reasons that will be contextualized later in this chapter as we explore the case of Rabbit Island. But one simple solution is to define NBT more broadly as all PA visitation (Eagles *et al.*, 2002). The economic impact of NBT is acknowledged among the IUCN priorities as one of the fundamental aims of PA management, but visitors' opinions are not often reflected in IAS mitigation policies despite their importance as stakeholders (Koichi *et al.*, 2012). Partly, this reflects the designation criteria of most PAs based on static distribution of biodiversity without acknowledgement of climate change, anthropogenic influence and the proliferation of IAS (Scott *et al.*, 2002). Controversially, IAS can include formerly domesticated feral animals such as horses or rabbits that roam free (Notzke, 2016; Todd, 2018) but are often stigmatized by PA managers even though their familiarity and appeal to visitors can make them valuable tourism assets. To date, there has been little empirical research that investigates NBT in a holistic manner covering the pros and cons of IAS simultaneously, hence this chapter's focus on Japan's feral rabbits.

IAS as a double-edged sword for NBT: rabbits as friend or foe?

Free-ranging populations of European rabbits (*Oryctolagus cuniculus*) have been introduced all over the world in many different habitats. Once native to the Iberian Peninsula, western France and the northern Atlas Mountains, the species has been dispersed widely for farming and as pets, often with severe consequences for local biodiversity (Devillard *et al.*, 2008). Although rabbits are usually described as feral, this terminology and its 'wild' counterpart reflect inconsistent social constructs that compound the partisan invasive species debate (Notzke, 2013). In New Zealand, for example, the 19th-century introduction of aggressive IAS such as possum has resulted in radical eradication programmes that seek to protect ground-dwelling birds such as the iconic kiwi (Potts, 2009). Research from that country suggests that 'international visitors' perceptions and preferences of relevant species were more likely to be influenced by their physical appearance and cultural representation rather than invasive and endemic status' (Zhang *et al.*, 2021). The same paper demonstrated visitors' preference for the invasive rabbit over the endemic wētā in photo-elicitation techniques. Leaving aside beautiful butterflies or iconic apiaries, visitor perceptions of insects such as wētā are invariably negative despite the biodiversity crisis that sees entomological populations severely threatened (Small, 2012). This confirms the findings of Ohsawa and Jones (2017) who noted that even park rangers and conservation experts tended to allow the introduction of certain more aesthetically appealing non-native species (e.g. beech trees) but not other less desirable invaders (e.g. cockroaches). Fortunately for rabbits, they are usually 'portrayed as happy and endearing animals worldwide', including in the British picture-book *The Tale of Peter Rabbit*, or as 'companions of the Moon-god (goddess) in traditional Indian, Chinese and Japanese folklore' (George, 2006, cited in Zhang *et al.*, 2021). In an era of Instagram and other social media, it is no surprise that rabbits could be represented as cute and cuddly bunnies in certain national parks even as they are persecuted as pests in others[2], but what are the ramifications for NBT policy in PAs? Before we turn to the particular parameters of our case-study site, the next section will first provide a brief overview of national park management in Japan.

The Japanese Ministry of the Environment's Role in Managing National Parks

MOE's role as *laissez-faire* national park managers

Japan's network of national parks stands out among the country's PA networks, covering an area that accounts for almost 6% of the total landmass (Kato, 2008). Legal responsibility for these low-key, *laissez-faire* national parks falls to the Japanese Ministry of the Environment. A predecessor, the Environment Agency, was established in 1971 after a series of major industrial pollution outbreaks and subsequent court cases. The Agency gained ministry status in 2001 and today the MOE takes charge of Japan's 34 national parks. However, the MOE's legal mandate under the Nature Parks Act is undermined by a lack of land ownership. The MOE owns only 0.2% of all national parkland (Kato, 2008), and the park administration division has been further hamstrung by insufficient funding and a lack of human resources (Imura, 2005). The 300-plus rangers spread across the country pale in comparison to the territorial size of their administrative domains and the number of visitors that certain PAs attract (Kato, 2008). Instead, the MOE relies on legal 'sticks' and 'carrots' laid out by the Nature Parks Act to achieve conservation aims, along with long-standing environmental laws such as the Wildlife Protection and Hunting Law[3] and the new Invasive Alien Species Act passed in 2004 (Mito and Uesugi, 2004; MOE, 2015).

Although Japan's national parks represent the strictest conservation level of a three-tiered nature park network, many PAs must also make space for 'secondary nature', including forestry, fishing and farming landscapes. Partly, this reflects the realities of land ownership, as around a quarter (25.6%) of all national parkland is in private ownership, with a further 12.4% of the total owned by local government, including prefectures and municipalities. Some parks, such as Ise-Shima in Mie prefecture, are almost entirely privately owned while others, such as Daisetsuzan in Hokkaido, comprise mostly state-owned land.[4] If the latter resembles a Yellowstone-style 'set-aside' park, the former is closer to European 'co-management' models of *laissez-faire* national parks (Mose and Weixlbaumer, 2007). Like many European counterparts, Japanese parks seek to regulate the activities of different stakeholders via an administration system known as *chiikisei*, which implies 'park management by zoning and regulation' but was interpreted by Hiwasaki (2005) as 'multi-purpose parks'. Unlike in former colonial countries such as USA and Canada, where large areas have been set aside for PA purposes, the Japanese system, instead, seeks to superimpose conservation goals via a sliding scale of regulations on pre-existing land use and development mosaics, with planning permission levers to limit human impact (Kato, 2008). Nor do Japan's national parks consist solely of primeval nature in the core zone, but instead they accommodate varying degrees of cultural heritage such as temples and shrines, primary industries such as farming and fisheries, and sizeable resident populations living in or around the park boundaries. This socio-economic dimension to PA management helps explain the cultural context for many parks in Japan that are located in remote, less developed or depopulating areas that are in desperate need of NBT due to the lack of viable economic alternatives (Jones and Ohsawa, 2016).

MOE policy for IAS mitigation: the Japanese PA context

Although this chapter later considers the controversial possibility for invasive rabbits to be accepted as potential NBT resources, a more conventional conservation narrative posits IAS as a threat to PAs to be mitigated or eradicated wherever possible. On a global scale, the annual Convention of Parties (COP) is the governing body of the Convention on Biological Diversity (CBD) that seeks to promote biodiversity via various policies including those that set specific targets to counter the threat of IAS. In Japan, an IAS act was passed in 2004 in the run-up to the COP10 hosted by Aichi prefecture (Mito and Uesugi, 2004). Those authors' affiliation with the MOE offers insights into the national perspective on IAS, including three overlapping spheres of influence exerted on ecosystems; human health and welfare; and primary industries (Table 8.1).

Pacific islands such as those in and around the Japanese archipelago form a biodiversity hotspot home to many endemic organisms (MOE, 2022a), including rare birds such as Amami jay (*Garrulus lidthi*) and woodpecker (*Dendrocopos owstoni*) and other vertebrates such as Amami rabbit (*Pentalagus furnessi*) (Ishida *et al*., 2015). Unfortunately, where mammal predators such as mongoose (*Herpestes javanicus*) have been introduced to islands, they have severely threatened native populations or even driven them to extinction, exacerbated by the historical absence of mammalian carnivores on islands such as Amami and Okinawa (Yagihashi *et al*., 2021).

Eradication of the mongoose was thus identified as a priority management action when Yambaru, located in the northern part of Okinawa Island, was designated as the country's 33rd national park in 2016 (Kato, 2008). This was followed shortly after by Amami Guntō, designated as a national park in 2017. Together Yambaru and Amami form part of a new UNESCO world natural heritage site listed in 2021. The serial property covers 427 km^2 comprised of five areas spread across four different subtropical islands on an archipelago that extends from south-west Japan toward Taiwan. UNESCO's designation criteria 'X' reflects the area's rich biodiversity that underpinned the

Table 8.1. Three categories and examples of IAS and effects. (Modified from Mito and Uesugi, 2004.)

Impact categories	Japanese example of IAS' entry path	Knock-on effects and risks
Impacts on ecosystems	Java mongoose (*Herpestes javanicus*) intentionally introduced to tropical islands to counter yellow-spotted lance-head snakes (*Trimeresurus flavoviridis*) on sugar plantations	Decreased habitats for endangered native species such as Okinawa rails (*Rallus okinawae*)
Impacts on human health and welfare	Snapping turtles (*Chelydra serpentina*) introduced as pets but later released into the wild	Risks of biting and injuring humans. Competition with native turtles
Impacts on farming, forestry and fisheries	Young raccoons (*Procyon lotor*) deliberately introduced as pets but later released into the wild due to the adults' large size and aggressive temperament	Certain agricultural products have been decimated by raccoons.

listing. In particular, 'five mammal species, three bird species, and three amphibian species in the property have been identified globally as Evolutionarily Distinct and Globally Endangered species' (WHC, 2022). The site's biodiversity, and associated threats to it, can thus be considered an an integral part of the World Heritage listing process, which underlines the political importance of the mongoose removal campaign.

In short, the evidence of the IAS mitigation policy from the Amami islands might suggest a strict and comprehensive MOE strategy to limit the spread of non-native species or even eradicate IAS such as the mongoose. New PAs have been designated and mongooses captured using a combination of traps and fences maintained mostly by local volunteers known as Mongoose Busters. However, in the case of Rabbit Island, the IAS mitigation debate has followed an entirely different direction due to the socio-economic context that will be discussed next via an overview of the national park in which the island is located.

Case study site: Setonaikai National Park

The Setonaikai National Park was one of the first batch to be designated across Japan in 1934 and is also the largest if marine territory is included. However, in terms of land ownership, only 79 km^2 of the Setonaikai's total 670 km^2 (<12%) is owned directly by the state, with the majority, 483 km^2 (>72%), belonging instead to private landowners (Kato, 2008). Due to this fragmented mosaic of land title deeds, only 9.5 km^2 (1.4%) of the park is designated as a 'special conservation zone' (i.e. the area with strictest conservation regulations). Conversely, 227 km^2 (34%) of parkland is classified within 'ordinary' or buffer zones that place few meaningful restrictions on planning or development. The Setonaikai thus typifies the *laissez-faire* approach to the extent that many visitors or even local residents might not recognize its PA status. The park management plan also reflects the reality of an inland seascape dominated by manufacturing hubs such as Osaka, Kobe and Hiroshima, which also includes islands such as Teshima, plagued by industrial pollution (Takatsuki, 2003). The overall landscape is peri-urban, with sightseeing spots dotted across the park not limited to pristine natural environments but also boasting gigantic anthropogenic attractions such as Kanmon suspension bridge,[5] which links the strait between Honshu and Kyushu. Decades of economic development have shaped the Setonaikai seascape, and the 20th-century history of Ōkunoshima was more abruptly reinvented from a wartime military base to the Rabbit Island of contemporary touristic fame.

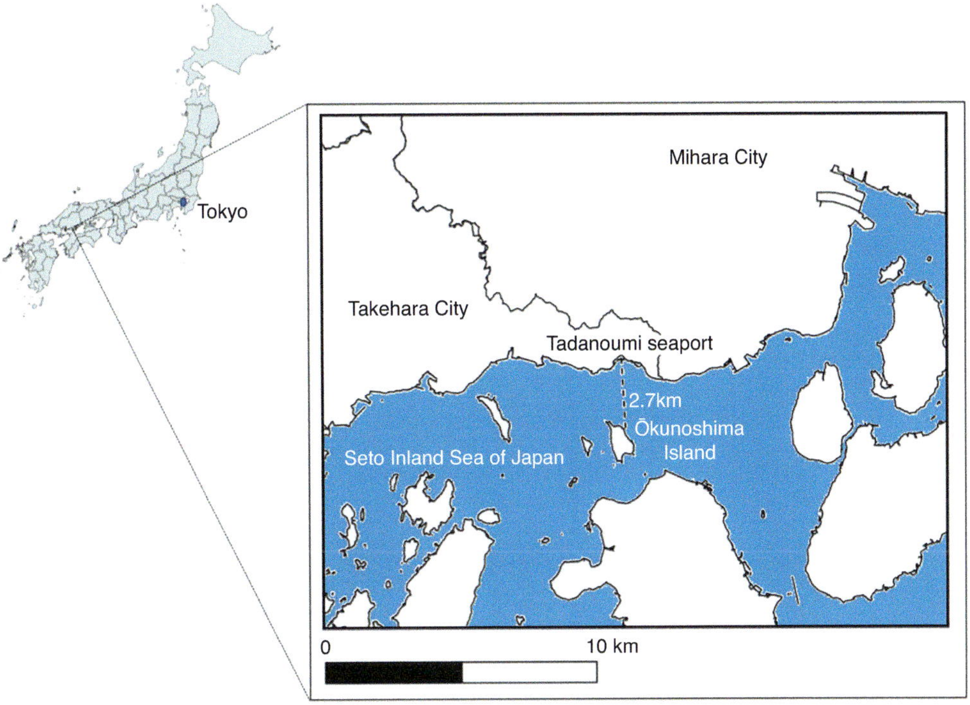

Fig. 8.1. Map of Ōkunoshima Island. (Modified from Usui, 2022.)

An Island Reinvented: From Nerve Gas to Nature-based Tourism

Ōkunoshima: from military base to Rabbit Island

Ōkunoshima is an island of approximately 0.7 km^2 in the Seto Inland Sea, located in Hiroshima prefecture, west Japan (Fig. 8.1). Due to the hundreds of introduced rabbits that roam freely, it is colloquially referred to as Rabbit Island. Ōkunoshima has had no permanent human residents since 1927, when the three households on the island were relocated to make way for an armaments silo (Takeda, 1987). Public access to the island was prohibited and Ōkunoshima was deleted from the map as a secret military base (Tomikawa, 2022). Altogether, the complex of over 100 factories and warehouses that covered most of the island continued to produce deadly nerve gases for 18 years (Maeda *et al.*, 1994; Sato, 2010).

After World War II, the military base was occupied by US troops until 1947, and in 1950 the island was designated within the Setonaikai National Park. However, due to the detection of toxic waste, public access was again restricted until 1957 to enable decontamination (Tomikawa, 2022). It was from 1960 onwards that Ōkunoshima's fortunes underwent a radical reversal as the island was actively redeveloped for leisure purposes. A resort hotel was constructed, one of the first in the country set up under the National Vacation Village Network (Takeo, 2010). The island's image was repurposed as a recreational hub replete with facilities such as a campsite, swimming pool, tennis courts and a fishing pond. A ropeway was also built to whisk tourists up to the top of the island, but this was later dismantled as Ōkunoshima suffered a subsequent stagnation in national park visitation that followed the bursting of Japan's economic bubble (Jones and Ohsawa, 2016). Meanwhile, a handful of architectural ruins from the poison gas facilities remained, and in 1988 a new museum opened to commemorate the island's dark history, attracting mostly male visitors interested in military tourism and schoolchildren on 'peace studies' field trips (Saito and Uodome, 2020; Tomikawa, 2022).

With overall tourist growth waning, rabbits emerged in the post-millennium as a new attraction, and by 2014 their growing exposure via social media resulted in increased numbers of international and domestic visitors (Usui *et al.*, 2018). The legend of how the rabbits were introduced has three main variants, with rabbits either escaping or released from a school, or the hotel, or left over from laboratories linked to the wartime nerve gas factory (DeMello, 2019). As the rabbit population proliferated, touristic images of visitors enjoying close contact with hundreds of cute bunnies began attracting a new profile including younger visitors and families (Fig. 8.2; Saito and Uodome, 2020). The exact number of tourists visiting Ōkunoshima is disputed – Tomikawa (2022) claims a peak of 400,000 in 2017 but ferry data suggest around 300,000 visitors in 2016 (Kunoshima Co. Ltd, 2019, cited in Usui, 2022). The exponential growth in overall numbers and increase in international tourists[6] also affected the rabbit population, which fluctuated broadly in line with the number of visitors to the island.

Nature-based tourism underlines familiar management challenges

National park management policies were underprepared for the post-millennial boom in tourists to Ōkunoshima (Usui, 2022). Today's tourists tend to engage in various interactions with rabbits, including feeding, petting, cuddling and holding them (Usui, 2021). The rabbits have become the island's predominant attraction, and most visitors partake in some form of hand feeding using vegetables or dedicated pellets brought especially from the mainland (Fig. 8.3; Usui and Funck, 2020). This feeding appears to have contributed to the explosion in rabbit births, with the population increasing threefold to over 920 recorded via manual count in 2018 (MOE, 2020a). The subsequent drop in tourists due to the COVID-19 pandemic led to a sharp decline in rabbit numbers, which were estimated at 360 in 2021. The population thus appears to have returned to a level similar to the 300 rabbits counted by the Visitor Centre staff in 2006. In the interim, increased numbers of rabbits caused concern about vegetation damage, transmission of infectious diseases (MOE, 2020a) and rabbits' own wellbeing (DeMello, 2019). Yet, paradoxically, the abundant presence of rabbits is the principal factor that attracts tourists to the island, as it provides a unique experience (DeMello, 2019; Usui, 2021).

Ōkunoshima's management dilemma is epitomized by the alien species law that categorizes the rabbits as an IAS that should require urgent eradication measures (MOE, 2018). Yet in a classic nature-based tourism conundrum, rabbits simultaneously represent the 'golden goose' that can offer added value to Ōkunoshima's otherwise limited portfolio of visitor attractions. Due to the unspoken economic dependence on the rabbits for tourism, neither MOE rangers nor the Visitor Centre staff intervene to regulate the hand feeding, and, as a result, tourists continue to bring food to the island and feed the rabbits freely. Officials, instead, pursue a passive, *laissez-faire* policy of 'demarketing' that seeks only to rein in rampant feeding by certain visitors who dump excess food on the island, thereby attracting pests such as rats, crows or even wild boar that swim to Ōkunoshima from surrounding islands (MOE, 2020a).

Discussion

As in many countries, administration of Japan's national parks remains a balancing act between conservation and development objectives. This chapter focused on the Setonaikai National Park that epitomizes many of the trade-offs between conservation and NBT exemplified by the controversial case of Rabbit Island. Unlocking the potential of such *laissez-faire* parks requires a co-management approach (Hiwasaki, 2005). Any sudden policy change, including limits to the number of tourists or rabbits, could spark an adverse reaction from local businesses. Implementation thus hinges on central and local government agencies but must also make space for local residents, landowners, tour operators and nature conservation organizations.

Fig. 8.2. Ōkunoshima attracts more female, family and international visitors. (Photo: T. Jones)

'Future Design' workshops and a new management system

It was precisely this kind of cross-cutting partnership that was envisaged when the MOE organized a series of 'Future Design' workshops from October 2019 to January 2020 (MOE, n.d.). Two of the workshops were opened to the general public and one of the authors (R.U.) participated as an observer. During one of the workshops, the MOE officer made it clear that there were no immediate plans to limit the island's rabbits or make them a target for eradication, which seemed to be the central concern for some participants. Instead, the primary purpose of the workshops was to gather information about the problems the island experienced, including not only rabbits but also a debate on how to improve the exhibits of the Visitor Centre (MOE, 2020b).

Fig. 8.3. Hand feeding by tourists affects the rabbit population. (Photo: T. Jones)

At the same time, it was also noted that the rabbits are designated as an IAS by the MOE. Parallel legislation akin to the Commonwealth Environment Protection and Biodiversity Conservation Act 1999 in Australia lists the effects of feral rabbits – such as land degradation, soil erosion and damage to crops – as a 'threatening process' for the environment (Mito and Uesugi, 2004). Invasive rabbits are also believed to contribute to the decline of native plant and animal species, sparking wildlife management efforts aimed at population control in Australia, including rabbit-proof fences, poisons, and even pathogens such as myxomatosis.

Ōkunoshima is yet to formally contemplate any such comprehensive counter-strategies, although the 0.7 km^2 island is located within a Level Two conservation zone of Setonaikai

National Park on land that falls under the direct jurisdiction of the MOE. Instead, the current management emphasis remains rooted in recreation and the island continues to host amenities such as a hotel, a tennis court, a campsite and an outdoor swimming pool that opens during the summer peak season. These facilities reflect the island's postwar reinvention as a hub for NBT, yet this management style has proved problematic to reconcile with conservation goals as typified by the debate over the rabbits. To reconsider the conservation/development trade-off, the MOE organized the series of 'Future Design' workshops as a 'round-table' forum to equilaterally discuss the tourism/IAS dilemma. However, the workshops were conducted as an information-collection and networking exercise that stopped short of a new direction for the management of Ōkunoshima, instead calling for volunteers to participate in rabbit-related stewardship activities (MOE, 2022b).

IAS as a potential drawcard for NBT?

In this current era of heightened recognition of biodiversity, and the eminent threats it faces, can IAS be considered a legitimate attraction for NBT? Based on empirical evidence from PAs around the world, instances of IAS such as feral horses being viewed positively already exist (Notzke, 2016). There are also cases of global appeal such as the blooming lupin weeds that attract visitors to New Zealand's South Island in spring (Williams, 2019). Our chapter investigated NBT management on Rabbit Island by drawing on the meeting minutes, PDF slides and other records of the workshops, triangulated with key-person interviews and ethnographic insights. The ramifications for PA management in Japan's national parks are raised and some of the possible future design solutions for Ōkunoshima discussed, including a limit on the number of tourists and/or rabbits on the island. The spectrum of potential management responses could also include a ban on feeding, but tourists tend to favour feeding opportunities, so stricter regulation could be a more feasible first step. Even this would be anathema to a cottage tourism industry that has sprung up around the rabbits. The island's nickname, its social media presence and even the logo displayed in the ferry windows all bear testimony to the bunnies' irresistible pull.

Post-pandemic pause for thought

A few months after the MOE meeting in January 2020, Japan closed its borders due to the coronavirus outbreak. The all-out ban on international arrivals, combined with the restrictions placed on domestic travel, ensured that visitor numbers to the island dropped dramatically (personal communication). The rabbit population has subsequently plummeted by two-thirds, but it is unclear if the reduced visitor footfall offers only a temporary respite to the dilemma of Rabbit Island, or if it could continue in the medium to long term. Either way, the study of ecosystem dynamics demands a holistic approach that also considers other invasive organisms.

The MOE's *laissez-faire*, low-key stance on Ōkunoshima's rabbits stands in stark contrast to their policy for other IAS, notably the mongoose on Amami. Together with the Okinawa prefectural government, the MOE launched a mongoose removal project (2000–2004) in Yambaru prior to the Invasive Alien Species Act enacted in 2005 (Yagihashi *et al.*, 2021). It is no coincidence that Yambaru was being groomed for national park designation and listed in 2021 as a UNESCO natural world heritage site. A mitigation plan was developed and thorough counter-measures implemented including control fences and kill-traps. In the case of Amami, regular monitoring suggests that a total eradication of mongooses may have already been achieved, although conclusive evidence remains elusive. Eradication of the mongoose from Amami would represent a significant success story for conservation biologists, but the presence of other threats such feral cats (*Felis silvestris*) and goats (*Capra hircus*) cannot be ignored (Ishida *et al.*, 2015). Meanwhile, more research is needed to understand why certain IAS should be eradicated in one PA context while becoming a tourism drawcard in another.

Notes

[1] Native to other parts of Japan, but not to the northern island of Hokkaido.
[2] For example, approximately 300,000 European rabbits have recently been eradicated from Macquarie Island, a UNESCO World Heritage site in Australia (Kean, 2022).
[3] Japan's Wildlife Protection and Hunting Law was originally enacted in 1918 but was amended in 2002.
[4] 96.1% of Ise-Shima National Park is privately owned due to the extensive Shinto shrine complex of Ise Jingu, while 94.7% of Daisetsu is owned by branches of government including the Forestry Agency.
[5] When it opened in 1973, the 1068 m suspension bridge was said to be the longest in Asia.
[6] The official number of inbound travellers from overseas increased from just 378 in 2013 to 17,125 in 2015 (Takehara City, 2017, cited in Usui *et al.*, 2018).

Acknowledgements

The authors would like to thank the Japan Institute of Tourism Research for their FY2020 research grant.

References

Azara, I., Michopoulou, E., Clarke, A., Niccolini, F. and Taff, B.D. (2018) *Tourism, Health and Wellbeing in Protected Areas*. CAB International, Wallingford, UK and Boston, Massachusetts. DOI: 10.1079/9781786391315.0000.

Bremner, A. and Park, K. (2007) Public attitudes to the management of invasive non-native species in Scotland. *Biological Conservation* 139(3–4), 306–314. DOI: 10.1016/j.biocon.2007.07.005.

CBD (Convention on Biodiversity) Secretariat (2015). Available at: www.cbd.int/convention/ (accessed 16 December 2022).

DeMello, M. (2019) The rabbits of Okunoshima: how feral rabbits alter space, create relationships, and communicate with people and each other. In: Middlehoff, F., Schonbeck, S., Borgards, R. and Gersdorf, C. (eds) *Texts, Animals, Environments: Zoopoetics and Ecopoetics*. Rombach Druck-und Verlagshaus, Berlin, pp. 231–240.

Devillard, S., Aubineau, J., Berger, F., Léonard, Y., Roobrouck, A. *et al.* (2008) Home range of the European rabbit (Oryctolagus cuniculus) in three contrasting French populations. *Mammalian Biology* 73(2), 128–137. DOI: 10.1016/j.mambio.2007.01.003.

Dudley, N. (2008) *Guidelines for Applying Protected Area Management Categories*. IUCN, Gland, Switzerland. DOI: 10.2305/IUCN.CH.2008.PAPS.2.en.

Eagles, P.F.J., McCool, S.F. and Haynes, C.D. (2002) *Sustainable Tourism in Protected Areas: Guidelines for Planning and Management (No. 8)*. IUCN, Gland, Switzerland. DOI: 10.1079/9780851995892.0000.

Emel, J. (1995) Are you man enough, big and bad enough? Ecofeminism and wolf eradication in the USA. *Environment and Planning D: Society and Space* 13(6), 707–734. DOI: 10.1068/d130707.

Fraser, A. (2006) Public attitudes to pest control: a literature review. Department of Conservation Research and Development Series no. 227. Department of Conservation, Wellington. Available at: https://www.doc.govt.nz/documents/science-and-technical/drds227.pdf (accessed 13 October 2023).

George, P.A. (2006) *East Asian Literatures: Japanese, Chinese and Korean: An Interface with India*. Northern Book Centre, Delhi.

Hiwasaki, L. (2005) Toward sustainable management of national parks in Japan: securing local community and stakeholder participation. *Environmental Management* 35(6), 753–764. DOI: 10.1007/s00267-004-0134-6.

Imura, H. Schreurs, M.A. (ed.). (2005) *Environmental Policy in Japan*. Edward Elgar Publishing, Cheltenham, UK.

Ishida, K., Murata, K., Nishiumi, I., Takahashi, Y. and Takashi, M. (2015) Endemic Amami Jay, invasive Small Indian Mongoose, and other alien organisms: a new century investigation of island aliens towards improved ecosystem management. *Journal of Ornithology* 156(S1), 209–216. DOI: 10.1007/s10336-015-1250-1.

IUCN (International Union for Conservation of Nature) (2004) The history of the International system of protected area management categories. *Parks* 14(3), 4–14.

Jones, T.E. and Ohsawa, T. (2016) Monitoring nature-based tourism trends in Japan's national parks: mixed messages from domestic and inbound visitors. *PARKS* 22(1), 25–36. DOI: 10.2305/IUCN.CH.2016.PARKS-22-1TJ.en.

Jones, T.E., Bui, H.T. and Apollo, M. (2021) Mountainous protected areas & nature-based tourism in Asia. In: Jones, T.E., Bui, H.T. and Apollo, M. (eds) *Nature-Based Tourism in Asia's Mountainous Protected Areas*. Springer, Cham, Switzerland, pp. 3–25. DOI: 10.1007/978-3-030-76833-1.

Kato, M. (2008) *National Park System of Japan*. Kokon Shoin, Tokyo.

Kean, Z. (2022) From 300,000 rabbits to none: a Southern Ocean island is reborn. *The Guardian*, 10 August. Available at: https://www.theguardian.com/environment/2022/aug/10/rats-and-rabbits-invasive-species-macquarie-island-southern-ocean-aoe (accessed 28 June 2023).

Koichi, K., Cottrell, A., Sangha, K.K. and Gordon, I.J. (2012) Are feral pigs (*Sus scrofa*) a pest to rainforest tourism? *Journal of Ecotourism* 11(2), 132–148. DOI: 10.1080/14724049.2012.703672.

Maeda, A., Awaya, Y., Ihara, Y., Yoshida, Y., Ishioka, S. *et al*. (1994) A study of autopsied lung cancer cases among retired workers of the Okuno-jima poison gas factory. *Japanese Journal of Lung Cancer* 34(4), 525–530. DOI: 10.2482/haigan.34.525.

Mito, T. and Uesugi, T. (2004) Invasive alien species in Japan: the status quo and the new regulation for prevention of their adverse effects. *Global Environmental Research* 8(2), 171–193.

MOE (Ministry of the Environment) (2015) A list of alien species. Available at: https://www.env.go.jp/nature/intro/2outline/ list/fuka_animal.pdf (accessed 13 October 2023).

MOE (Ministry of the Environment) (2018) List of regulated living organisms under the Invasive Alien Species Act [Animal Kingdom]. Available at: https://www.env.go.jp/content/900452906.pdf (accessed 28 June 2023).

MOE (Ministry of the Environment) (2020a) *Ōkunoshima mirai zukuri wāku shoppu (Ōkunoshima future-design workshop)*. Chugoku-Shikoku Regional Environment Office.

MOE (Ministry of the Environment) (2020b) *Ōkunoshima to watashitachi no korekara* (Ōkunoshima Island and our future). Available at: https://chushikoku.env.go.jp/content/900128027.pdf (accessed 13 October 2023).

MOE (Ministry of the Environment) (2022a) Yambaru National Park: characteristics. Available at: https://www.env.go.jp/en/nature/nps/park/yambaru/point/index.html (accessed 28 June 2023).

MOE (Ministry of the Environment) (2022b) Okunoshima Mirai Zukuri Sapōtā boshū shimasu (Calling for Ōkunoshima Future Design supporters). Available at: https://chushikoku.env.go.jp/pre_2022/post_134.html (accessed 28 June 2023).

MOE (Ministry of the Environment) (n.d.) Ōkunoshima Miraizukuri Workshop. Available at: https://chushikoku.env.go.jp/nature/mat/post_21.html (accessed 28 June 2023).

Mose, I. and Weixlbaumer, N. (2007) A new paradigm for protected areas in Europe. In: Mose, I. (ed.) *Protected Areas and Regional Development in Europe: Towards a New Model for the 21st Century*. Routledge, London, pp. 3–20.

Notzke, C. (2013) An exploration into political ecology and nonhuman agency: the case of the wild horse in western Canada. *The Canadian Geographer* 57(4), 389–412. DOI: 10.1111/cag.12028.

Notzke, C. (2016) Wild horse-based tourism as wildlife tourism: the wild horse as the other. *Current Issues in Tourism* 19(12), 1235–1259. DOI: 10.1080/13683500.2014.897688.

Ohsawa, T. and Jones, T.E. (2017) How can protected area managers deal with nonnative species in an era of climate change? *Natural Areas Journal* 37(2), 240–253. DOI: 10.3375/043.037.0213.

Pellizzaro, P.C., Hardt, L.P.A., Hardt, C., Hardt, M. and Sehli, D.A. (2015) Stewardship and management of protected natural areas: the international context. *Ambiente & Sociedade* 18(1), 19–36. DOI: 10.1590/1809-4422ASOC509V1812015en.

Potts, A. (2009) Kiwis against possums: a critical analysis of anti-possum rhetoric in Aotearoa New Zealand. *Society & Animals* 17(1), 1–20. DOI: 10.1163/156853009X393738.

Saito, A. and Uodome, Y. (2020) A study of maintaining the natural environment and sightseeing in Ohkunoshima – animal and human symbiosis. *Bulletin of Institute for Interdisciplinary Studies of Culture, Doshisha Women's College of Liberal Arts* 37, 146–153. [In Japanese].

Sandlund, O.T., Schei, P.J. and Viken, A. (1999) Introduction: the many aspects of the invasive alien species problem. In: Sandlund, O.T., Schei, P.J. and Viken, A. (eds) *Invasive Species and Biodiversity Management*. Kluwer Academic, Dordrecht, The Netherlands. DOI: 10.1007/978-94-011-4523-7.

Sato, R. (2010) *Sea-dumped Chemical Weapons: Japan*. Global Green, Washington, DC. USA. Available at: https://static1.squarespace.com/static/5548ed9034b0b0a763d0e704/t/55548e55e4b0cef71ee e4bd8/1431604821083/publication-114-1.pdf
Schüttler, E., Rozzi, R. and Jax, K. (2011) Towards A societal discourse on invasive species management: a case study of public perceptions of mink and beavers in Cape Horn. *Journal for Nature Conservation* 19(3), 175–184. DOI: 10.1016/j.jnc.2010.12.001.
Scott, D., Malcolm, J.R. and Lemieux, C. (2002) Climate change and modelled biome representation in Canada's national park system: implications for system planning and park mandates. *Global Ecology and Biogeography* 11(6), 475–484. DOI: 10.1046/j.1466-822X.2002.00308.x.
Sharp, R.L., Larson, L.R. and Green, G.T. (2011) Factors influencing public preferences for invasive alien species management. *Biological Conservation* 144(8), 2097–2104. DOI: 10.1016/j.biocon.2011.04.032.
Small, E. (2012) The new Noah's Ark: beautiful and useful species only. Part 2. The chosen species. *Biodiversity* 13(1), 37–53. DOI: 10.1080/14888386.2012.659443.
Stolton, S. (2010) Protected areas: linking environment and well-being. In: Stolton, S. and Dudley, N.B. (eds) *Arguments for Protected Areas*. Routledge, London, pp. 22–33.
Stolton, S., Dudley, N.B., Çokçalışkan, A., Hunter, D. and Ivanić, K.Z. (2015) Values and benefits of protected areas. In: Worboys, G.L., Lockwood, M., Kothari, A., Feary, S. and Pulsford, I. (eds) *Protected Area Governance and Management*. Australian National University Press, pp. 145–168. DOI: 10.22459/PAGM.04.2015.
Takatsuki, H. (2003) The Teshima Island industrial waste case and its process towards resolution. *Journal of Material Cycles and Waste Management* 5(1), 26–30. DOI: 10.1007/s101630300005.
Takeda, E. (1987) *Chizu kara kesareta shima – Ōkunoshima dokugasu kōjō [The island that was removed from the map – Ōkunoshima Island's poisonous gas factory]*. Domesu Publishing Inc. [In Japanese].
Takeo, S. (2010) Hiroshima Heiwa no Tabi [peace journey in Hiroshima]. *PRIME* 32, 131–135.
Todd, T.R. (2018) *Pigs of Paradise: The Story of the World-famous Swimming Pigs*. Skyhorse Publishing, New York.
Tomikawa, K. (2022) Transition of tourist attractions and tourist behaviors on rabbit-Island, Okunoshima. *The Journal of Island Studies* 23(1), 1–13. DOI: 10.5995/jis.23.1.1.
Usui, R. (2021) Feral animals as a tourism attraction: characterizing tourists' experiences with rabbits on Ōkunoshima Island in Hiroshima, Japan. *Current Issues in Tourism* 26(16), 1–16. DOI: 10.1080/13683500.2021.1978950.
Usui, R. (2022) Does the path toward resilience go beyond conservation at Ōkunoshima Island? *Geographical Sciences* 76(3), 131–139.
Usui, R. and Funck, C. (2020) Rabbits in the wild: close encounters on an equal footing. In: Bertella, G. (ed.) *Wildlife Tourism Futures: Encounters with Wild, Captive and Artificial Animals*. Channel View Publications, Clevedon, UK, pp. 24–39.
Usui, R., Wei, X. and Funck, C. (2018) The power of social media in regional tourism development: a case study from Ōkunoshima Island in Hiroshima, Japan. *Current Issues in Tourism* 21(18), 2052–2056. DOI: 10.1080/13683500.2017.1372393.
Valentine, P. (1992) *Nature-based Tourism*. Belhaven Press, London.
Weaver, D.B. (2001) *The Encyclopedia of Ecotourism*. CAB International, Wallingford, UK. DOI: 10.1079/9780851993683.0000.
WHC (World Heritage Convention) (2022) Amami-Oshima Island, Tokunoshima Island, northern part of Okinawa Island, and Iriomote Island. Available at: https://whc.unesco.org/en/list/1574/ (accessed 28 June 2023).
Williams, A. (2019) Lupin-spotting in New Zealand, cited in Zhang *et al.* (2021). Available at: https://www.dangerous-business.com/2014/12/lupin-spotting-new-zealand/ (accessed 17 October 2023).
Yagihashi, T., Seki, S.I., Nakaya, T., Nakata, K. and Kotaka, N. (2021) Eradication of the mongoose is crucial for the conservation of three endemic bird species in Yambaru, Okinawa Island, Japan. *Biological Invasions* 23(7), 2249–2260. DOI: 10.1007/s10530-021-02503-w.
Zhang, G., Higham, J.E.S. and Albrecht, J.N. (2021) Ecological restoration in Aotearoa New Zealand: contrasting tourist conservation narratives. *Tourism Management Perspectives* 37, 100761. DOI: 10.1016/j.tmp.2020.100761.

9 What Pushes Nature Enthusiasts? A Five-country Comparison

Thomas Bausch[1,2]* and Verena Tauber[2]
[1]Department of Tourism, Munich University of Applied Sciences; [2]Competence Center Tourism and Mobility, Free University of Bozen-Bolzano, Italy

Abstract
Travel push factors are manifold and individual. Sun and warmth, being close to nature, relaxing or actively exploring a destination are just some of many motives pushing people to travel. Some factors focus on socializing, while others are of a more intrinsic nature. To understand the entire set of push factors of travellers with high nature affinity, the results from a five-country study (CH, DE, FIN, IT, USA), with 1531 participants, are presented. Twenty-three basic push factors were derived from literature. Seven higher-level push factors were found using a principal component analysis. Their importance for nature lovers is compared to travellers with low or medium nature affinity. Moreover, a typology of travellers with high nature affinity is presented. Finally, the chapter makes recommendations for product development and communication at destination level, considering the heterogeneity of nature enthusiasts.

Introduction

Activities in nature, enjoying the atmosphere of pure nature, or observing flora and fauna in its natural environment are, for many travellers, a must during a vacation. The interest in vacations where nature is an integral part of the trip has seen a steady increase over the last decades. For example, in the German travel market about 13.9% of travellers classified their main vacation trip as 'nature-vacation' in 1999. Twenty years later, the same survey revealed a level of 21.3% for the pre-COVID-19 travel year, 2019 (statistics based on data from German travel analysis 2000 and 2020). During the pandemic, a further increase in outdoor and nature-oriented travel could be observed due to several phenomena. Craig (2021) reported a significant increase in the sector of camping and 'glamping' in the USA as people wanted to maintain social distancing. Lebrun *et al.* (2021) found a strong increase in domestic tourism to natural parks in France as international travelling was prohibited and people tried to avoid crowded places. Schlemmer and Schnitzer (2021) observed a boom in ski touring in the Alps as in many regions the ski resorts and ropeways were closed. Some authors assume that such new experiences might have a long-term effect not only on the demand side but also the supply side (Stankov *et al.*, 2020) and, this way, nature-based tourism sites might see a further growth of demand. Others are more sceptical and assume that once the pandemic is over, travellers will return to their old habits (Hall *et al.*, 2020; Tauber and Bausch, 2022).

The market of nature-based tourism is, as for all types of tourism, the result of choices made by the travellers. A significant share of travellers have an affinity to nature and select their holiday destination because of expectations

*Corresponding author: thomas.bausch@hm.edu

 Nature-based Tourism and Wellbeing: Impacts and Future Outlook (eds F. Niccolini *et al.*)
DOI: 10.1079/9781800621411.0009

of natural or nature-based elements that satisfy their preferred vacation desires. Thus, the natural elements of a destination become a part of the destination-choice process. Woodside and Lysonski (1989) defined in their model of traveller destination choice two main blocks of factors that lead to the formation of the destination choice sets (see also Decrop, 2010): the marketing variables at the destination side and the traveller variables at the consumer side. The traveller variables characterize the travellers concerning their motivations and expectations, which are the result of past travel experiences, personal situation (e.g. life cycle, income, age), lifestyle and value system, which play an important role during the decision-making process (Cohen *et al.*, 2014). The combination of the traveller's individual profile with the marketing variables forms the choice set, including destinations that are considered for a trip and those that are excluded. Finally, affective associations, destination preferences as well as situational variables influence the destination choice.

In this chapter we focus on the question of what distinguishes travellers with a high nature affinity from those with a medium or low one. More specifically, we look at their travel push factors. All models for travel decision-making (Sirakaya and Woodside, 2005) assume that a travel decision is based on an individual mix of different expectations or motives of the traveller and not on one single factor alone, e.g. the natural environment of the destination. Destination choice is based on a complex combination of factors pushing the traveller towards a set of destinations during the process of decision-making (Smallman and Moore, 2010). Therefore, we first give a short overview of literature discussing travel push factors. Then we present an empirical study from five countries (CH, DE, FIN, IT, USA) analysing the following questions:

- Which travel push factors are of outstanding importance for people with high nature affinity?
- Which combinations of travel push factors are found most frequently in the group of travellers with high nature affinity?
- Do we see intercultural differences of travel push factors and personal values between travellers from different countries?

The chapter will close with conclusions for marketers and managers of destinations with a high potential for nature-based tourism.

Travel Push Factors in Literature

The reasons people engage in leisure activities have been a major focus of tourism research. Since motivation is one of the major driving forces behind every human behaviour, it is seen as a crucial factor in tourism (Crompton, 1979). Motivation to travel is the 'set of needs and attitudes which predisposes a person to act in a specific touristic goal-directed way' (Pizam *et al.*, 1979). It is believed that motivations are largely constant, fundamental human traits (Manfredo *et al.*, 1996). However, they might vary from trip to trip depending on time, money, travel partner, etc. But, generally, they do not change fundamentally, or only slowly over time. In rare instances, motivations can shift irregularly due to, for example a disruptive event such as childbirth, a new relationship or sickness. Moreover, motivations are not to be considered as separate entities, as they complement each other, and thus a single motivation cannot be used to ascribe causation to a certain behaviour (Crompton, 1979).

Psychographic personality types of vacationers' travel patterns and preferences differ in main motivations to travel (Plog, 1974). On the basis of in-depth interviews, Crompton (1979) identified nine travel motivations and classified seven of them as socio-psychological (neither related to nor derived from a particular destination's attributes) and two of them as cultural (they can be related to a destination). The leisure motivation scale has been applied to tourism by Ryan and Glendon (1998) in the UK, who found four motivational factors: social dimension, relaxation dimension, intellectual dimension, and competence-mastery dimension. Another scale was proposed by Manfredo *et al.* (1996). Their REP scale (recreation experience preference scale) focuses on the desired outcomes of travel that help people achieve certain psychological and physical objectives. They created a comprehensive list of leisure benefits through a meta-analysis from studies conducted between 1975 and 1979 in an American setting.

Similarly, Moscardo *et al.* (1996) linked travel benefits with activities and destination choice and found eight travel benefit factors.

The distinction between push and pull factors has also been accepted in literature. Push factors refer to the traveller as subject and discuss the elements that influence the decision to travel (e.g. escape, nostalgia); pull factors are elements of the touristic product (e.g. attractions or services), attracting travellers with a profile of push factors that match these elements. This way, push factors are temporally antecedent to pull factors (Dann, 1977). In tourism research, benefits were most often associated with travellers' motivations (Frochot and Morrison, 2000), and for a destination to continue to have some pulling power on a person its perceived pull factors must be in line with the motivations or push factors of the traveller (Uysal *et al.*, 2009) and the expected rewards (benefits gained).

Focusing on motivations and related travel push factors, Bieger and Laesser (2002) segmented the Swiss market using ten different motivations that resulted in four clusters (compulsory travel, cultural hedonism, family travel, and personality marketing), while Lee *et al.* (2002) distinguished six push motivational factors (escape, novelty, relaxation, travel brag, hedonism and family) for German travellers going overseas. A German travel analysis (FUR, 2014) uses 29 items to find the most important travel motivations of Germans and it cites, among others, sun and warmth, distance from daily routine, and relaxation as major travel motivations. Pesonen (2012) distinguished four segments of Finnish rural tourists that differed in terms of travel motivations, namely social travellers, wellbeing travellers, home region, and family travellers. Dong *et al.* (2013) segmented American tourists to a rural area in Pennsylvania and named five factors of benefits sought for visiting that area, namely personal growth and escape, nature and rural exploration, relaxation, social bonding seekers, and family fun, while Loker and Perdue (1992) named only four benefit factors for Americans in North Carolina. However, there is a lack of a universally accepted scale to measure tourist motivation, and authors tend to create their own lists and scales to measure it, adapting scales from previous publications and adding or deleting items to fit their purpose.

Methodology

The presented results are taken from a study that was conducted in five countries (no. of participants: DE, 272; CH, 254; IT, 252; FIN, 253; USA, 500) in spring 2021 about sustainability and sustainable tourism. The study and the participants' recruitment were conducted by the commercial online panel provider Norstat. The participants had to be frequent travellers who had left on vacations lasting at least 5 days each year for the preceding 5 years. The survey contained questions about the participants' socio-demographic data, including gender, age, education level, household size, kids in household, living in rural or metropolitan area, employment status, income, and zip code for regional distribution. To ensure the representivity of the survey, quota-sampling was applied. The quotas were based on the above-listed socio-demographic items and statistics taken from the census of the national statistical offices of each of the five countries.

The study included a list of travel push factors/motivations and recreational benefits that participants had to rate according to the importance of their last holiday on a six-point Likert scale ranging from very important to not important at all. As there is no universally accepted scale to measure tourist motivation available in literature, a list consisting of 23 items was created by the authors based on available scales in literature for the participating countries, combining and merging similar factors and motivations. This study uses items published and used by Loker and Perdue (1992), Manfredo *et al.* (1996), German travel analysis (FUR, 2014), Pesonen (2012) and Bieger and Laesser (2002). All items consisted of a title and a short description; see, for example, item two in Table 9.1, 'discovery', with the description 'discovery of unknown; other countries and cultures; adventure, thrill, experiencing excitement'.

Another section of the questionnaire included the shortened version of the Portrait Values Questionnaire (Schwartz, 2001), as it is a widely accepted measure for values across nations. The PVQ consists of 21 items and was included using the method and answer scale proposed by Schwartz. Participants had to rate

Table 9.1. Push factors derived from literature.

Push factor	Description
New experiences	Having many new experiences, new impressions
Discovery	Discovery of unknown; other countries and cultures; adventure, thrill, experiencing excitement
Socializing	Meeting and talking to new people; observing other people; making contact with locals
Entertainment and fun	Having fun; enjoying entertainment and partying; being together with people who enjoy the same things
Pleasure	Enjoying life with all senses; getting pampered; indulging in luxury for some days
Family, partner and friends	Having time for and sharing time with family, partner and friends; doing things with members of my group
Sun and warmth	Enjoying sun, warmth; sunbathing and suntan
Relaxation	Just relaxing, chilling, lounging, doing nothing
Tension release	Distance from daily routine; avoidance of everyday responsibilities
Mental distance	Giving your mind a rest; slowing down; getting into other thoughts
Health and physical fitness	Maintaining or improving my health status; keeping me physically fit; exercise and feeling good after being physically active
Nature and scenery	Being close to nature; enjoying pristine nature; viewing beautiful scenery; observing and studying nature
Physical activities	Being physically active; doing sports; outdoor activities such as hiking, biking, fishing, hunting
Personality	Rediscovering myself; gaining self-fulfilment and self-confidence; introspection and mindfulness
Enrichment	Enriching perspective on life; broadening personal horizons; taking new or other perspectives
Recognition	Recognition; sharing my trip on social media; visiting and sharing places friends haven't been; gaining prestige
Spirituality	Development and growth of personal spiritual values; reflecting on religious or other spiritual values
Nostalgia	Bringing back and reflecting on pleasant memories; thinking about good times had in the past; visiting familiar places
Interpersonal escape	Experiencing tranquillity, solitude, peace; being on one's own away from people and crowds
Learning and abilities	Learning something; development of skills and abilities; testing abilities; being creative
Autonomy and leadership	Being independent and free; having no obligations; doing things own way; being in control of things that happen

Continued

Table 9.1. Continued

Push factor	Description
Safety and social security	Being safe; avoiding the unexpected; being with others if you need them; to be near considerate people
Motion and flow	Being in motion; being on the move; getting around; going with the flow

Table 9.2. Socio-demographic data of the sample.

		DE (272)	IT (252)	CH (254)	FIN (253)	USA (500)	Total (1531)
Gender	Male	50.0%	52.4%	51.2%	43.1%	47.0%	47.7%
	Female	49.3%	47.6%	48.4%	56.1%	52.0%	51.0%
	Diverse	0.7%	0.0%	0.4%	0.8%	1.0%	1.3%
Age group	18–29	21.7%	19.8%	25.2%	27.3%	27.2%	25.0%
	30–44	28.7%	23.4%	27.2%	28.1%	37.4%	29.6%
	45–65	49.6%	56.7%	47.6%	44.7%	35.4%	45.3%
College degree	Yes	39.7%	55.2%	46.5%	32.0%	51.2%	44.9%
	No	60.3%	44.8%	53.5%	68.0%	48.8%	55.1%
Household size	1 person	41.9%	16.7%	27.6%	38.7%	19.4%	30.4%
	2 persons	33.1%	34.9%	37.0%	36.0%	32.4%	33.3%
	3 or more persons	25.0%	48.4%	35.4%	25.3%	48.2%	36.3%
Children under 14	Yes	16.9%	17.9%	24.4%	21.7%	31.6%	22.4%
	No	83.1%	82.1%	75.6%	78.3%	68.4%	77.6%
Living area	Metropolitan area/large city	69.5%	61.1%	58.3%	78.3%	57.8%	65.2%
	Medium/small town or rural area	30.5%	38.9%	41.7%	21.7%	42.2%	34.8%
Monthly pre-tax household income	Below €2000/ CHF 3000 / $3000	30.5%	36.5%	20.1%	24.5%	26.0%	27.3%
	Up to €4000/ CHF 6000 /$6000	34.9%	41.7%	41.7%	37.6%	35.4%	36.6%
	More than €4000/CHF 6000 / $6000	34.6%	21.8%	38.2%	37.9%	38.6%	36.1%

the similarity of themselves to short descriptions of persons on a six-point Likert scale ranging from 'very much like me' to 'not like me at all'.

The questions and items were translated from German into English, Italian and Finnish and the translations were subsequently cross-checked by and discussed with native speakers. Results were then analysed quantitatively using SPSS 25. Table 9.2 presents the socio-demographic structure of the sample.

We used two items to classify study participants to be of high, medium or low nature

Table 9.3. Nature affinity as composed of push factor 'nature and scenery' (I1) and personal value 'care for nature' (I2)

(I2) Care nature (PVQ) / (I1) Importance Nature	Very much like me	Like me	Somewhat like me	A little like me	Not like me	Not like me at all
Very important Important	High affinity (675)					
Somewhat important Somewhat unimportant			Medium affinity (711)			
Not important Not important at all			Low affinity (145)			

Table 9.4. Nature affinity by country.

	Country					
Nature affinity	CH	DE	FIN	IT	USA	All
High	42.5%	44.5%	36.3%	48.9%	46.2%	44.1%
Medium	48.4%	47.8%	53.0%	44.4%	42.4%	46.4%
Low	9.1%	7.7%	10.7%	6.7%	11.4%	9.5%

affinity. Both items were measured on a six-point Likert scale. The first item (I1) was the push factor that asked about the importance of nature during a vacation (see Table 9.1) – 'being close to nature, enjoying pristine nature, view beautiful scenery, observing and studying nature'. This was measured from 'very important' to 'not at all important'. As the second item (I2), we used a universalism value item from Schwartz's (2001) Human Value Scale – 'He/she strongly believes that people should care for nature. Looking after the environment is important to him/her'. We chose this second item as values influence behaviour (Schwartz and Cieciuch, 2021). Environmentalism or environmentally significant behaviour is based on altruistic values, such as 'respecting the earth, harmony with other species' and 'protecting the environment, preserving nature' (Stern *et al.*, 1999; Stern, 2000). Hence, we define nature affinity not only by the importance of nature as push factor but also by the universalism value describing the appreciation and protection for nature and environment. As mentioned above, the classifications were made as follows:

High: I1: ((very important) OR (important)) AND I2: ((very much like me) OR (like me))

Medium: I1: ((somewhat important) OR (somewhat unimportant)) AND I2: at least (a little like me)) OR I1: (at least (somewhat unimportant)) AND I2: ((somewhat like me) OR (a little like me))

Low: All other

Table 9.3 exemplifies the classification for nature affinity at the levels high, medium, and low. Out of the 1531 study participants, 675 were classified as high, 711 as medium and 145 as low.

Findings

In a first step, the defined three groups with high, medium and low nature affinity were compared concerning socio-demographic characteristics. No significant structural differences could be found for gender, age, education level or income (cross-tabs with chi-square independency test). When looking for structural differences between the share of the three groups in the five countries (see Table 9.4), some differences became visible, especially for Finland, which has a slightly lower share of high nature-oriented people. However, these differences are not significant when using a chi-square independency test at 5% level. Thus, comparing the three groups' socio-demography and group structure across the countries was disregarded for further analysis.

To analyse the travel push factors from Table 9.1, a factor analysis using principal component analysis (PCA) was done. As the 'nature and scenery' push factor was used to define the three levels of nature affinity, this one was excluded from the PCA. A solution with seven factors explained 65.9% of the total variance. The communalities of the 22 remaining travel push factors ranged from 0.56 to 0.86, which indicates that at least half of the variance from each push factor was used for the extracted

Table 9.5. Results from PCA and group comparison using ANOVA.

Travel factors		PCA factor score average			ANOVA	
PCA factor	Travel push factors	High	Medium	Low	F	Sig.
Healthy activities	Physically active; sports and outdoor activities; maintaining and improving health; staying fit	–0.25	0.09	0.72	67.21	<0.001
Introspection	Tranquility; solitude; rest for mind; rediscovering oneself; feeling safe near to considerate people; development of personal spiritual values	–0.28	0.16	0.55	62.50	<0.001
Discovery	Discover the unknown; have new experiences; enriching perspective; new people; learning and skills	–0.24	0.12	0.56	50.61	<0.001
Escape	Distance from daily routine and everyday responsibilities; being independent and free	–0.20	0.13	0.30	27.60	<0.001
Personal relations	Having time for and sharing time with family, partner and friends; doing things with members of my group	–0.10	0.06	0.16	6.59	0.001
Relax	Enjoying sun; relaxing and doing nothing; enjoying life with all senses indulging in luxury for some days	–0.07	0.05	0.09	3.29	0.037
Socializing and fun	Recognition on social media; having fun and party; pleasant memories; being in motion; going with the flow	–0.02	0.00	0.09	0.68	0.503

solution. After a varimax rotation, the seven factors could be well interpreted (see Table 9.5). The seven factors then were compared for the three groups high, medium and low nature affinity using an analysis of variance. The ANOVA revealed significant group differences for six of the seven factors. As the original scale of the push factors ranged from 1 – very important to 6 – not important at all, negative average values of the factor scores indicate a tendency to the direction of 'very important' while positive values indicate a tendency to the direction of 'not important at all'.

Looking at the ANOVA results in Table 9.5, the strongest differences between the three nature affinity groups can be observed for the three factors *healthy activities*, *introspection* and *discovery*. For the group with high nature affinity, *introspection* has a slightly higher importance than *healthy activities*. The average importance for *discovery* ranks similarly to *healthy activities*. Moreover, the groups differ strongly in the factors *discovery* and *escape*. Lower, but still significant, differences can be found for the factors *personal relations* and *relax*, while the factor *socializing and fun* does not differ significantly across the three groups.

Even though the three groups show significantly different levels of importance for the extracted PCA travel factors and are well separated, the three groups cannot be assumed to be homogenous in themselves. Therefore, in a further step, the group of study participants with high nature affinity was analysed concerning subgroups using cluster analysis. First, hierarchical clustering with Ward method, Euclidean distance and elbow criterion was applied and a five-clusters solution could be found. After, the centroids of this solution were used for a second clustering by K-Means algorithm. Seventy-nine per cent of the cluster members remained in their original cluster from the initial hierarchical clustering. These participants can be seen as the core members of each cluster. To describe these five subgroups of travellers with high nature affinity, the cluster centroids of the core groups were used. Table 9.6 shows the differences of the five clusters for all seven PCA travel push factors from the centroid of all remaining cases in the core clusters. Negative values indicate a higher interest in a PCA factor by the members of a cluster compared to all members; positive values show a lower interest.

Based on these results the role of nature during a vacation can be interpreted for the five clusters as follows:

1. nature as a possibility to escape from daily routine and obligations without one's family/partner or peer group (9.4%)
2. nature as place to share time with family/partner – low interest in discovery (10.5%)
3. nature as place to share time with family/partner – discovery and introspection (19.7%)
4. nature as scenery for socializing and fun, healthy activities and relaxing (39.3%)
5. nature as scenery to relax and share time with family/partner – wish for discovery but low interest in physical activities or introspection (21.2%)

Table 9.6. Five-cluster solution group high nature affinity.

	Cluster					
	1	2	3	4	5	
Share of cluster	9.4%	10.5%	19.7%	39.3%	21.2%	Centroid all
PCA travel factor	Difference from centroid all					
Healthy activities	–0.07	0.33	–0.12	–0.40	0.73	–0.22
Introspection	0.29	–0.18	–0.32	–0.27	0.76	–0.30
Discovery	0.46	1.96	–0.39	–0.16	–0.51	–0.27
Escape	–0.61	–0.20	0.44	0.13	–0.28	–0.18
Personal relations	2.14	–0.45	–0.43	0.00	–0.33	–0.06
Relax	0.72	–0.05	0.74	–0.34	–0.35	–0.10
Socializing and fun	0.59	0.07	1.02	–0.83	0.29	–0.03

Table 9.7. Country differences for high nature affinity group.

	CH	DE	FIN	IT	USA	Mean all	F	Sig.
Socializing and fun	0.45**	0.47**	−0.03	0.00	−0.50**	−0.02	28.05	<0.001
Introspection	−0.40**	−0.25	0.33**	−0.32	−0.47**	−0.28	16.16	<0.001
Personal relations	−0.01*	0.17**	−0.48**	0.06**	−0.21**	−0.10	8.96	<0.001
Escape	−0.24	−0.36**	−0.31**	−0.18	−0.07**	−0.20	2.97	0.019
Healthy activities	−0.34*	−0.30	−0.08**	−0.15**	−0.30	−0.25	1.69	0.150
Discovery	−0.20	−0.29	−0.13**	−0.40**	−0.20	−0.24	1.33	0.258
Relax	−0.06	−0.17**	0.08**	−0.03	−0.11	−0.07	1.15	0.332

Finally, we analysed the differences between the five countries of the group with high nature affinity of the study. Table 9.7 shows the average factor scores for the five countries. Again, a negative value of an average factor score indicates the direction towards 'very important' while positive values point towards 'not important at all'. Two types of tests were used to assess significant differences between the countries. ANOVA was used to test differences between the five countries generally. In addition, t-tests were done to look for significant deviations from the total average value, country. Significant deviations of average values at 1% level are marked in the table by two asterisks; those at 5% level by one asterisk.

Looking at the ANOVA results, the strongest differences in country comparison can be found for the factor *socializing and fun*. This factor stands out for US travellers as very important for their vacation while it plays a minor role for Swiss and German travellers. Looking at the country values of the original travel push factors (Table 9.1) that have a high loading in the PCA on the factor *socializing and fun*, we detected that *having fun, enjoying entertainment and partying, being together with people who enjoy the same things* plays a very important role for US participants, while *recognition, sharing my trip on social media, visiting and sharing places friends haven't been, gaining prestige* is of secondary importance for them. Concerning the factor *introspection*, Finnish travellers show a much lower interest compared to all other countries in *experiencing tranquility, solitude, peace, being on one's own away from people and crowds*. In contrast, Finnish travellers assign an evident importance to personal relations. Even though the ANOVA does not show significant differences for the factors 'discovery' and 'relax', the detailed analysis using t-tests reveals some further country-specific characteristics. Italians are above-averagely interested in *discovery*, whereby *meeting and talking to new people, observing other people, making contact with locals* and *learning something, development of skills and abilities, testing abilities, being creative* play a central role. The factor 'relax' seems to be only of average importance for Finnish travellers, but the detailed analysis shows that the aspect *enjoying sun, warmth, sunbathing and suntan* is much below the average of the other countries while *just relaxing, chilling, lounging, doing nothing* is slightly above.

Conclusions

In our study we defined three groups of travellers with high, medium and low nature affinity using a nature-oriented travel push factor in combination with a nature and environment-oriented social value. First, we compared the three groups on the basis of socio-demographic variables. Surprisingly, we did not find significant differences. Nature affinity is not a question of gender, age, education level or income. Therefore, traditional market segmentation approaches based on socio-demography will fail for nature-based destinations. We then compared the three groups based on seven general travel push factors derived by a PCA factor analysis. Thereby, the crucial role of 'healthy activities' and 'introspection' became visible. Nature is not the primary focus of people who are drawn towards a nature-based location by one or both factors. The natural space is used for outdoor activities such as hiking, biking or swimming, or to find a calm atmosphere for solitude and reflection. Nature is reduced to a means to an end, a mere resource, and is not the subject of study or exploration. The same can be assumed for the factor 'escape', as nature is just a counter-world of ordinary and everyday life. Of course, most nature-based destinations promote their natural potential towards travellers with these three general push factors. However, this also, potentially, attracts travellers with lower nature affinity who place little value on the protection of nature or natural resource conservation.

Only for the push factors that were bundled in the factor 'discovery' is a connection of travellers' interests with nature as the central object likely: discovering new things, learning something, broadening personal horizon, or having new impressions. Observing flora and fauna, listening to a ranger or guide explain the particularity of a natural element and meeting farmers who do landscape conservation, are just a few examples of many of what discovery might entail. Managers from nature-based destinations must be aware that this necessitates the creation of specific products and related services that allow the travellers to immerse themselves in nature in the role of a discoverer.

The cluster analysis of the group of travellers with high nature affinity backs up these basic findings. Only two out of five groups attach particular importance to discovery. Only one of these two groups wishes to be physically active, emphasizing the importance of visitor centres in nature and national parks as well as an offer of short and easy-to-access experience trails. Such installations allow those who are unable (e.g. due to a disability) or unwilling to venture out into the rough, wild nature, to fulfil their desire for discovery as well.

Examining the seven general push factors for the five countries in the group of travellers with high nature affinity, the factor 'socializing and fun' showed the most significant differences. Travellers from the US love to be entertained while German and Swiss travellers tend to place less emphasis on this factor, even though they enjoy it as well. Swiss and German nature lovers seem to see nature as something that has to be protected and should not be used for entertainment. In discussions regarding the future development of nature-based destinations in Europe, the term 'disneyization' is common (N.N, 2003), since it is commonly used as an euphemism for what should not happen in any case. Excessive entertainment is associated with over-exploitation and crowding, degrading the experience in nature-based destinations.

Many nature-based destinations report an over-visiting of natural places of unique attractiveness. Destination managers identified social media picture posting as a main driver, as 'Instagramable' places are shared by social media influencers. In our study, participants generally assigned a low or medium priority to the specific travel push factor 'recognition, sharing my trip on social media, visiting and sharing places friends haven't been, gaining prestige'. However, we found a significantly greater share of travellers in the group with high nature affinity to whom this push factor is important or very important (34.1%). In this field, further research is needed to develop destination strategies to reduce pressure caused by social media.

However, destination managers should not rule out *a priori* the possibility of connecting discovery and entertainment in a sustainable way as social media can be used to increase awareness for the delicate ecosystems of nature-based destinations.

References

Bieger, T. and Laesser, C. (2002) Market segmentation by motivation: the case of Switzerland. *Journal of Travel Research* 41(1), 68–76. DOI: 10.1177/004728750204100110.

Cohen, S.A., Prayag, G. and Moital, M. (2014) Consumer behaviour in tourism: concepts, influences and opportunities. *Current Issues in Tourism* 17(10), 872–909. DOI: 10.1080/13683500.2013.850064.

Craig, C.A. (2021) Camping, glamping, and coronavirus in the United States. *Annals of Tourism Research* 89, 103071. DOI: 10.1016/j.annals.2020.103071.

Crompton, J.L. (1979) Motivations for pleasure vacation. *Annals of Tourism Research* 6(4), 408–424. DOI: 10.1016/0160-7383(79)90004-5.

Dann, G.M.S. (1977) Anomie, ego-enhancement and tourism. *Annals of Tourism Research* 4(4), 184–194. DOI: 10.1016/0160-7383(77)90037-8.

Decrop, A. (2010) Destination choice sets: an Inductive longitudinal approach. *Annals of Tourism Research* 37(1), 93–115.

Dong, E., Wang, Y., Morais, D. and Brooks, D. (2013) Segmenting the rural tourism market: the case of Potter County, Pennsylvania, USA. *Journal of Vacation Marketing* 19(2), 181–193. DOI: 10.1177/1356766712471231.

Frochot, I. and Morrison, A.M. (2000) Benefit segmentation: a review of Its applications to travel and tourism research. *Journal of Travel & Tourism Marketing* 9(4), 21–45. DOI: 10.1300/J073v09n04_02.

FUR (2014) *Abschlussbericht zu dem Forschungsvorhaben: Nachfrage für Nachhaltigen Tourismus im Rahmen der Reiseanalyse (Final report on the research project: Demand for sustainable tourism in the context of travel analysis)* (p.77). Bundesministerium für Umwelt, Naturschutz, Bau und Reaktorsicherheit (BMUB). Available at: https://www.bmu.de/fileadmin/Daten_BMU/Download_PDF/Tourismus_Sport/nachhaltiger_tourismus_nachfrage_bericht_bf.pdf (accessed 28 June 2023).

Hall, C.M., Scott, D. and Gössling, S. (2020) Pandemics, transformations and tourism: be careful what you wish for. *Tourism Geographies* 22(3), 577–598. DOI: 10.1080/14616688.2020.1759131.

Lebrun, A.-M., Su, C.-J. and Bouchet, P. (2021) Domestic tourists' experience in protected natural parks: a new trend in pandemic crisis? *Journal of Outdoor Recreation and Tourism* 35, 100398. DOI: 10.1016/j.jort.2021.100398.

Lee, G., O'Leary, J.T., Lee, S.H. and Morrison, A. (2002) Comparison and contrast of push and pull motivational effects on trip behavior: an application of a multinomial logistic regression model. *Tourism Analysis* 7(2), 89–104. DOI: 10.3727/108354202108749970.

Loker, L.E. and Perdue, R.R. (1992) A benefit-based segmentation of a nonresident summer travel market. *Journal of Travel Research* 31(1), 30–35. DOI: 10.1177/004728759203100107.

Manfredo, M.J., Driver, B.L. and Tarrant, M.A. (1996) Measuring leisure motivation: a meta-analysis of the recreation experience preference scales. *Journal of Leisure Research* 28(3), 188–213. DOI: 10.1080/00222216.1996.11949770.

Moscardo, G., Morrison, A.M., Pearce, P.L., Lang, C.-T. and O'Leary, J.T. (1996) Understanding vacation destination choice through travel motivation and activities. *Journal of Vacation Marketing* 2(2), 109–122. DOI: 10.1177/135676679600200202.

N.N (2003) 'Disneysierung' der Schweiz wird es nicht geben. *Tourist Ausria International* 1650, 8.

Pesonen, J.A. (2012) Segmentation of rural tourists: combining push and pull motivations. *Tourism and Hospitality Management* 18(1), 69–82. DOI: 10.20867/thm.18.1.5.

Pizam, A., Neumann, Y. and Reichel, A. (1979) Tourist satisfaction. *Annals of Tourism Research* 6(2), 195–197. DOI: 10.1016/0160-7383(79)90146-4.

Plog, S.C. (1974) Why destination areas rise and fall in popularity. *Cornell Hotel and Restaurant Administration Quarterly* 14(4), 55–58. DOI: 10.1177/001088047401400409.

Ryan, C. and Glendon, I. (1998) Application of leisure motivation scale to tourism. *Annals of Tourism Research* 25(1), 169–184. DOI: 10.1016/S0160-7383(97)00066-2.

Schlemmer, P. and Schnitzer, M. (2021) Research note: ski touring on groomed slopes and the COVID-19 pandemic as a potential trigger for motivational changes. *Journal of Outdoor Recreation and Tourism* 41, 100413. DOI: 10.1016/j.jort.2021.100413.

Schwartz, S.H. (2001) A proposal for measuring value orientations across nations (Ch. 7). In: European Social Survey, City University, London (ed.) *European Social Survey Core Questionnaire Development* (Ch. 7). Available at: http://www.europeansocialsurvey.org/docs/methodology/core_ess_questionnaire/ESS_core_questionnaire_human_values.pdf (accessed 28 June 2023).

Schwartz, S.H. and Cieciuch, J. (2021) Measuring the refined theory of individual values in 49 cultural groups: psychometrics of the revised portrait value questionnaire. *Assessment* 29(5), 1005–1019. DOI: 10.1177/1073191121998760.

Sirakaya, E. and Woodside, A.G. (2005) Building and testing theories of decision making by travellers. *Tourism Management* 26(6), 815–832. DOI: 10.1016/j.tourman.2004.05.004.

Smallman, C. and Moore, K. (2010) Process studies of tourists' decision-making. *Annals of Tourism Research* 37(2), 397–422. DOI: 10.1016/j.annals.2009.10.014.

Stankov, U., Filimonau, V. and Vujičić, M.D. (2020) A mindful shift: an opportunity for mindfulness-driven tourism in a post-pandemic world. *Tourism Geographies* 22(3), 703–712. DOI: 10.1080/14616688.2020.1768432.

Stern, P.C., Dietz, T., Abel, T.D., Guagnano, G. and Kalof, L. (1999) A value-belief-norm theory of support for social movements: the case of environmentalism. *Human Ecology Review* 6(2), 18.

Stern, P.C. (2000) Toward a coherent theory of environmentally significant behavior. *Journal of Social Issues* 56(3), 407–424. DOI: 10.1111/0022-4537.00175.

Tauber, V. and Bausch, T. (2022) Will COVID-19 boost sustainable tourism: wishful thinking or reality? *Sustainability* 14(3), 1686. DOI: 10.3390/su14031686.

Uysal, M., Li, X. and Sirakaya-Turk, E. (2009) Push–pull dynamics in travel decisions. *Handbook of Hospitality Marketing Management* 28.

Woodside, A.G. and Lysonski, S. (1989) A general model of traveler destination choice. *Journal of Travel Research* 27(4), 8–14. DOI: 10.1177/004728758902700402.

10 The Sustainability of Scuba Diving Tourism: Principles and Implications

Daniela Marzo* and Iacopo Cavallini
Department of Economics and Management, University of Pisa, Italy

Abstract
Sustainable scuba diving tourism can be studied through the lens of systemic perspective that highlights the intersections between environmental, socio-cultural and economic aspects. Specifically, the key issue is how sustainable diving tourism can help restore and maintain coastal and marine ecosystems by managing the interdependent goals of recreation and resource protection. The establishment of marine protected areas is broadly considered as a milestone for the quality of underwater sites and for marine resources conservation, as well as one of the key strategies that contribute to pushing the diving system towards a path of sustainability. The systemic approach used to conceptualize the sustainable diving tourism demand advances the idea that the management of a tourism scuba diving destination and its outcomes are also related to the study of the multiple relationships between stakeholders to ensure participation in the decision-making process for the sustainability of marine resources.

Introduction

The debate on sustainable tourism development has become an urgent priority today, especially after the COVID-19 pandemic. Ecotourism, nature-based tourism and heritage tourism come from the management perspective of sustainable tourism (Butler, 1991; Zolfani *et al.*, 2015) and promote a win-win scenario (Ardoin *et al.*, 2015) in which visitors are guaranteed a quality experience, the living standards for local communities are improved, and an ecologically oriented and profit-generating activity is promoted (Lim and McAleer, 2005).

Over 20 years after its theorization, the concept of sustainable tourism does not seem obsolete at all; on the contrary, it is capable of continuously enriching the scientific debate, especially due to the implications related to the theme of global warming and climate change.

This chapter aims to explore how a diver-oriented tourism strategy would contribute to the sustainable development of a tourist destination.

Starting from the idea of sustainable tourism, the chapter focuses on scuba diving tourism, a particular nature-based segment that has significantly increased in recent years. The main characteristics of sustainable diving tourism are analysed through a systemic approach and then the strategies that contribute to pushing the diving system towards a path of sustainability are outlined.

*Corresponding author: daniela.marzo@ec.unipi.it

DOI: 10.1079/9781800621411.0010

Sustainability in Tourism

As in other sectors of the economy (Meadows *et al.*, 1972), the sustainable approach has emerged in the tourism sector, primarily in response to the negative effects of human activities, with the purpose of promoting a balance between the conservation of ecosystems and social and economic development. Several authors (Bramwell and Lane, 1993; Koens *et al.*, 2009; Mondino and Beery, 2018; Wondirad *et al.*, 2020) highlighted how the negative impact of mass tourism on the environment and society has generated a new awareness regarding the need to adopt tourism development strategies that respect the natural environment and the social wellbeing of local communities.

Moving from the paradigm of sustainable development, introduced globally in 1987 with the Brundtland Report (WCED, 1987), the purpose of sustainable tourism is to seek a balance between the protection of biodiversity and natural resources, the safeguarding of the values and identity of local communities, and the promotion of economic benefits through the improvement of the quality of life of local populations (Liu, 2003) of both developed and developing countries (Swarbrooke and Horner, 2004; Mitchell and Hall, 2005). In this view, sustainable tourism, as Bramwell and Lane (2011) argued, can be considered the practical application of the idea of sustainable development to the tourism sector, with a focus on the social and economic wellbeing of local communities and on the recreational experience of tourists, for both present and future generations.

The growing debate within the academic community on sustainable tourism has received a strong impetus also thanks to the numerous international initiatives, such as those promoted by the World Tourism Organization (WTO), in order to spread a more forward-looking vision of tourism that promotes the intrinsic value of natural and cultural heritage. The 1995 World Conference on Sustainable Tourism was one of the first initiatives aimed at developing a systemic framework on the principles for sustainable tourism, which led to the drafting of the Charter of Lanzarote (Holden, 1995). A further contribution was made in 2015 with the promotion of the UN Sustainable Development Goals (SDGs), subsequently, in 2016, with the ratification of the Paris Agreement, and then again with the designation by the United Nations of 2017 as the International Year of Sustainable Tourism.

The growing interest for the sustainability of tourism has led to the development of different meanings of the notion, which differ in placing a greater emphasis on environmental protection or economic development issues (Bramwell and Lane, 2011). Many economists have described an economic growth that does not have irreversible negative effects on the environment (Goodland and Ledec, 1987; Solow, 1991; Turner, 1993; Beltratti, 1997), while environmentalists have pushed for a vision of sustainable tourism, with a responsible use of natural resources, which does not cause the deterioration of the environment (Rubenstein, 1993). Hunter (2002) argued that the idea of an 'equilibrium' could be used to mask economic growth as the primary goal. Dávid (2011) underlined the complexity of the notion of sustainable tourism in terms of an interactive and symbiotic connection between tourists, local communities, business companies, tourist attractions and nature.

According to Liu (2003), sustainable tourism embraces all types of tourism that contribute to sustainable development. There are many forms of 'alternative' tourism, compatible with the principles of sustainability, capable of meeting the needs of a specific tourist target. All these forms of sustainable tourism share a common objective: to promote respect for the environment and local communities.

On the other hand, Butler (1999) observed that the perspective of sustainable tourism aims not only at ensuring the development of 'cultural and environment-oriented' forms of niche tourism but also at making current forms of mass tourism as sustainable as possible. In this regard, Clark (2014) highlighted how sustainable tourism should not necessarily be considered as a positive alternative to mass tourism. According to this author, in fact, sustainability should be seen as an objective towards which every type of tourism (mass or niche) should work. A fundamental prerequisite for the promotion of a sustainable approach is that tourism has to be based on planning and management tools capable of driving appropriate behaviour, one that is compatible with respect for the environment and local identities.

The marine and coastal environment plays a leading role in the development of the tourism sector in many countries. In fact, there are many recreational activities linked to the coastal and marine environment, such as swimming, sunbathing, bathing, aquatic activities such as water sports (sailing, surfing, canoeing, etc.) or activities where people are in a more direct contact with marine resources, such as snorkelling and diving. Natural resources and the beauty of the marine and coastal areas attract many visitors, causing, especially at certain periods of the year, a great pressure on the environment and ecosystems, due to multiple factors such as nautical traffic, a greater consumption of water, increased waste production, and increased CO_2 emissions from air, road and sea transport. Scuba diving is one of the coastal and marine tourism segments that have expanded to a certain extent in recent years. The WTO estimated that there were 6 million certified divers already in 2001, about a third of whom came from Europe. The richness of marine biodiversity and the beauty of the seabed make the underwater recreational experience increasingly requested by tourists.

The Pillars of Sustainable Diving Tourism

According to the World Tourism Organization (UNWTO), scuba diving tourism can be defined as travelling to destinations with the main purpose of participating in scuba diving recreational activities.

Scuba diving has become a key segment of the tourism market and a multi-billion-dollar industry (Lucrezi *et al.*, 2017), in addition to being one of the fastest-growing recreational activities (Hunt *et al.*, 2013). The most popular scuba diving sites are located in tropical and subtropical regions, such as Australia, Egypt, South Africa and Thailand.

In recent years, many researchers have conducted studies on scuba diving tourism, which gave insight into various topics and issues that go from the divers' impacts on the marine environment to the divers' motivations, satisfaction and safety.

An interesting body of research has explored scuba diving from a systemic perspective to understand the relationships between the various elements of the system and to explore how to improve a sustainable management of these activities (Dimmock and Musa, 2015). A systemic approach is useful to understand the complexity of the issues around the scuba diving segment and the inter-relations between humans and their environments.

Moving from this approach, the study of scuba diving tourism and its sustainability could be developed through the analysis of the key elements that characterize the demand for scuba diving tourism (Fig. 10.1). This approach helps clarify the major issues to be tackled and the key priorities to be pursued for a sustainable local development. The marine environment is the core element of the framework, as it represents the main motivation for diving tourists (Dearden *et al.*, 2007; Uyarra *et al.*, 2009). The second element regards the evaluation of the socio-economic impacts (as well as the associated benefits) of the scuba diving tourism at local level (Dicken, 2014; Saayman and Saayman, 2018). Finally, the cultural dimension encompasses a number of aspects related to how divers perceive the underwater world (Lindsey and Holmes, 2002; Dodds *et al.*, 2010). The three elements identified in this framework are inter-related and equally concur to characterize the scuba diving tourism demand. This background information helps ensure that tourist destinations are managed sustainably (Uyarra *et al.*, 2009) and supports an equitable development path (Grilli *et al.*, 2021). Understanding the systemic dimensions of sustainable scuba diving tourism can help build the profile of a tourist destination and manage its related development, planning and governance issues (Dodds *et al.*, 2010). In addition, a better understanding of the pillars of sustainable scuba diving tourism from a systemic point of view can provide a broad knowledge to reduce negative impacts on the coastal and marine environment, generate local wellbeing, satisfy recreational underwater experiences for tourists and produce economic benefits. According to Lucrezi *et al.* (2017), the assessment of the sustainability of scuba diving tourism requires

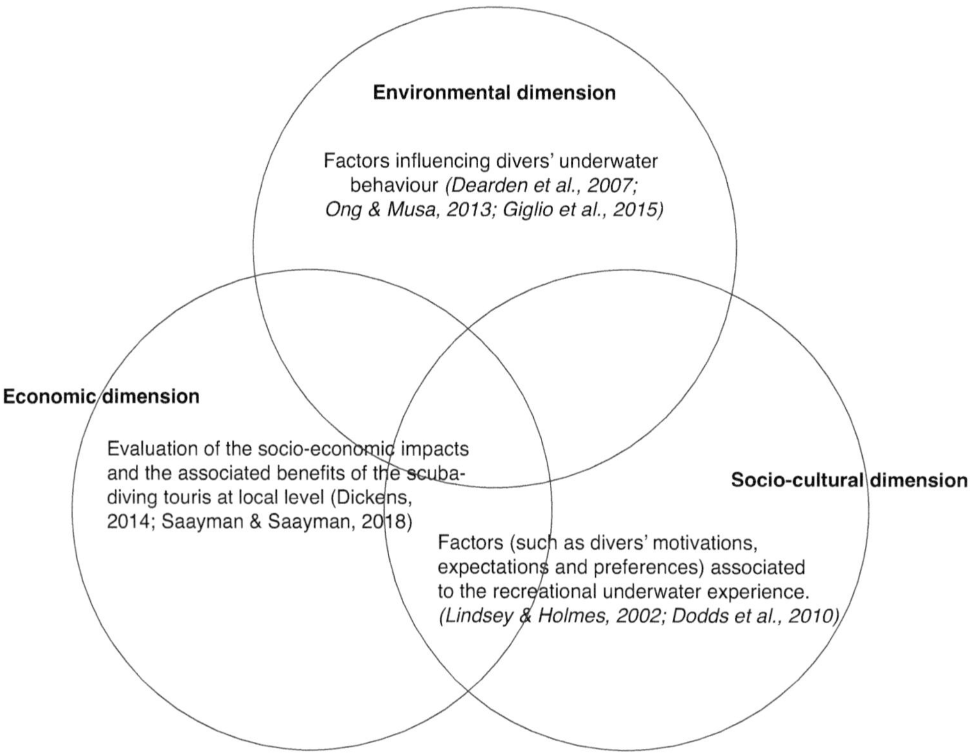

Fig. 10.1. The key scuba diving demand dimensions.

a holistic view to understand the intersections between environmental, socio-cultural and economic aspects.

The environmental dimension

Although scuba diving tourism is commonly described as a 'niche tourism', and often as a sustainable form of tourism, it can actually impact the natural assets, fish populations and habitats. The intensive recreational pressure of diving on certain tourist destinations can negatively affect the marine environment, causing damage to corals, disturbing marine life and causing marine pollution with boat traffic, thus generating environmental degradation, with the consequent dissatisfaction of divers (Wongthong and Harvey, 2014). A particularly prolific research branch has explored the factors that influence divers' underwater behaviour (Thapa *et al.*, 2005; Dearden *et al.*, 2007; Worachananant *et al.*, 2008; Ong and Musa, 2013; Lucrezi *et al.*, 2013, 2021; Giglio *et al.*, 2015; Hammerton and Bucher, 2015; Hayes *et al.*, 2017). Most of these studies provide strategies that help minimize man-induced pressures and protect the marine environment. Davis and Tisdell (1995) suggested that strict carrying capacity plans should be implemented to mitigate environmental damages and the overcrowding of dive sites.

The economic dimension

The economic approach aims to measure the benefits associated with the environment and the related tourism activities. Many tourist destinations promote their marine resources with the objective to become popular scuba diving destinations. The socio-economic perspective highlights the idea of scuba diving tourism as a source of revenue (Giddy and Rogerson, 2018), especially in developing countries, where it is a critical segment for tourism economies (Giddy and Rogerson, 2018). In fact, diving tourism often contributes to the economic development of the coastal communities that surround dive sites (Musa and Dimmock, 2012). In addition, the study of the economic dimension is useful for tourism planning and management (Saayman and Saayman, 2018). Many studies highlighted the economic benefits derived from scuba diving activities (Arin and Kramer, 2002; Asafu-Adjaye and Tapsuwan, 2008; Casey *et al.*, 2010; Thur, 2010; Vianna *et al.*, 2012; Schuhmann *et al.*, 2013, 2019; Saayman and Saayman, 2018). Some authors focused on the analysis of tourist expenditure (Craggs and Schofield, 2006; Kruger *et al.*, 2010; Saayman *et al.*, 2011), while others studied travel motivation factors (Saayman and Saayman, 2009; Kruger *et al.*, 2013).

All these research studies that demonstrated the economic benefits of scuba diving contribute, above all, to identifying successful financial strategies for the conservation and management of marine and coastal areas.

The socio-cultural dimension

The socio-cultural aspect of scuba diving tourism consists in the study of the divers' profile through information about their motivations, expectations and preferences associated with the recreational underwater experience. This kind of knowledge can provide an explorative idea about tourists' inclination towards environmental conservation (Lindsey and Holmes, 2002; Dodds *et al.*, 2010), contributing to both the sustainable management and the strategic planning of a tourist destination.

A broader literature has dealt with this topic, with studies on scuba divers' motivations, perceptions and preferences in several diving destinations around the world (Williams and Polunin, 2000; Mundet and Ribera, 2001; Musa, 2002; Thapa *et al.*, 2005; Maccarthy *et al.*, 2006; Musa *et al.*, 2006; Dearden *et al.*, 2007; Gössling et al., 2008; Uyarra *et al.*, 2009; Ong and Musa, 2011; Ong and Musa, 2012; Edney, 2012; Paterson *et al.*, 2012; Lucrezi *et al.*, 2013; Dicken, 2014; Giglio *et al.*, 2015; Lucrezi *et al.*, 2019).

A whole branch of studies (Mundet and Ribera, 2001; Musa *et al.*, 2006; Giglio *et al.*, 2015) revealed an association between scuba diving experiences and divers' preferences and motivations. The authors' contribution is related to the assessment of divers' motivations, preferences, expectations and related implications for sustainable tourism management (Davis and Tisdell, 1996; Williams and Polunin, 2000; Gössling et al., 2008; Uyarra *et al.*, 2009; Edney, 2012; Giglio *et al.*, 2015; Lucrezi *et al.*, 2019).

In most of these studies, researchers found that, for experienced divers, the naturalistic features of the marine environment are the most important drivers for their choice of a dive site. Giglio *et al.* (2015) explored the preferences and perceptions of divers visiting Brazil's coral reefs and found that experienced divers were more interested in cryptic and small fish, while novices preferred the megafauna. Some authors explored the different attributes of the scuba diving experience that affect divers' satisfaction (Mundet and Ribera, 2001; Musa, 2002; Maccarthy *et al.*, 2006; Musa *et al.*, 2006; Paterson *et al.*, 2012). Musa (2002) highlighted that the abundance and variety of marine species, the accessibility of dive sites and the social aspect of sharing the underwater experience are the major attributes that positively influence divers' satisfaction.

Tourism Planning: Strategies for a Sustainable Diving Tourism

The growing demand for new forms of tourism in contact with nature has increased the importance of marine protected areas (MPAs) as popular sea tourism destinations, where special regulations and careful management principles mitigate the potential negative impact of tourism on the natural environment. Without the implementation of plans, regulations and management policies, the economic purposes of tourism can have negative effects on ecosystems. In places where sustainable tourism strategies are planned and implemented, such as in many

MPAs, tourism can encourage the conservation of the natural and cultural heritage.

Therefore, sustainable tourism principles have found a fertile soil in marine protected areas (MPAs) (Dikou, 2011), the richness and biodiversity of which are preserved and improved and attract a large number of tourists, thus creating an interdependence between the visitors' experience and the conservation strategy of the MPAs. The creation of an MPA is broadly considered a flagship for the quality of underwater sights and for the conservation of marine resources (Dicken, 2014). MPAs are useful tools for the conservation of biodiversity because they exclude or control consumptive uses of marine resources and enhance non-consumptive uses (Lucrezi *et al.*, 2017). The establishment of an MPA is generally accompanied by careful planning aimed at avoiding unsustainable tourism pressure (Lindsey and Holmes, 2002) and enhancing social acceptance. A key starting phase in an MPA tourism planning process oriented towards sustainability should be the study of the tourist system that revolves around the targeted marine and coastal resources and its environmental and economic impact on the ecosystem.

In the context of an MPA, divers are often the main or, at least, a significant category of ecotourists (Lucrezi *et al.*, 2017) as they contribute to generating economic benefits, especially in areas characterized by peripheral, monostructured economies (Garrod and Gössling, 2008).

Many of the most popular diving destinations are part of MPAs and, in some cases, the existence of an MPA increases visitors' satisfaction (Davis and Tisdell, 1995). The institutional purposes of MPAs enucleate the values of sustainability and the mission of conservation for present and future generations, thus providing contexts for experimenting with new models for a sustainable management of tourism resources.

MPAs encourage diving and snorkelling activities, as they are more focused on the natural heritage and share a sensitivity for the conservation of ecosystems, often promoting awareness and education strategies for tourists and economic operators to mitigate the impacts that also a niche tourism can have on an ecosystem. Information regarding divers' preferences and motivations, their expenditure, environmental awareness, and commitment towards conservation, are important for the development of effective sustainable tourism strategies.

The use of this information, *inter alia*, may help identify preferred diving sites to manage use conflicts or reduce overcrowding. Understanding divers' preferences and motivations can also assist tourism operators in developing multiple communication strategies to attract a broader range of tourists and charge a fee to raise funds for conservation programmes. As suggested by several scholars (Cater, 2008; Townsend, 2008; Ong and Musa, 2012; Giglio *et al.*, 2018; Lucrezi *et al.*, 2018), the link between the attractiveness of the diving sector and the fascination of learning about underwater life can promote education on the conservation of underwater ecosystems, and consequently a more sustainable tourism development. The educational approach is even more important for those tourist destinations that are in the process of institutionalizing an MPA (Lindsey and Holmes, 2002), as additional awareness-raising initiatives may be included as part of the standard training of recreational and professional scuba divers. Pre- and post-dive briefings are excellent opportunities to emphasize the educational message, making divers aware of the positive and negative impact their behaviour has on the marine environment (Townsend, 2008; Giglio *et al.*, 2018; Lucrezi *et al.*, 2018).

Implications for a Sustainable Scuba Diving Tourism System

Sustainable scuba diving improves the conservation of ecosystems while generating income and supporting the development of local communities, especially in the context of MPAs, where multiple players (with different perceptions and needs) interact with each other in a complex manner. Sustainability is a fundamental prerequisite for a successful tourism planning and development (Byrd, 2007). Achieving sustainable outcomes with effective sustainable tourism strategies requires the implementation of an approach that encompasses the views of the different stakeholders involved in the scuba diving tourism system, such as divers, the host community and the tourism industry (Dimmock and Musa, 2015). According to Imran *et al.* (2014), tourism creates benefits for local communities and generates potential positive attitudes towards ecosystem conservation principles.

A sustainable scuba diving tourism is an opportunity for opening a constructive dialogue between the various stakeholders on the different perspectives of tourism development from the economic, social and environmental points of view. For this reason, the stakeholder theory defined by Freeman in 1984 introduced an interesting approach for the study, organization and management, both strategically and operationally, of scuba diving tourism activities according to the principles of sustainability. The stakeholder theory (Freeman, 1984; Wondirad *et al.*, 2020) helped scholars identify the key players that operate in a given tourism destination and the role they can play in the decision-making process (Dodds *et al.*, 2010).

The involvement of all the stakeholders in the development of sustainable tourism strategies allows for the identification of new perspectives and scenarios associated with the local context, thus ensuring that the priorities of the various players and their opinions are taken into consideration in the planning and management of the tourist destination (Castellani and Sala, 2010). Imran *et al.* (2014) stressed the importance of recognizing and taking into account the perceptions of stakeholders of environmental and sustainable tourism development issues.

The systemic view, combined with the stakeholder approach, helps identify important associations and relationships that impact tourism development processes and drives towards sustainable outcomes (Dimmock and Musa, 2015).

As shown in Fig. 10.2, the scuba diving tourism system proposed by Dimmock and Musa (2015) and redefined by Lucrezi *et al.* (2017) encompasses the marine environment and a variety of stakeholders, such as the divers themselves, the scuba diving industry (i.e. diving operators, the hospitality industry, the retail and transport sectors) and the host community (including residents, as well as local authorities and policy-makers). A tourism system can be defined as 'sustainable' if it promotes satisfactory and economically viable recreational experiences for tourists, while preserving marine resources and meeting the needs of local communities (Lucrezi *et al.*, 2017). The systemic approach reinforces the interdependence and

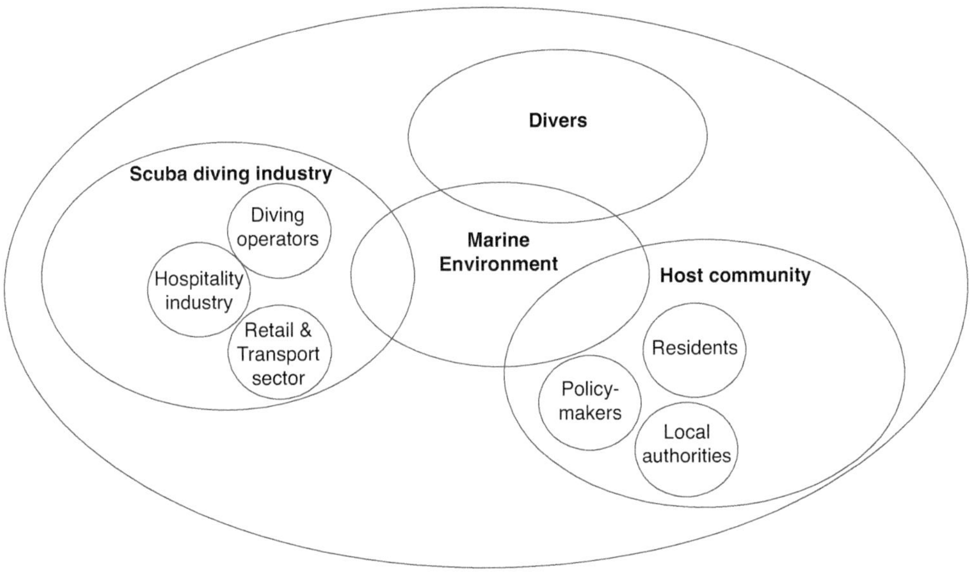

Fig. 10.2. The key dimensions of scuba diving tourism system. (Reinterpretation from Dimmock and Musa, 2015; Lucrezi *et al.*, 2017.)

integration between all these factors, while promoting each stakeholder's participation in the management and decision-making process (Dimmock and Musa, 2015). The involvement of stakeholders is fundamental in the designation, implementation and management phases of MPAs. The first step is to identify and characterize the stakeholders in order to decide which ones should be involved and how. Walton *et al.* (2013) stated the importance of understanding not only who these stakeholders are but also what kind of relationships and attitudes they have towards the MPA.

The 17th SDG ratified at the Rio+20 Conference in 2012, 'Partnership for the Goals', also focuses on the need to promote inclusive partnership strategies, 'built upon principles and values, and upon a shared vision and shared goals, placing people and the planet at the centre'.

Conclusions

One of the key aspects that emerged from the discussion reported in this chapter is the need to maintain the attractiveness of the marine ecosystem as a starting point for a sustainable development not only of the diving industry but also of an entire tourist destination. MPAs, as well as most marine and coastal contexts, are highly complex systems affected by strong anthropogenic pressures that can compromise their biological equilibrium. Even a niche tourism can often generate significant negative impacts on ecosystems, while, conversely, mass tourism, if based on the values of sustainability, can mitigate impacts on the marine and coastal environment. For this reason it is fundamental for a governance structure to play a role that, through careful regulation and planning, embeds the vision of sustainability in the strategies implemented for tourism management.

The establishment of some form of resource conservation programme (including MPAs) should be accompanied by a preliminary background analysis of the tourism system so that strategies and plans can be shaped to promote the vision of a more equitable balance between environmental, economic and tourist interests for the purpose of preserving the marine environment. In light of the above considerations and of the recent repercussions of the COVID-19 pandemic on tourism in general, the institutions involved in the governance of MPAs should address local challenges by creating a system aimed at enhancing and engaging the local ecotourism demand rather than only tackling global conservation issues.

References

Ardoin, N.M., Wheaton, M., Bowers, A.W., Hunt, C.A. and Durham, W.H. (2015) Nature-based tourism's impact on environmental knowledge, attitudes, and behavior: a review and analysis of the literature and potential future research. *Journal of Sustainable Tourism* 23(6), 838–858. DOI: 10.1080/09669582.2015.1024258.

Arin, T. and Kramer, R.A. (2002) Divers' willingness to pay to visit marine sanctuaries: an exploratory study. *Ocean & Coastal Management* 45(2–3), 171–183. DOI: 10.1016/S0964-5691(02)00049-2.

Asafu-Adjaye, J. and Tapsuwan, S. (2008) A contingent valuation study of scuba diving benefits: case study in Mu Ko Similan Marine National Park, Thailand. *Tourism Management* 29(6), 1122–1130. DOI: 10.1016/j.tourman.2008.02.005.

Beltratti, A. (1997) Growth with natural and environmental resources. In: Carraro, C. and Siniscalco, D. (eds) *New Directions in the Economic Theory of the Environment*. Cambridge University Press, Cambridge, UK. DOI: 10.1017/CBO9780511560019.

Bramwell, B. and Lane, B. (1993) Sustainable tourism: an evolving global approach. *Journal of Sustainable Tourism* 1(1), 1–5. DOI: 10.1080/09669589309450696.

Bramwell, B. and Lane, B. (2011) Critical research on the governance of tourism and sustainability. *Journal of Sustainable Tourism* 19(4–5), 411–421. DOI: 10.1080/09669582.2011.580586.

Butler, R.W. (1991) Tourism, environment, and sustainable development. *Environmental Conservation* 18(3), 201–209. DOI: 10.1017/S0376892900022104.

Butler, R.W. (1999) Sustainable tourism: a state-of-the-art review. *Tourism Geographies* 1(1), 7–25. DOI: 10.1080/14616689908721291.

Byrd, E.T. (2007) Stakeholders in sustainable tourism development and their roles: applying stakeholder theory to sustainable tourism development. *Tourism Review* 62(2), 6–13. DOI: 10.1108/16605370780000309.

Casey, J.F., Brown, C. and Schuhmann, P. (2010) Are tourists willing to pay additional fees to protect corals in Mexico? *Journal of Sustainable Tourism* 18(4), 557–573. DOI: 10.1080/09669580903513079.

Castellani, V. and Sala, S. (2010) Sustainable performance index for tourism policy development. *Tourism Management* 31(6), 871–880. DOI: 10.1016/j.tourman.2009.10.001.

Cater, C. (2008) Perceptions of and interactions with marine environments: diving attractions from great whites to pygmy seahorses. In: Garrod, B. and Gössling, S. (eds) *New Frontiers in Marine Tourism: Diving Experiences, Sustainability, Management*. Elsevier, Oxford, UK, pp. 49–64.

Clark, J. (2014) A framework of approaches to sustainable tourism. In: Williams, S. (ed.) *Tourism: Critical Concepts in the Social Sciences. Vol. 3. Tourism, Development and Sustainability*. Routledge, London.

Craggs, R. and Schofield, P. (2006) *Expenditure Segmentation and Visitor Profiling: Regenerating the Quays in Salford, UK*. University of Salford, UK.

Dávid, L. (2011) Tourism ecology: towards the responsible, sustainable tourism future. *Worldwide Hospitality and Tourism Themes* 3(3), 210–216. DOI: 10.1108/17554211111142176.

Davis, D. and Tisdell, C. (1995) Recreational scuba-diving and carrying capacity in marine protected areas. *Ocean & Coastal Management* 26(1), 19–40. DOI: 10.1016/0964-5691(95)00004-L.

Davis, D. and Tisdell, C. (1996) Economic management of recreational scuba diving and the environment. *Journal of Environmental Management* 48(3), 229–248.

Dearden, P., Bennett, M. and Rollins, R. (2007) Perceptions of diving impacts and implications for reef conservation. *Coastal Management* 35(2–3), 305–317. DOI: 10.1080/08920750601169584.

Dicken, M. (2014) Socio-economic aspects of the Sodwana Bay SCUBA diving industry, with a specific focus on sharks. *African Journal of Marine Science* 36(1), 39–47. DOI: 10.2989/1814232X.2014.893257.

Dikou, A. (2011) Ecotourism in marine protected areas: development, impacts and management – a critical review. In: Krause, A. and Weir, E. (eds) *Ecotourism: Management, Development and Impact*. Nova Science, New York, pp. 1–42.

Dimmock, K. and Musa, G. (2015) Scuba Diving Tourism System: A framework for collaborative management and sustainability. *Marine Policy* 54, 52–58. DOI: 10.1016/j.marpol.2014.12.008.

Dodds, R., Graci, S.R. and Holmes, M. (2010) Does the tourist care? A comparison of tourists in Koh Phi Phi, Thailand and Gili Trawangan, Indonesia. *Journal of Sustainable Tourism* 18(2), 207–222. DOI: 10.1080/09669580903215162.

Edney, J. (2012) Diver characteristics, motivations, and attitudes: Chuuk Lagoon. *Tourism in Marine Environments* 8(1/2), 7–18. DOI: 10.3727/154427312X13262430523983.

Freeman, R.E. (1984) *Strategic Management: A Stakeholder Approach*. Pitman, Boston, Massachusetts.

Garrod, B. and Gössling, S. (eds) (2008) *New Frontiers in Marine Tourism: Diving Experiences, Sustainability, Management*. Elsevier, Oxford, UK.

Giddy, J.K. and Rogerson, C.M. (2018) Tracking SCUBA diving adventure tourism in South Africa. *EuroEconomica* 1(37), 47–62.

Giglio, V.J., Luiz, O.J. and Schiavetti, A. (2015) Marine life preferences and perceptions among recreational divers in Brazilian coral reefs. *Tourism Management* 51, 49–57. DOI: 10.1016/j.tourman.2015.04.006.

Giglio, V.J., Luiz, O.J., Chadwick, N.E. and Ferreira, C.E.L. (2018) Using an educational video-briefing to mitigate the ecological impacts of scuba diving. *Journal of Sustainable Tourism* 26(5), 782–797. DOI: 10.1080/09669582.2017.1408636.

Goodland, R. and Ledec, G. (1987) Neoclassical economics and principles of sustainable development. *Ecological Modelling* 38(1–2), 19–46. DOI: 10.1016/0304-3800(87)90043-3.

Gössling, S., Linden, O., Helmersson, J., Liljenberg, J. and Quarm, S. (2008) Diving and global environmental change: a Mauritius case study. In: Garrod, B. and Gössling, S. (eds) *New Frontiers in Marine Tourism: Diving Experiences, Sustainability, Management*. Elsevier, Oxford, UK, pp. 93–113.

Grilli, G., Tyllianakis, E., Luisetti, T., Ferrini, S. and Turner, R.K. (2021) Prospective tourist preferences for sustainable tourism development in small Island developing States. *Tourism Management* 82, 104178. DOI: 10.1016/j.tourman.2020.104178.

Hammerton, Z. and Bucher, D. (2015) Levels of intervention: reducing SCUBA-diver impact within subtropical marine protected areas. *Journal of Ecotourism* 14(1), 3–20. DOI: 10.1080/14724049.2015.1073738.

Hayes, C.T., Baumbach, D.S., Juma, D. and Dunbar, S.G. (2017) Impacts of recreational diving on hawksbill sea turtle (*Eretmochelys imbricata*) behaviour in a marine protected area. *Journal of Sustainable Tourism* 25(1), 79–95. DOI: 10.1080/09669582.2016.1174246.

Holden, A. (1995) The world conference on sustainable tourism, Lanzarote, 24–29 April 1995. *Journal of Sustainable Tourism* 3(3), 173–174. DOI: 10.1080/09669589509510722.

Hunt, C.V., Harvey, J.J., Miller, A., Johnson, V. and Phongsuwan, N. (2013) The Green Fins approach for monitoring and promoting environmentally sustainable scuba diving operations in South East Asia. *Ocean & Coastal Management* 78, 35–44. DOI: 10.1016/j.ocecoaman.2013.03.004.

Hunter, C. (2002) Aspects of the sustainable tourism debate from a natural resources perspective. In: Harris, R., Griffin, T. and Willams, P. (eds) *Sustainable Tourism: A Global Perspective*. Routledge, London.

Imran, S., Alam, K. and Beaumont, N. (2014) Environmental orientations and environmental behaviour: perceptions of protected area tourism stakeholders. *Tourism Management* 40, 290–299. DOI: 10.1016/j.tourman.2013.07.003.

Koens, J.F., Dieperink, C. and Miranda, M. (2009) Ecotourism as a development strategy: experiences from Costa Rica. *Journal of Environment, Development and Sustainability* 11(6), 1225–1237. DOI: 10.1007/s10668-009-9214-3.

Kruger, M., Saayman, M. and Ellis, S.M. (2010) Does loyalty pay? First-time versus repeat visitors at the national arts festival. *Southern African Business Review* 14(1), 79–104. DOI: 10.4102/ac.v14i1.220.

Kruger, M., Viljoen, A. and Saayman, M. (2013) Who pays to view wildflowers in South Africa? *Journal of Ecotourism* 12(3), 146–164. DOI: 10.1080/14724049.2013.871286.

Lim, C. and McAleer, M. (2005) Ecologically sustainable tourism management. *Environmental Modelling & Software* 20(11), 1431–1438. DOI: 10.1016/j.envsoft.2004.09.023.

Lindsey, G. and Holmes, A. (2002) Tourist support for marine protection in Nha Trang, Vietnam. *Journal of Environmental Planning and Management* 45(4), 461–480. DOI: 10.1080/09640560220143503.

Liu, Z. (2003) Sustainable tourism development: a critique. *Journal of Sustainable Tourism* 11(6), 459–475. DOI: 10.1080/09669580308667216.

Lucrezi, S., Ferretti, E., Milanese, M., Sarà, A. and Palma, M. (2021) Securing sustainable tourism in marine protected areas: lessons from an assessment of scuba divers' underwater behaviour in non-tropical environments. *Journal of Ecotourism* 20(2), 165–188. DOI: 10.1080/14724049.2020.1856122.

Lucrezi, S., Saayman, M. and Merwe, P.V.D. (2013) Managing diving impacts on reef ecosystems: analysis of putative influences of motivations, marine life preferences and experience on divers' environmental perceptions. *Ocean & Coastal Management* 76, 52–63. DOI: 10.1016/j.ocecoaman.2013.02.020.

Lucrezi, S., Milanese, M., Markantonatou, V., Cerrano, C., Sarà, A. *et al.* (2017) Scuba diving tourism systems and sustainability: Perceptions by the scuba diving industry in two Marine Protected Areas. *Tourism Management* 59, 385–403. DOI: 10.1016/j.tourman.2016.09.004.

Lucrezi, S., Milanese, M., Sarà, A., Palma, M., Saayman, M. *et al.* (2018) Profiling scuba sivers to assess their potential for the management of temperate marine protected areas: a conceptual model. *Tourism in Marine Environments* 13(2-3), 85–108. DOI: 10.3727/154427318X15225542424207.

Lucrezi, S., Milanese, M., Cerrano, C. and Palma, M. (2019) The influence of scuba diving experience on divers' perceptions, and its implications for managing diving destinations. *PLOS ONE* 14(7), e0219306. DOI: 10.1371/journal.pone.0219306.

Maccarthy, M., O'neill, M. and Williams, P. (2006) Customer satisfaction and scuba-diving: some insights from the deep. *The Service Industries Journal* 26(5), 537–555. DOI: 10.1080/02642060600722841.

Meadows, D.H., Meadows, D.L., Randers, J. and Behrens, W.W. (1972) *The Limits to Growth*. Universe Books, New York.

Mitchell, M. and Hall, D. (2005) Rural tourism as sustainable business: key themes and issues. In: Hall, D., Kirkpatrick, I. and Mitchell, M. (eds) *Rural Tourism and Sustainable Business*. Channel View Publications, Clevedon, UK, pp. 3–16. DOI: 10.21832/9781845410131.

Mondino, E. and Beery, T. (2018) Ecotourism as a learning tool for sustainable development: the case of monviso transboundary biosphere reserve, Italy. *Journal of Ecotourism* 18(2), 1–15. DOI: 10.1080/14724049.2018.1462371.

Mundet, L. and Ribera, L. (2001) Characteristics of divers at a Spanish resort. *Tourism Management* 22(5), 501–510. DOI: 10.1016/S0261-5177(01)00016-4.

Musa, G. (2002) Sipadan: a SCUBA-diving paradise: an analysis of tourism impact, diver satisfaction and tourism management. *Tourism Geographies* 4(2), 195–209. DOI: 10.1080/14616680210124927.

Musa, G. and Dimmock, K. (2012) Scuba diving tourism introduction to special issue. *Tourism in Marine Environments* 8(1-2), 1–5. DOI: 10.3727/154427312X13262430523947.

Musa, G., Kadir, S.L.S.A. and Lee, L. (2006) Layang Layang: an empirical study on scuba divers' satisfaction. *Tourism in Marine Environments* 2(2), 89–102. DOI: 10.3727/154427306779436273.

Ong, T.F. and Musa, G. (2011) An examination of recreational divers' underwater behaviour theories. *Current Issues in Tourism* 14(8), 779–795. DOI: 10.1080/13683500.2010.545370.

Ong, T.F. and Musa, G. (2012) Examining the influences of experience, personality and attitude on SCUBA divers' underwater behaviour: a structural equation model. *Tourism Management* 33(6), 1521–1534. DOI: 10.1016/j.tourman.2012.02.007.

Ong, T.F. and Musa, G. (2013) Responsible underwater behaviour. In: Musa, G. and Dimmock, K. (eds) *Scuba Diving Tourism*. Routledge, New York, pp. 117–133.

Paterson, S., Young, S., Loomis, D.K. and Obenour, W. (2012) Resource attributes that contribute to nonresident diver satisfaction in the Florida Keys, US. *Tourism in Marine Environments* 8(2), 47–60. DOI: 10.3727/154427312X13262430524108.

Rubenstein, D.I. (1993) Science and the pursuit of a sustainable World. *Ecological Applications* 3(4), 585–587. DOI: 10.2307/1942090.

Saayman, M. and Saayman, A. (2009) Why travel motivations and socio-demographics matter in managing a National Park. *Koedoe* 51(1), 1–9. DOI: 10.4102/koedoe.v51i1.381.

Saayman, M. and Saayman, A. (2018) Are there economic benefits from marine protected areas? An analysis of scuba diver expenditure. *European Journal of Tourism Research* 19, 23–39. DOI: 10.54055/ejtr.v19i.323.

Saayman, M., Van der Merwe, P. and Rossouw, R. (2011) The impact of hunting for biltong purposes on the SA economy. *Acta Commercii* 11(1), 1–12. DOI: 10.4102/ac.v11i1.143.

Schuhmann, P.W., Casey, J.F., Horrocks, J.A. and Oxenford, H.A. (2013) Recreational SCUBA divers' willingness to pay for marine biodiversity in Barbados. *Journal of Environmental Management* 121, 29–36. DOI: 10.1016/j.jenvman.2013.02.019.

Schuhmann, P.W., Skeete, R., Waite, R., Lorde, T., Bangwayo-Skeete, P. *et al.* (2019) Visitors' willingness to pay marine conservation fees in Barbados. *Tourism Management* 71, 315–326. DOI: 10.1016/j.tourman.2018.10.011.

Solow, R.M. (1991) *Sustainability: An Economist's Perspective*. Marine Policy Center, Woods Hole Oceanographic Institution, Massachusetts, USA.

Swarbrooke, J. and Horner, S. (2004) *Consumer Behavior in Tourism*. Butterworth-Heinemann, Burlington, Massachusetts.

Thapa, B., Graefe, A.R. and Meyer, L.A. (2005) Moderator and mediator effects of SCUBA diving specialization on marine-based environmental knowledge-behavior contingency. *The Journal of Environmental Education* 37(1), 53–67. DOI: 10.3200/JOEE.37.1.53-68.

Thur, S.M. (2010) User fees as sustainable financing mechanisms for marine protected areas: an application to the Bonaire National Marine Park. *Marine Policy* 34(1), 63–69. DOI: 10.1016/j.marpol.2009.04.008.

Townsend, C. (2008) Interpretation and environmental education as conservation tools. In: Garrod, B. and Gössling, S. (eds) *New Frontiers in Marine Tourism: Diving Experiences, Sustainability, Management*. Elsevier, Oxford, UK, pp. 189–200.

Turner, R.K. (1993) Sustainability: principles and practice. In: R.K, T. (ed.) *Sustainable Environmental Economics and Management*. Belhaven Press, London.

Uyarra, M.C., Watkinson, A.R. and Côté, I.M. (2009) Managing dive tourism for the sustainable use of coral reefs: validating diver perceptions of attractive site features. *Environmental Management* 43(1), 1–16. DOI: 10.1007/s00267-008-9198-z.

Vianna, G.M.S., Meekan, M.G., Pannell, D.J., Marsh, S.P. and Meeuwig, J.J. (2012) Socio-economic value and community benefits from shark-diving tourism in Palau: A sustainable use of reef shark populations. *Biological Conservation* 145(1), 267–277. DOI: 10.1016/j.biocon.2011.11.022.

Walton, A., Gomei, M. and Di, G. (2013) *Stakeholder Engagement: Participatory Approaches for the Planning and Development of Marine Protected Areas*. World Wildlife Fund for Nature and NOAA-National Marine Sanctuary Program, Rome.

WCED (World Commission on Environment and Development) (1987) *Our Common Future*. Oxford University Press, Oxford, UK.

Williams, I.D. and Polunin, N.V.C. (2000) Differences between protected and unprotected reefs of the western Caribbean in attributes preferred by dive tourists. *Environmental Conservation* 27(4), 382–391. DOI: 10.1017/S0376892900000436.

Wondirad, A., Tolkach, D. and King, B. (2020) Stakeholder collaboration as a major factor for sustainable ecotourism development in developing countries. *Tourism Management* 78, 104024. DOI: 10.1016/j.tourman.2019.104024.

Wongthong, P. and Harvey, N. (2014) Integrated coastal management and sustainable tourism: a case study of the reef-based SCUBA dive industry from Thailand. *Ocean & Coastal Management* 95, 138–146. DOI: 10.1016/j.ocecoaman.2014.04.004.

Worachananant, S., (Bill) Carter, R.W., Hockings, M. and Reopanichkul, P. (2008) Managing the impacts of SCUBA divers on Thailand's Coral Reefs. *Journal of Sustainable Tourism* 16(6), 645–663. DOI: 10.1080/09669580802159677.

Zolfani, S.H, Sedaghat, M., Maknoon, R. and Zavadskas, E.K. (2015) Sustainable tourism: a comprehensive literature review on frameworks and applications. *Economic Research* 28(1), 1–30. DOI: 10.1080/1331677X.2014.995895.

11 Sustainable Tourism Development through Shadow Destinations: Case Studies from Central Sweden

Anna Sörensson[1,2]* and Ulrich Schmudde[1]
[1]Mid Sweden University, Östersund, Sweden; [2]Inland Norway University of Applied Science, Norway

Abstract
This chapter aims to provide in-depth knowledge of how nature-based tourism destinations could expand and develop sustainability through shadow destinations. The analysis uses two case studies from central Sweden to show how shadow destinations could contribute to a more sustainable tourism development. Interviews were conducted with different stakeholders, such as destination marketing organizations, tourism companies and public actors. The results show that a shadow destination can be developed along with the main destination by networking with the destination marketing organization. This study's practical implications indicate that a shadow destination should work with the main tourist attractions or destinations to expand the tourists' perception of the 'destination'.

Motivations of Nature-based Tourism

The popularity of tourism centred around nature has grown and it is crucial to examine the relationship between recreation and natural resources (Fossgard and Fredman, 2019). In the field of tourism, 'nature-based tourism' encompasses all types of tourism where the natural environment serves as the main draw or location (Taczanowska *et al.*, 2019). Nature is an important part of tourism development, and businesses that rely on natural environments are growing in popularity. To ensure that tourists and nature-based industries can grow, it is crucial to safeguard the surrounding environment, like rivers, lakes, forests, beaches and mountains (Lundberg and Fredman, 2012). However, tourism literature has made significant progress in addressing why people flock to nature-based locations in such enormous numbers (Lundberg and Fredman, 2012) but has not thoroughly addressed shadow destinations. For example, Baloglu and Uysal (1996) used psychological theory to discuss tourists' motivations through the push-and-pull perspective. Push factors are the internal motivations or drivers that prompt tourists to engage in activities outside their homes to satisfy their desires (Gnoth, 1997). Pull factors, in contrast, are what draw tourists to a particular destination. These factors are further explained in modern psychology using the terms 'instincts', 'drivers' and 'needs' (Gnoth, 1997; Tasci and Ko, 2017). These factors then explain the motivations and drivers of people's travels (Witt and Wright, 1992).

In contrast to push factors, pull factors are external (Kliem, 2003); they are the result of the environment. Destinations appeal to travellers

*Corresponding author: anna.sorensson@miun.se

 Nature-based Tourism and Wellbeing: Impacts and Future Outlook (eds F. Niccolini *et al.*)
DOI: 10.1079/9781800621411.0011

based on their offerings (Mundt, 2012). Once travellers have decided they want to travel, pull factors impact their decision concerning their destination (Witt and Wright, 1992). Pull factors may be substituted to compare with push factors (Kliem, 2003) as the preferability of the surroundings can change depending on the individual and the perception of the destination's offerings. Thus, for destinations that are not as well known, the concept of 'shadow destinations' is crucial as it aims to understand the pull factors that attract different tourist groups.

Existing research shows that shadow destinations are present in the peripheral perception of a well-established, main attraction-focused destination with a strong and recognizable brand (Brambini and Vang, 2013). These destinations have received little attention from researchers despite the fact that they could be developed and contribute to sustainable tourism development. Many stakeholders, such as the tourism industry, development organizations and municipal officials are taking steps to promote tourism in certain areas and make natural sites attractive tourist destinations (Winter *et al.*, 2019). One of the quickest-expanding segments of the tourism industry is nature-based tourism, with many public lands and protected areas becoming key tourist destinations. The large number of tourists flocking to a small area of land may have a negative impact on the area. Therefore, the need for a more sustainable development in the region becomes of interest, making studies on the contributions of shadow destinations noteworthy. By having tourists visit a larger area and not just the 'hot spot' or main destination, municipalities can ensure that tourism does not damage important natural resources, and therefore may contribute to the tourism development.

Previous research has shown that some places are impacted by a shadow effect (Hudman and Jackson, 2003). This means that destinations that are geographically close to well-known places are affected by the 'rain shadow'. There are places that get less rain due to precipitation since it is affected by the mountains or the wind. This metaphor is used for destinations where some destinations are geographically affected by the shadow effect (rain shadow). In tourism, this means that some places are more attractive to tourists compared to the place close by. Some destinations are more attractive because they are established destinations with strong brands among tourists. Tourists therefore stay longer at the well-known place rather than the shadow destination (Hudman and Jackson, 2003, p. 31).

Tourists visiting nature-based destinations might also be interested in exploring the surrounding area. Tourists often view a destination as a whole, even though it includes many components (Haugland *et al.*, 2011). Determining a destination's geography is challenging, because in the minds of tourists, it is not clear where a destination really starts and ends. A shadow destination located near a well-known destination can take advantage of the main destination's strong brand and attract more tourists (Ashton, 2014). A destination with a clear identity and image has a strong brand. Tourists feel a sense of recognition and familiarity when visiting such a place. Studies have indicated that creating tourist routes, such as wine, scenic or culinary routes, can be an effective way to attract tourists to regions near a specific attraction (Schlesinger *et al.*, 2020). Basing ourselves on this analysis, we would like to contribute to explore knowledge on how nature-based tourism destinations could expand and develop sustainably through shadow destinations. The analysis uses two case studies to show how shadow destinations could contribute to more sustainable tourism development.

Methodology and the Swedish Cases

This study uses a qualitative case study approach. Two cases (destinations) were chosen in central Sweden, discussed below. They face similar challenges; the difference is that one (Duved/Åre) is primarily a winter destination while the other (Härnösand/Höga Kusten) is primarily a summer destination. They are both among the largest tourist destinations in Sweden. The destinations aim to become year-round ones but currently have only one peak season when most tourists visit. Fig. 11.1 shows a map of Sweden and the location of the two destinations; these are explained in more detail below.

The researchers conducted interviews with the destination management organization (DMO), tourism companies and stakeholders from the public government in each location. A total of 18 respondents were interviewed.

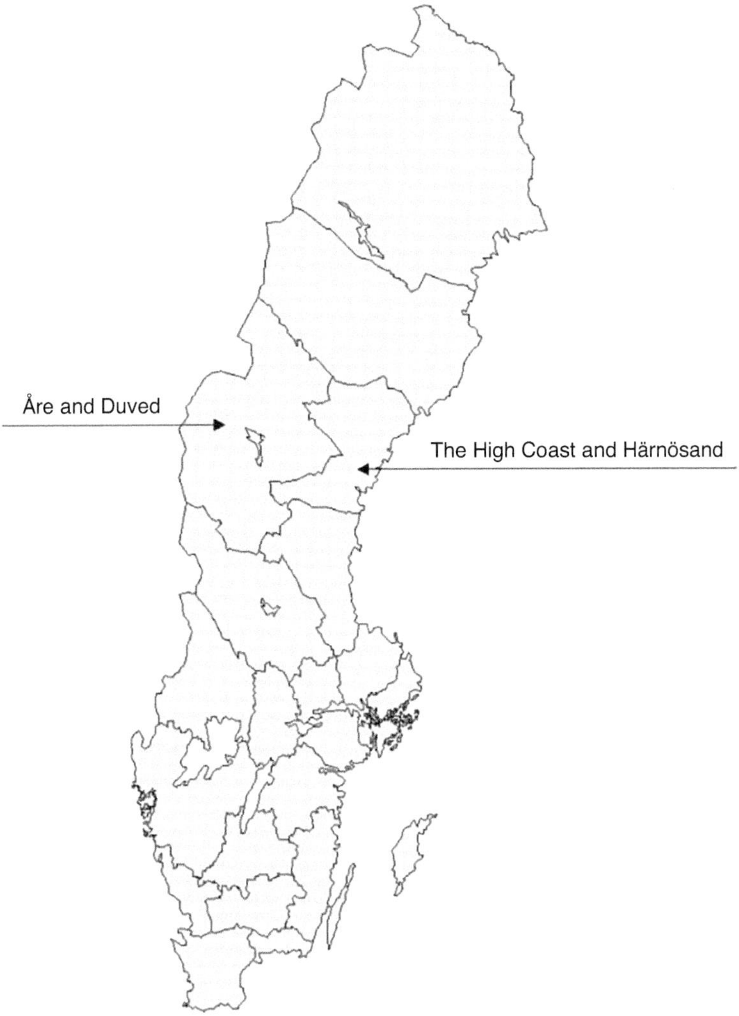

Fig. 11.1. Map of Sweden, with the two cases highlighted. (Based on SCB, 2023.)

The interview guide was constructed based on previous research and had questions on tourism development (Hudman and Jackson, 2003; Brambini and Vang, 2013), the role of the brand, image, identity (Aaker, 2014); and the role of tourist routes (Sims, 2009).

Åre and Duved

Åre is comprised of old agricultural villages that have become one of Europe's top tourist destinations. It is Scandinavia's alpine capital and a gathering point for year-round activities. It is a Swedish mountain gem with an international atmosphere. It offers world-class skiing and cycling together with a wide range of activities for a holiday full of experiences, regardless of when tourists choose to travel. Visitors can hike on the mountain trails of Åreskutan and its surrounding mountains, go dog sledding on snow-covered mountain expanses, and enjoy good food from Åre's renowned restaurants. Åre is perhaps best known as a winter resort, with ski slopes, mountains and skiing to suit everyone. It offers several children's areas where both old and young skiers can learn to ski. The mountain of Åreskutan and the surrounding mountains offer close to 90 pistes of all degrees of difficulty and 48 lifts that allow visitors to travel seamlessly. Visitors can also go cross-country skiing on several miles of cross-country tracks for both classic skiing and skating and find their favourite piste in the digital trail guide. Åre also offers ski arenas that annually become the playing field for the world elite in alpine skiing, freestyle and cross-country skiing. It has a strong tradition of hosting events like Alpine WC (https://aresweden.com, 2022).

Duved is located in the westernmost part of Åredalen and is just 8 km away from the centre of Åre. It is included in the destination of Åre, but the lift system is not connected to Åre's lift system. This means that Duved is partly its own destination and can be defined as a shadow destination.

The Åre destination offers a well-known brand where everything from the homes to the lift tickets are more expensive than in Duved. Duved is part of the destination's brand, and visitors can buy lift tickets that include skiing in both Åre and Duved. During the weeks with the highest number of tourists, such as Christmas, New Year, Easter and school holidays, the skiing queues are noticeably longer in Åre compared to Duved. Many tourists crowd into a small area, which does not create a positive impression of the destination. One solution would be to get more tourists to choose the ski system in Duved to spread out the visitors and reduce the pressure in Åre. The municipality of Åre has attempted to encourage tourists to visit the whole area by introducing uniform signs in the area. They believe this can create the impression of a singular destination, thus spreading the number of visitors.

The High Coast and Härnösand

The High Coast is a UNESCO World Heritage Site and Sweden's most well-known tourist destination. It boasts the world's longest shoreline and the most impressive land elevation resulting from the retreat of glaciers within its borders about 10,000 years ago. Wildlife includes bears, eagles and whales, which can frequently be seen in the region. The highest recorded land uplift can be found in Skuleberget, with the previous coastline now standing 286 m above sea level, as reported by the High Coast in 2019. This is a rare occurrence on the planet and showcases the remarkable land uplift that took place in the region after each Ice Age. The region houses Sweden's tallest island, Mjältön, which stands 236 m above the sea; it also has the deepest part of the Baltic Sea, Ulvödjupet, measuring 293 m deep. The High Coast had around 542,000 visitors staying overnight in 2017.

South of the World Heritage Site is located Härnösand, which is a city of 18,000 inhabitants. The total population of the municipal region is around 25,000. King Johan III founded the city in 1585 and it was created as the county capital of Västernorrland, which is located in central Sweden, and borders three counties, including the Baltic Sea's Gulf of Bothnia. Västernorrland is Sweden's sixth-largest county, covering 21,685 km^2, and it is home to 5% of the country's overall population. The county is made up of seven municipalities; Härnösand is one of these municipalities, located 23 km from the High Coast by the E4 road. Despite its various attractions, such as automotive, boat and county museums, many travellers neglect it as a destination.

Härnösand has not yet reached its full potential as a tourist destination despite there being the UNESCO High Coast World Heritage Site in its nearby region. Härnösand's municipality has stated that the High Coast is of significance for the brand image as an internationally recognized site which also influences the city of Härnösand (Tillväxtstrategi Härnösand, 2017). It is fully understood that Härnösand and the nearby region is of importance for the tourism industry in the region. Research has shown that many tourism destinations around the globe have problems with over-tourism. There are places that are gateways into nature areas like heritage sites, national parks as well as wilderness areas.

Challenges Faced by Popular Nature-based Destinations

The main challenge these two cases face is that there are too many tourists at certain times, which creates issues, such as parking and congestion.

Duved and Åre

The ski area in Åre and Duved is owned by SkiStar, which is a listed company that owns several destinations in the Nordic region. SkiStar had a very profitable 2022, with a record number of visitors. It reported the best half-year results ever, in terms of both visitor and sales records, an increase of 106% compared to the same period the previous year. Visitor numbers in all of SkiStar's destinations have broken visitor records because of lifted restrictions after the COVID-19 pandemic, which has created high demand for mountain vacations. It is not just the school holiday weeks that have had a record number of visitors; people have also gone to the mountains in the periods between the holidays (Dagens PS, 2022). The interest in spending Easter in the mountains was greater than ever in 2022, and several ski facilities in Jämtland County exceeded visitor records. Thus, Åre had more visitors in 2022 than before the pandemic. On Good Friday alone, over 18,000 skiing visitors visited Åre, the highest number ever (SVT, 2022a).

Åre has also witnessed rapid development, and the increased number of visitors requires modern lifts with greater capacity. Old lifts must be replaced with more modern ones. A completely new lift will be built, and it will be weatherproof, as larger lifts (VM8) often stop due to strong winds, cutting off access to three of the larger ski slopes. When this occurs, there is considerable congestion on the other slopes. The new lift will eliminate this problem (Jamtlandstidning, 2022). People are also increasingly using Duved and Tegefjäll, which are fantastic ski areas with ski lifts that cope with weather challenges better than the ski lift WC8 in Åre. This is a positive development as tourism destinations are used better, and guests get a significantly better experience than if everyone is in the same place (Freeride, 2022).

Härnösand and The High Coast

Skuleskogen's National Park is, together with the mountain of Skuleberget and the High Coast, among the top five most popular hiking areas in the High Coast. It has 3.5 miles of hiking trails. The Skuleskogen's National Park has three entrances – the south, north and west entrances. The most popular tours start from the south entrance, while the west entrance is suitable for shorter tours or as a starting point for longer hikes. Visits to the High Coast also broke new records during summer 2022. In June 2022, visitors increased by 34% compared to the same period the year before. These record numbers of guest nights are due to the *New York Times Magazine* ranking it as a top location, an increased general interest in the outdoors and a new strategy for the area for the future (Besöksliv, 2022).

High Coast experiences continue to attract a record number of visitors, and international guests have returned after the pandemic. The nature in this area continues to attract local tourists despite the fact that Swedes are now allowed to travel freely around the world after the lifting of the pandemic restrictions. A major challenge that the High Coast faces is parking for the increased number of tourists. The area has one of the most sought-after hiking trails in the county. The south entrance is facing a shortage of parking spaces due to the high volume of visitors currently. Twice as many parking spaces are needed than are available (SVT, 2020a).

In addition, residents of the small village of Käl have opposed tourism development as the south entrance of Skuleskogen's National Park goes straight through their village. An increasingly large stream of vehicles to the park has created significant traffic in the small village. The county board has proposed alternatives to resolve the situation. In ten years, vehicle traffic to the national park has tripled as people increasingly want to hike in Skuleskogen's National Park. Thus, this destination presents the pitfalls of tourist success. For instance, in summer 2020, just under 40,000 vehicles drove to and from the park (SVT, 2020b).

Discussion and Implications

This chapter aimed to explore in-depth how nature-based tourism destinations could expand in a sustainable way through shadow destinations. The two cases are examples of destinations with a large number of tourists in a small area. Expanding the area to include shadow destinations can decrease the pressure on the destination and provide tourists with a better experience. Several sub-areas can be promoted to encourage tourists decide to visit areas beyond the hotspot itself. Increasing the area of interest is not easy and requires that different stakeholders work together to attract tourists. The DMO can collaborate to highlight a larger area for tourists to visit that includes shadow destinations. One way to pull these tourists away from the hotspots is to show them that other attractions exist in the close-by area. Furthermore, it is quite important to have formal and informal networks to achieve sustainable development at shadow destinations. Hotspots, with their strong brands, have been hit hard during the pandemic since tourists have only been able to travel nationally due to lockdowns and travel bans; subsequently, these destinations have witnessed a large growth in the number of tourists. While these destinations want to promote the idea of untouched nature, tourists who visit are so numerous that they must stand in queues to enter.

Fig. 11.2 illustrates how the shadow destination can be used to widen the destination.

As previous research by Schlesinger *et al.* (2020) has shown, routes may play a key role in moving tourists around in the region. The Åre municipality has chosen to develop its own sign system that is uniform across the municipality. This step aims to have tourists perceive the destination as larger because they visually associate the main destination with the shadow destination. Furthermore, alternative routes can be promoted to make tourist experiences include more than just the hotspot. Furthermore, in both cases discussed above, tourist buses reduce personal vehicle transport, thus reducing the need for parking.

In conclusion, the cases show that it is possible to develop destinations and shadow destinations to create more sustainable tourist destinations and improve the tourists' experiences.

This study's practical implications show how sustainable tourism development can occur. The main destination may have issues due to an influx of a huge number of tourists who only go to the hot spot. By finding ways to expand the destination to include the shadow destination, municipalities can contribute to decreasing the number of tourists to the hotspot. It is also important to work with the DMO to expand this area since there are many different stakeholders with their own agenda, but the sustainability of the destination needs to be the main focus. By using a common sign system, it is easier for the tourists to see the main and shadow destinations as a singular destination. Another initiative destinations can pursue is to create routes that lead tourists to larger areas and not just the hotspot. Future research should focus on other places that struggle with similar challenges, to understand how to foster more sustainable tourism.

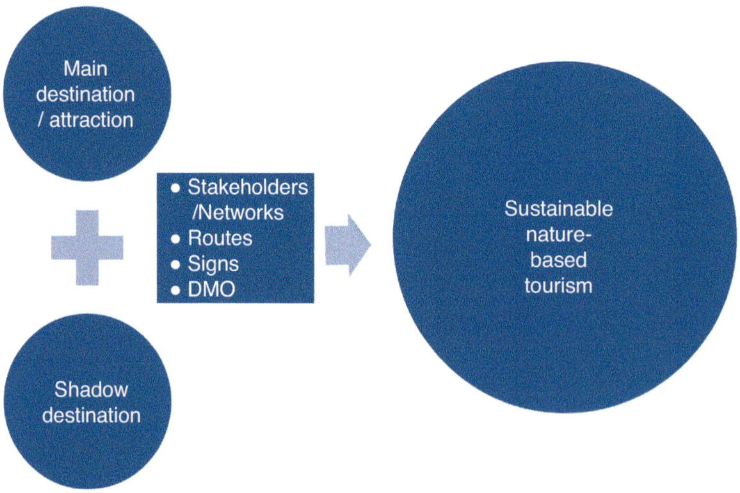

Fig. 11.2. Sustainable tourism development through shadow destinations.

References

Aaker, D. (2014) *Aaker on Branding: 20 Principles that Drive Success*. Morgan James Publishing, New York.

Ashton, A.S. (2014) Tourist destination brand image development – an analysis based on stakeholders perception: a case study from Southland, New Zealand. *Journal of Vacation Marketing* 20(3), 279–292. DOI: 10.1177/1356766713518061.

Baloglu, S. and Uysal, M. (1996) Market segments of push and pull motivations: a canonical correlation approach. *International Journal of Contemporary Hospitality Management* 8(3), 32–38. DOI: 10.1108/09596119610115989.

Besöksliv (2022) Toppresultat för Höga Kusten. Available at: https://www.besoksliv.se/nyheter/toppresultat-for-hoga-kusten/ (accessed 29 August 2022).

Brambini, A. and Vang, J. (2013) Policy measures for creating an integrated and brand-focused regional innovation system in tourism in a shadow destination: insights from Pisa's destination development strategy. *International Journal of Business and Globalisation* 10(2), 194–219. DOI: 10.1504/IJBG.2013.052251.

Dagens, P.S. (2022) SkiStar slår rekord – fler besökare än före pandemin. Available at: https://www.dagensps.se/bors-finans/skistar-slar-rekord-fler-besokare-an-fore-pandemin/ (accessed 29 August 2022).

Fossgard, K. and Fredman, P. (2019) Dimensions in the nature-based tourism experiencescape: an explorative analysis. *Journal of Outdoor Recreation and Tourism* 28, 100219. DOI: 10.1016/j.jort.2019.04.001.

Freeride (2022) Skistars Chef: Åre Hogst Prio är Att 'Vädersäkra' Vårt Skidsystem. Available at: https://www.freeride.se/skistars-chef-are-hogst-prio-ar-att-vadersakra-vart-skidsystem/ (accessed 29 August 2022).

Gnoth, J. (1997) Tourism motivation and expectation formation. *Annals of Tourism Research* 24(2), 283–304. DOI: 10.1016/S0160-7383(97)80002-3.

Haugland, S.A., Ness, H., Grønseth, B.O. and Aarstad, J. (2011) Development of tourism destinations: an integrated multilevel perspective. *Annals of Tourism Research* 38(1), 268–290.

Hudman, L. and Jackson, R. (2003) *Geography of Travel and Tourism*, 4th edn. Thomson Learning, London.

Jamtlandstidning (2022) SkiStar planerar nya liftar inför nästa säsong. Available at: https://jamtlandstidning.se/besoksnaring-naringsliv-naringsliv/skistar-planerar-nya-liftar-infor-nasta-sasong/132274 (accessed 29 August 2022).

Kliem, T. (2003) Reisemotive, Reiseverhalten und Wahrnehmungen deutscher Touristen in Norwegen als Grundlage der Entwicklung neuer Konzepte für die norwegische Tourismuswirtschaft (Travel motives, travel behaviour and perceptions of German tourists in Norway as a basis for the development of new concepts for the Norwegian tourism industry), Dissertation, Faculty of Social Sciences, University of Duisburg-Essen.

Lundberg, C. and Fredman, P. (2012) Success factors and constraints among nature-based tourism entrepreneurs. *Current Issues in Tourism* 15(7), 649–671. DOI: 10.1080/13683500.2011.630458.

Mundt, J.W. (2012) *Tourismus (Tourism), 4. überarbeitete und ergänzte Auflage*, 4th edn. De Gruyter, Oldenbourg, Germany.

SCB (2023) Statistics Sweden. Available at: https://www.scb.se/hitta-statistik/regional-statistik-och-kartor/regionala-indelningar/kartor-over-indelningar/ (accessed 17 April 2023).

Schlesinger, W., Cervera-Taulet, A. and Pérez-Cabañero, C. (2020) Exploring the links between destination attributes, quality of service experience and loyalty in emerging mediterranean destinations. *Tourism Management Perspectives* 35, 100699. DOI: 10.1016/j.tmp.2020.100699.

Sims, R. (2009) Food, place and authenticity: local food and the sustainable tourism experience. *Journal of Sustainable Tourism* 17(3), 321–336. DOI: 10.1080/09669580802359293.

SVT (2020a) Stor parkeringsbrist vid skuleberget i kramfors. Available at: https://www.svt.se/nyheter/lokalt/vasternorrland/parkeringskaos-pa-skuleberget-i-kramfors (accessed 29 August 2022).

SVT (2020b) Så ska bybornas trafikkaos lösas – här är nya vägalternativen. Available at: https://www.svt.se/nyheter/lokalt/vasternorrland/skuleskogsvagen-3 (accessed 29 August 2022).

SVT (2022a) Skistar: 'Det har varit en rekordsäsong. Available at: https://www.svt.se/nyheter/lokalt/jamtland/skistar-det-har-varit-en-rekord-sasong (accessed 29 August 2022).

Taczanowska, K., González, L.-M., García-Massó, X., Zięba, A., Brandenburg, C. *et al.* (2019) Nature-based tourism or mass tourism in nature? segmentation of mountain protected area visitors using self-organizing maps (SOM). *Sustainability* 11(5), 1314. DOI: 10.3390/su11051314.

Tasci, A.D. and Ko, Y.J. (2017) Travel needs revisited. *Journal of Vacation Marketing* 23(1), 20–36. DOI: 10.1177/1356766715617499.

Tillväxtstrategi Härnösand (2017). Available at: https://harnosand.se/download/18.4b6c551c1679dec5d0ab4854/1578656626659/Tillv%C3%A4xtstrategi%20H%C3%A4rn%C3%B6sand%20kortversion.pdf (accessed 16 October 2023).

Winter, P.L., Selin, S., Cerveny, L. and Bricker, K. (2019) Outdoor recreation, nature-based tourism, and sustainability. *Sustainability* 12(1), 81. DOI: 10.3390/su12010081.

Witt, C.A. and Wright, P. ;L (1992) Tourist motivation: life after Maslow. In: Johnson, P. and Thomas, B. (eds) *Choice and Demand in Tourism*. CAB International, Wallingford, UK, pp. 33–35.

12 The Future of (Slow) Travel: Tourist Attitudes as a Guide to Sustainable Tourism Development

Isabelle Adler[1] and Kim Hartmann[2*]
[1]University of Lleida, Lleida, Spain; [2]University of Strathclyde, Strathclyde Business School, Glasgow, UK

Abstract
The role of sustainability in travellers' vacation choices is a much-discussed topic. This chapter explores the potential of applying slow travel principles to mass tourism. The sustainability attitude, past and future travel behaviour, as well as willingness to pay, on the demand side, is examined to identify potential starting points towards more sustainable holiday offers without minimizing their desirability. Based on a literature review and a quantitative study among German travellers, concrete recommendations for the nature-based transformation of tourism offers are presented. Activity-centred slow travel components, such as intense engagement with local culture and the natural environment, are highly relevant to most travellers. In contrast, travellers are opposed to slower transportation to and from destinations. Bundling slow travel activities with fast transportation options facilitates a more sustainable mass tourism leading to not only better tourist choices and more sustainable economies but also higher revenues due to travellers' willingness to pay sustainability surcharges.

Introduction and Problem Statement

During the COVID-19 pandemic, the travel and tourism industry experienced an unprecedented crisis affecting both the demand and supply side. Travel regulations and restrictions resulted in a focus both on local destinations and short travel distances. Thus, the use of transport with lower carbon emissions increased. In the context of sustainable tourism, these developments fostered new opportunities as more travellers indulged in sustainable travel behaviour, many for the first time (Simpson *et al.*, 2008, p. 297). However, this behaviour opposes the traditional self-concept of the tourism industry. The quote 'Everyone wants to get back to nature, just not on foot', by German writer Werner Mitsch (cited in WWF, 2014) underlines the growing societal awareness of climate change, yet it also illustrates that only a minority of people are prepared to restrain themselves. This also applies to the discrepancy between environmental awareness and actual travel behaviour of tourists. Tourism is often accompanied by negative impacts on the environment, especially when it is carried out on a large scale or affects ecologically sensitive areas (Schmied, 2004). One example is the share of climate change attributed to air travel, which

*Corresponding author: kim.hartmann@strath.ac.uk

 Nature-based Tourism and Wellbeing: Impacts and Future Outlook (eds F. Niccolini *et al.*)
DOI: 10.1079/9781800621411.0012

has become an issue of concern in recent years. It is important not to close one's eyes to the harm done to the ecosystem and local society, but to ensure that a trip leaves the smallest possible ecological footprint and preserves the cultural identity of a place.

The concept of 'slow travel' is a response to the carbon-intensive travel practices that shape tourism. Slow travel summarizes travel behaviour characterized by a synergy between experiential aspects of vacationing and discourse on the environment, particularly climate change. The term slow travel has, consequently, been used to describe a new form of holiday that is intended to establish a more sustainable future for tourism. It combines low-carbon transportation and the pleasure of travel with an intense engagement with local cuisine, culture and community (Dickinson *et al.*, 2011).

The aim of this chapter is to explore the potential of slow travel as a manifestation of holiday design and the relevance of sustainability in holiday choices. Thereby, new practices relating to nature-based tourism will be established. The demand side and its attitudes will be evaluated from the perspective of sustainable tourism development to serve as a guide in this regard. Recommendations for action are derived in order to allow tourism companies to properly respond to consumer demand, to build sustainable economies, and to foster a competitive advantage within the increasing competition in the tourism industry.

The Slow Travel Phenomenon

Sustainable tourism

The tourism industry represents one of the largest and fastest-growing industries in the world. Tourism and related economic activities generated 10% of World GDP in 2019 and were responsible for the creation of one in four jobs worldwide (WTTC, 2020). In 1950 there were 25 million international tourist arrivals worldwide; by 1990 that had grown to 435 million. Almost 30 years later, in 2018, the number had grown more than threefold, reaching 1.442 billion, with tourist arrivals projected to reach 1.8 billion by 2030 (The World Counts, 2022).

As a result of the tourism industry's significant growth during the last half-century, the concept of sustainable tourism development emerged with the aim of reducing the negative impacts of mass tourism. A multitude of tourism forms in various types of destinations, including mass tourism, as well as niche segments, can benefit from sustainable tourism development guidelines to counter the negative impacts resulting from unmanaged tourism (Marson, 2011; Novelli *et al.*, 2022). Correspondingly, destinations that neglect the principles of sustainable tourism development run the risk of destroying the very basis of their own profile and unique selling propositions such as natural landscapes. The UNWTO's (1997) Agenda 21 fortified the claims of sustainable development by stating that 'the only correct way to develop society is to alleviate poverty, strive for a society of social justice and conserve natural resources'. Therefore, sustainable development gives guidance for developing tourism practice based on environmental, social and economical aspects.

Within the relevant literature, researchers define sustainable tourism differently and do not agree on universal approaches or definitions. Exemplary definitions range from 'tourism that takes full account of its current and future economic, social, and environmental impacts, addressing the needs of visitors, the industry, the environment, and host communities' (UNWTO, 2017) to 'a management tool that has been co-opted by the tourism industry to maintain the status quo and legitimize the expansion of the industry' (Goessling *et al.*, 2013, p. 243). However, an area of agreement among researchers can be found regarding the key aspects of sustainable tourism, which can be divided into (i) the natural perspective with emerging concepts such as ecotourism, green tourism and slow tourism; and (ii) the social perspective tightly linked to the terms responsible and inclusive tourism (Roblek *et al.*, 2021). Summarizing, the UNWTO provides an all-encompassing definition for sustainable tourism development: 'Tourism that takes full account of its current and future economic, social and environmental impacts, addressing the needs of visitors, the industry, the environment and host communities' (UNEP & UNWTO, 2005).

Defining slow travel

The emerging trend of slow travel has arisen in response to carbon-intensive and fast-paced modes of travel. Although it has not yet gained attention on a large scale, the potential of slow travel as a climate change mitigation and adaptation strategy for tourism is substantial (Goessling *et al.*, 2013; Lukic *et al.*, 2018). This is particularly due to lower carbon emissions as a result of the chosen mode of transport. To date, no universal definition of slow travel has been agreed upon, but many authors see mobility (especially transportation to and from a destination) as a characteristic feature of the concept (Goessling *et al.*, 2013; Lozano-Oyola *et al.*, 2019). As a matter of differentiation, Ceron and Dubois (2007) argue that slow tourism primarily comprises an experience associated with slow food and long-term visits, whereas slow travel is associated with journey-centred tourism (e.g. by boat or train), in which the distance covered shapes unique experiences and is only available when undergoing the time-intense journey. According to Caffyn (2012), advocates of slow travel concentrate primarily on the journey itself, experiencing a place by travelling slowly through its environment. This type of travel meets the needs of vacationers who want to move on rather than stay in one place. However, this is more difficult to implement in remote areas, such as small islands (Conway and Timms, 2010). A vacation with numerous brief and/or pointless pauses at a variety of locations, as in cruise tourism, could be a result of focusing too much on the voyage and moving on. In this sense, slow travel is closely linked to the interests of mobility-centred studies. While modes of transportation such as planes, boats and vehicles are essential and convenient for travelling to faraway places or remote slow travel destinations, they are rarely environmentally friendly (Serdane, 2020). According to a study by McDonald *et al.* (2015), sustainable consumers opt to fly because of its convenience and/or low price, even though they know about aviation's negative environmental impact. Resultingly, even though air travel contributes significantly to climate change, slow travellers do not forgo travelling by minimizing the number of long-distance trips and thus lowering aviation pollution. Especially regarding the transport options, Dickinson and Lumsdon (2010) separate tourists into 'hard' and 'soft' slow travellers. The former are ecologically conscious persons whose conduct is similar whether at home or on vacation and who refuse to fly unless it is unavoidable. The latter are concerned about the environment yet use a variety of forms of transportation during their trips, including flights. According to Lumsdon and McGrath (2011), slow travel is strongly associated with its nature as the antithesis to speed. When defining it, reference is made, in particular, to tourism activities that can be considered slow travel, e.g. travelling off the beaten path, learning about the culture of the destination, using alternative modes of transport, or enjoying the journey as much as the destination (Lumsdon and McGrath, 2011). In terms of activities at the destination and localness, several studies emphasize that slow travel delivers all-encompassing and prolonged travel experiences rather than rapid and meaningless short excursions (Dickinson and Lumsdon, 2010), as well as a focus on sense of place and the meaning of travel through exploration and learning (Oh *et al.*, 2016). The more the aforementioned components prevail, the slower a vacation is considered. According to Conway and Timms (2010), slowing down while travelling also promotes inclusivity in the area and prevents the leaking of foreign capital.

A pragmatic approach towards an all-encompassing definition adopts the prevalent elements in order to be effective in practice – for destinations, tourism companies as well as travellers. In general, slow travel can be characterized by four main features: (i) slowness and the value of time; (ii) localness and activities at the destination; (iii) mode of transportation and experience of the journey; and (iv) environmental awareness (Lumsdon and McGrath, 2011). Gardner (2009) adds to this understanding by stating that 'Slow travel is a state of mind. It's about having the courage not to go the way of the crowd.' Resultingly, slow travel is about taking a personalized approach to one's journey and having more intense experiences, even if they take more effort. In general, the core ingredients of slow travel as a sustainable form of travel focus on the overall tourist experience, including manifold different aspects, reflecting both the duration of the voyage and the experience of the place (Oliveira, 2020). One key characteristic of slow travel is the traveller investing time to relish the trip and engage intensively with the destination as well as its culture and people. In addition, environmental concerns are of key importance and – given the international pressure to reduce carbon emissions – an important component to promote if this form of travel is to significantly contribute to reducing tourism's CO_2 impact (Dickinson *et al.*, 2011). Furthermore, slow travel involves a more autonomous and non-mass way of organizing travel. Thus, it implies more flexibility and participation of tourists. In addition, the authenticity of the experience at the destination plays an essential role, often pronounced by respect for local communities and their particular social and cultural characteristics (Kondoudaki *et al.*, 2017).

From the traveller's point of view, many factors influence the individual's desire to engage in slow travel options, e.g. spending power, cultural background, geographic diversity, tourism offerings, local weather, and the traffic connection of the hometown. Regarding this, it has further been argued that 'people's desire to "slow down" may be a consequential reflection of their mental resistance to the ever-accelerating pattern of lifestyle' (Oh *et al.*, 2016, p. 205). Even though the discussion regarding the distinctness of definitions is highly relevant, it can be noted that all perspectives have a specific attitude towards tourism and the planet, which often involves a reaction – and commonly aversion – against mass tourism. For the purpose of this chapter, slow travel is considered as a phenomenon that coexists with faster forms of tourism. In this sense, the slow mode of travel has been viewed as a desirable alternative that directs tourist behaviour towards maximizing opportunities to achieve the goals of sustainable tourism development. For the further course of this chapter, slow tourism is explored based on two of the characteristics specified by Lumsdon and McGrath (2011): localness and activities at the destination, and mode of transportation in combination with the actual experience of the journey.

Distinction from mass tourism

As previously indicated, growing numbers of travellers related to mass tourism can result in negative impacts on the local community and the environment. In contrast, slow travel can serve as an antidote to the harmful influence of overtourism by standing apart from the constructed experience of mass tourism, which negates the essential meaning and essence of tourism and transforms destinations into standardized products (Kondoudaki *et al.*, 2017). In order to clearly differentiate between both forms of travel, their respective characteristics are juxtaposed in Table 12.1. Summarizing, slow travel can be seen as a response to the negative aspects of mass tourism, especially regarding the chosen mode of transport and the nature of activities and experiences at the destination.

Empirical Analysis of Tourist Attitudes Regarding Slow Travel

Current situation and research aim

In order to tailor sustainable tourism offers and communication measures adequately to travellers' demands, it is essential to recognize the needs of the target group as well as general social trends. Accordingly, this section provides theoretical evidence on the status quo of sustainable travel behaviour.

Nowadays, moments in which people take time to recharge their own energy levels – and not just their phones – are becoming increasingly rare. Even on weekends or holidays, it has become a norm to be 'always on'. Every third employee is under permanent stress and can't disconnect properly. Hence, the increasing speed of life leads to high pressure, a hectic pace and sensory overload. Consequently, the desire for rest is great (Weidner, 2015). People are seeking more and more ways to escape the fast pace of life by engaging in slower forms of consumption. Deceleration leads to a change of lifestyle towards self-awareness and ecological consciousness, as well as an inward focus on needs. With travellers being exhausted by overcrowding, standardized sights and a sea of smartphones, the counter-movement of slow travel is currently gaining momentum (Oh *et al.*, 2016; Oliveira, 2020).

Overall, it appears that even though awareness of ecological problems and attitudes towards

Table 12.1. Comparison of mass tourism and slow travel.

Mass tourism	Slow travel
Fast travel	Low-speed travel
Airplane as prevalent transport mode	Alternative transport mode with low CO_2 emissions
Visit to numerous tourist attractions	Appreciation of the local character of the place
Quantification of visited areas	Qualification of the tourist experience
Passive tourist	Active tourist
Standardized experiences	Authentic experiences
Standardized meals	Local and traditional food
Impersonal encounters and little contact with the destination and locals	Intensive contact with the destination and genuine communication with locals
Accommodation in large, commercial resorts or hotels	Accommodation in smaller hotels or inns
Group options and lack of flexibility	Individuality and flexibility
Omnipresence of communication technologies	Elimination of communication technologies

Note: Own elaboration based on Kondoudaki *et al.* (2017)

environmental protection and more sustainable consumer behaviour have increased in society, this awareness often fails to influence the actual behaviour of tourists while on vacation. Travellers show high levels of environmental awareness, but they hardly accept any behavioural restrictions. This phenomenon, known as the attitude-behaviour gap, can be seen as an 'inconsistency of sustainability-inclined tourists' (Nikolić *et al.*, 2021, p. 4). Nevertheless, studies confirm that half of tourists would be willing to pay an environmental surcharge (e.g. a set amount per day of stay in the destination) provided that the amount is not perceived as excessive (Freyer, 2011). Resultingly, the willingness to pay a surcharge represents the very nature of the attitude-behaviour gap, as travellers would rather spend an additional fee for environmental protection, and thus have a clear conscience, than change their actual behaviour. However, there are singular studies pointing out that some travellers would be willing to engage in sustainable travel choices and even pay more for these offers than for a standardized mass experience (Pulido-Fernández and López-Sánchez, 2016).

The overall rise of climate-friendly attitudes is likely to become even more significant in the future and result in more sustainable choices across a variety of consumption contexts. It needs to be noted that this development is prone to a combination of multiple influencing factors when it comes to an individual's travel behaviour, e.g. higher climate-friendly awareness, decreasing spending power, and rising transportation costs (which potentially lead to less air travel) (Rein and Strasdas, 2017). Although environment- and climate-friendly holiday offers already exist, they are predominantly niche offers. In order to successfully increase the market share of sustainable travel options, the focus should therefore lie on an environmentally and socially compatible transformation of mass-market offers, as their suppliers cater to a large number of travellers. In general, travel products are only successful if they meet the needs and wishes of potential clients in the best possible way. In this respect, conventional providers are often guided by general market trends. Further, the average size of the potential target group has to be big enough to ensure economic success. Consequently, the following sections set off to analyse whether, how, in what form and at what price slow travel vacation options can be marketed to a wider audience. Only a differentiated knowledge of tourists' expectations allows the development of more suitable trips and prompts tourism companies to take the guiding principle of sustainability into account when developing their services (Schmied *et al.*, 2009).

For now, most slow travel studies seem to overlook the perspective of 'normal' travellers.

The current state of research is mainly characterized by specific destinations (Meng and Choi, 2016; Lin *et al.*, 2020), qualitative interviews (Lumsdon and McGrath, 2011; Oliveira, 2020) and samples with an affinity towards sustainability (Dickinson *et al.*, 2011; Piispa, 2021). Consequently, this study followed a quantitative approach to add a new perspective to the field. The research aim was to add a new level of understanding to the literature about travellers' preferences and expectations concerning slow travel, the ongoing attitude-behaviour gap and the barriers related to this mode of tourism.

Methods

Based on the reviewed literature, this explorative study set out to (i) measure the influence of sustainability attitude on previous slow travel activities, previous slow travel transportation, future slow travel behaviour, and willingness to pay a sustainability surcharge; and (ii) explore the phenomenon of the attitude-behaviour gap (Nikolić *et al.*, 2021). The survey was programmed and administered via Qualtrics' online version of survey software. All data were collected through this online survey. The survey link was distributed in January and February 2021 within the personal network of the first author of this chapter via social media platforms (such as LinkedIn and Instagram) as well as via an e-mail to the student mailing list of the International School of Management (ISM). The sample was set to German tourists, as Germany ranks among the three countries with the highest tourism spending worldwide but is rather under-represented in tourism research in contrast to the two preceding nations, namely China and the US (UNWTO, 2019). The German sample included a total of 165 participants. The average age was 42.38 years (SD = 18.42), with 47.9% of participants being female and 52.1% male.

The variables – (i) sustainability attitude, (ii) previous slow travel activities, (iii) previous slow travel transportation, and (iv) future slow travel behaviour – were each measured with four to ten items, on a five-point Likert scale from 1 = strongly disagree to 5 = strongly agree. There are many scales which measure sustainability attitude or comparable constructs. However, they are often rather generic and/or too long for the purpose of this study (e.g. the EAI; Milfont and Duckitt, 2010). Therefore, a scale measuring sustainability attitude was created, focusing on aspects such as energy saving, (regional) shopping and everyday mobility. Slow travel variables and items were loosely based on Caffyn's (2012) slow travel elements and split up into transport-related ('previous slow travel transportation') and activity-related ('previous slow travel activities') variables, following the slow travel definition outlined above ('Defining slow travel'). Considering the debate in previous studies (Dickinson and Lumsdon, 2010; Lumsdon and McGrath, 2011; Oh *et al.*, 2016) about whether transportation mode should be considered as a main characteristic of slow travel, this study examined it separately to measure potential differences in past behaviour regarding the transportation to and from a destination in contrast to behaviour at a destination. The variable future slow travel behaviour merged both perspectives to allow for a combined outlook on the travellers' future travel intentions and characteristics. The variable willingness to pay a sustainability surcharge was measured using a slider from 0 to 100 (as a percentage of travel price) and was loosely adapted from Netemeyer *et al.* (2004).

Table 12.2 provides a summary of all research variables, their corresponding number of items, and levels of Cronbach's Alpha. Data analysis was performed with the statistical software IBM SPSS Statistics 27.

Results

The research results demonstrate that all variables share rather positive manifestations (higher values for the first four variables indicate agreement with sustainable attitudes/slow travel behaviour). The participants show a clear tendency towards sustainable attitudes (M = 3.94, SD = 0.50) and have engaged rather strongly in slow travel activities during their past holidays (M = 3.42, SD = 0.61). In contrast, the mode of transport chosen in their past holidays was not clearly characterized by slow travel transport choices (M = 3.06, SD = 0.87).

Table 12.2. Overview of variables.

Variable	Range	# Items	α	Exemplary item
Sustainability attitude	1–5	10	.62	I deliberately buy aliments from the region.
Previous slow travel activities	1–5	9	.75	It is important for me to get to know a place well.
Previous slow travel transportation	1–5	4	.74	I try to avoid modes of transport with a high environmental impact, such as air travel.
Future slow travel behaviour	1–5	10	.77	Making the arrival or departure a central part of my holiday.
Willingness to pay a sustainability surcharge	0–100	1		What percentage of the holiday price would you be willing to pay extra for a sustainable holiday?

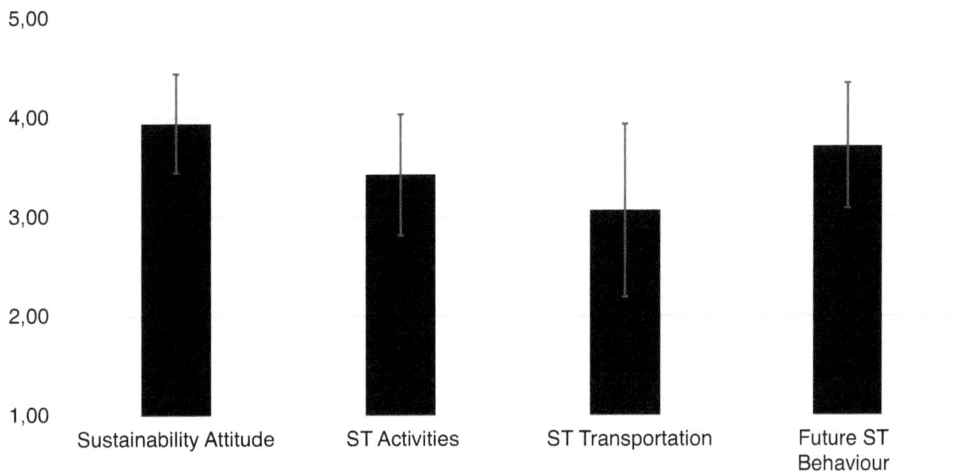

Fig. 12.1. Descriptive statistics of independent and dependent variables.

The participants share an openness towards slow travel in their future holidays (M = 3.71, SD = 0.63) and are willing to pay a sustainable surcharge of more than 20% of the initial price of the holiday (M = 21.13, SD = 14.69).

Resultingly, the differences between independent variables (sustainability attitude) and dependent variables suggest the presence of an attitude-behaviour gap, as demonstrated in Fig. 12.1 for better visual comparability. Sustainability attitude was significantly higher than corresponding slow travel variables, namely previous slow travel activities ($t_{(164)} = 10.86$, $p_{2\text{-tailed}} < .001$), previous slow travel transportation ($t_{(164)} = 15.34$, $p_{2\text{-tailed}} < .001$) and future slow travel behaviour ($t_{(164)} = 5.33$, $p_{2\text{-tailed}} < .001$).

Next, a multiple regression analysis was performed to measure the relationship between sustainability attitude (IV) and the four slow travel variables (DV). Gender and age were included as control variables. Table 12.3 presents the positive effects that sustainability attitude has on the four dependent variables.

Table 12.3. Regression model.

		Slow travel activities	Slow travel transportation	Future slow travel behaviour	Willingness to pay sustainability surcharge
Sustainability attitude	b	0.49***	0.91***	0.61***	8.86***
Gender	b	0.02	0.02	−0.10	2.00
Age	b	−0.00	−0.00	−0.01**	−0.05
Overall model	R^2	.17***	.28***	.31***	.09**

Note: *p<.05; **p<.01; ***p<.001; Gender: O=Female, 1=Male

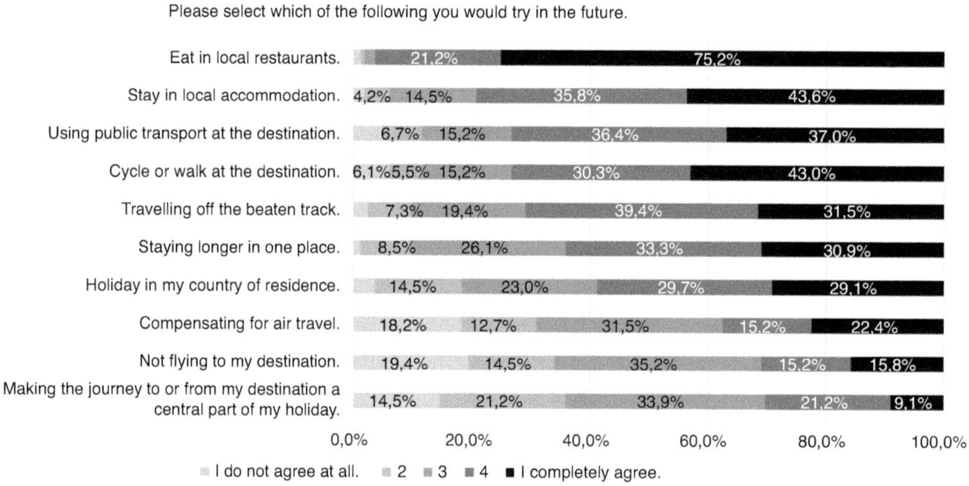

Fig. 12.2. Descriptive statistics of future slow travel behaviour.

All these effects have proved to be significant. Further, the model's coefficients of determination (R^2) range between medium and large effect sizes. So, it can be said that sustainability attitude is a good predictor for slow travel attitude and behaviour. However, sustainability attitude seems to be higher than respective slow travel variables and 'only' accounts for 9%–31% of variance in slow travel variables, thus indicating an attitude-behaviour gap.

In addition, the ten items measuring the variable future slow travel behaviour were analysed separately to obtain in-depth insights concerning the participants' openness towards slow travel in the future. As shown in Fig. 12.2, participants are most interested in slow travel activities at the destination. The highest-ranking aspects of slow travel are eating in local restaurants, staying in local accommodation and using public transport at the destination. However, a critical stance persists towards using slow and time-intense means of transportation to the destination. This is underlined by the lowest-ranking aspects of slow travel: making the journey a central part of the holiday and not flying to a destination.

Lastly, the survey included common barriers hindering participants from engaging in slow travel in the past and in the future. The results expose overall travel time (62.4% agreement), inconvenience related to transfers, e.g. between trains (52.2% agreement), potential for delays and missed connections (45.5% agreement), and the need to carry luggage (40.0% agreement) as the key barriers when considering slow travel options.

Discussion

This chapter set out to expand the knowledge on travellers' attitudes towards slow travel, the ongoing attitude-behaviour gap, and the barriers related to this mode of tourism and its principles. This research aims to provide the structure for the following discussion.

Slow travel: traveller attitudes

According to the findings of this study, certain slow travel principles, such as staying longer in a place or engaging intensively with the culture, are already unconsciously considered by the majority of respondents while other principles (e.g. regarding transportation) are difficult to implement due to time restraints. Resultingly, slow travel activities at the destination are not only essential for slow travel (Serdane, 2020) but can also serve as desirable starting points towards more sustainable vacations apart from 'hard' slow travel (Dickinson and Lumsdon, 2010). This assumption results from this study's findings and supports Manthiou *et al.* (2022) stating that slow and fast travel practices can coexist within the same vacation. The hypothesis that slow travel is more suitable for advanced capitalist countries with good infrastructure may be confirmed to a certain extent as well, since a lack of connectivity is a barrier for alternative forms of travel. The findings of this study are in line with other studies claiming that slow travel involves holiday type as well as travel mode choices (Dickinson *et al.*, 2011). Therefore, it could be worthwhile for tourism businesses, such as tour operators or travel agencies, to bundle slow travel activities at the destination with convenient and fast transportation options. Thus, the development of a more sustainable 'mass tourism' would be advanced and made accessible and attractive to broader target groups.

As in previous studies, the results confirm that some travellers are willing to pay price premiums for travel options that are more sustainable. The average accepted surcharge (21.13%) is higher than in comparable studies, which stated that those travellers 'willing to pay more would not pay more than 10% above the cost of travel' (Pulido-Fernández and López-Sánchez, 2016, p. 13). If tourism businesses were willing to claim sustainability surcharges in the future, the additional proceeds could be invested in local sustainability initiatives and in offsetting travel-related emissions. As the data provided evidence that only a third of participants are willing to compensate their air travel emissions, it appears reasonable for tour operators to be in charge of ensuring adequate offsetting schemes for carbon emissions.

Attitude-behaviour gap

The findings indicate that the attitude-behaviour gap plays a relevant role in the slow travel context. Thus, existing studies are confirmed (e.g. Juvan and Dolnicar, 2014). Nevertheless, it needs to be noted that the observed attitude-behaviour gaps within this study are particularly more pronounced for slow travel transportation related to arrival at and departure from a destination. Despite the suggestion that growing interest in sustainability would increase the popularity of slow travel, the results of this study show that people would not be willing to give up flights, as they are often justified as the only viable method of transportation to most vacation destinations, confirming earlier findings about reluctance to abandon air travel (e.g. Dickinson and Lumsdon, 2010; Serdane, 2020). The differentiation between 'hard' and 'soft' slow travellers from Dickinson and Lumsdon (2010) is supported, stating that 'soft' slow travellers engage in sustainable travel but will most likely take an airplane to reach a distant or remote region, which they will then explore on foot or bicycle. This can be explained by the increased travel time and cost of alternative transportation, which were seen as big obstacles. In contrast, the gaps between attitude and actual behaviour are considerably smaller when it comes to slow travel activities at a destination. The smallest attitude-behaviour gap was reported for the dependent variable of future slow travel behaviour. This indicates that travellers intend to make more sustainable choices in their future travel planning compared to their past choices. However, it needs to be taken into account that this might (at least to some extent) be influenced by a prevalent social-desirability bias, which is particularly common in sustainability-related studies and describes the respondents' propensity to give answers that they believe will make them look good (i.e. to the researcher), concealing their true opinions (Nederhof, 1985; Krumpal, 2013).

Theoretical implications

Overall, the presented results function as an important and empirically based evaluation of travellers' perceptions of and attitudes towards more sustainable forms of travel. They provide a detailed overview of slow travel principles (not) applicable to mass tourism and the traveller's past and future behaviour concerning these principles. Thus, concrete tourism experiences in demand (slow activities) could be differentiated from those that would not yield significant interest of potential travellers (slow journeys to/from destinations). Existing slow travel studies were expanded by examining the main components of slow travel in a separated way, thus differentiating between the actual journey to or from a destination in comparison to the activities and behaviour *at* the destination. Further, many prior studies focused on niche samples (e.g. climate activists; Piispa, 2021) and their propensity to travel slow, or on specific destinations. In contrast, this study followed a general approach regarding the sample ('normal' travellers) and their attitudes towards slower travel options as well as not focusing on a specific destination. Lastly, the employed quantitative method adds to the field, which is otherwise shaped by conceptual and qualitative work.

The importance of slowness and the value of time, cultural immersion, environmental awareness, as well as mobility, were explored, taking into account past patterns and future intentions to provide insights regarding the drivers of travel choices. These additions have significant implications not only for how a slower mass tourism can be envisioned but also for how it must be appropriately marketed and operated. By serving as a guide for the development of more sustainable tourism offers, this study helps tourism scholars and practitioners respond to the challenges of mass tourism while shaping its future.

Lastly, the study highlights important opportunities that could be used to advance sustainable tourism research. Since the tourist experience includes multiple phases (e.g. pre-trip, on-trip, post-trip), it would be worthwhile to investigate the role of different slow travel characteristics across all touchpoints, and which characteristics travellers look for or expect at various stages of the experience. Future studies should set out to minimize the influence of social-desirability bias, e.g. by analysing official travel data from tour operators or tourism authorities. Transnational sampling strategies could be used to allow for comparisons between source markets. Furthermore, it would be of interest to explore if and how additional slow travel-induced revenues (e.g. sustainability surcharges) can and should be locally invested to facilitate successful sustainability initiatives taking into account different destinations and regions.

Conclusion

This chapter has shown that travel-related sustainability has become a matter of great importance and will continue and remain so in the future. In general, a trend towards 'fewer and better' experiences can be identified in the tourism sector. Throughout this chapter it has been demonstrated that progress is being made towards a more sustainable society and therefore a more sustainable tourism industry. People are more concerned about the environment, act more mindfully and strive for more authenticity. The tourism industry is shaped by an increase in competition as well as more and more interchangeable travel offers. Resultingly, those businesses daring to develop more ethical and sustainable services to respond to new and often unmet consumer demands will be able to create a competitive advantage. At present, there is an opportunity for tour operators to further develop the market for sustainable travel options, for example via the improving European rail network. Sustainable travel options could guide tourists towards mindful and environmentally sensitive travel behaviour.

However, it remains a challenge to combine the need to reduce the carbon footprint of a vacation and the focus on the actual travel experience as guiding principles of slow travel. It is advisable to focus future actions of tourism development on the quality of visits rather than the quantity of visitors, in order to improve travel experiences while preserving the character of a destination and its local communities. As demonstrated by the study, people are willing to pay more for sustainable tourism offers. In this context, tour operators that strive to protect and preserve the environment and local culture, as well as hotels with a responsible environmental and social concept, can certainly expect to benefit from the current and future rise of the sustainability-aware customer.

The limitations of these findings need to be pointed out. First, the actual past and future behaviour of participants is difficult to verify. Further, the results are not representative or transferable to other source markets and the survey was conducted during the COVID-19 pandemic. The crisis and the induced travel regulations and restrictions might have influenced individual responses. It is suggested that the results are complemented by comparable studies with a larger and/or more representative sample after the pandemic and across a variety of source markets.

To conclude, the concept of slow travel has gained momentum in the last decade and represents a type of holiday with the potential to minimize negative environmental and social impacts. As such, it is worth further investigation to explore how tourism can be continuously adapted or reshaped to provide enhanced tourist experiences in combination with reduced impacts on the local nature, people and culture. After all, the principles of slow travel can serve as a guiding light for the development towards more sustainable tourism offers and, resultingly, more sustainable tourist choices.

References

Caffyn, A. (2012) Advocating and implementing slow tourism. *Tourism Recreation Research* 37(1), 77–80. DOI: 10.1080/02508281.2012.11081690.

Ceron, J.P and Dubois, G. (2007) Limits to tourism? A back casting scenario for sustainable tourism mobility in 2050. *Tourism and Hospitality Planning & Development* 4(3), 191–209. DOI: 10.1080/14790530701736598.

Conway, D. and Timms, B.F. (2010) Re-branding alternative tourism in the caribbean: the case for 'slow tourism.' *Tourism and Hospitality Research* 10(4), 329–344. DOI: 10.1057/thr.2010.12.

Dickinson, J.E. and Lumsdon, L.M. (2010) *Slow Travel and Tourism*. Earthscan, London. DOI: 10.4324/9781849776493.

Dickinson, J.E., Lumsdon, L.M. and Robbins, D. (2011) Slow travel: issues for tourism and climate change. *Journal of Sustainable Tourism* 19(3), 281–300. DOI: 10.1080/09669582.2010.524704.

Freyer, W. (2011) *Tourismus: Einführung in die Fremdenverkehrsökonomie* [Tourism: Introduction to tourism economics] (10th edn). DeGruyter, Berlin.

Gardner, N. (2009) A manifesto for slow travel. *Hidden Europe Magazine* 25. Available at: https://www.hiddeneurope.eu/a-manifesto-for-slow-travel (accessed 18 October 2023).

Goessling, S., Hall, C.M. and Scott, D. (2013) *Tourism and Water: Interactions, Impacts, and Challenges*. Channel View Publications, Clevedon, UK.

Juvan, E. and Dolnicar, S. (2014) The attitude–behaviour gap in sustainable tourism. *Annals of Tourism Research* 48(2014), 76–95. DOI: 10.1016/j.annals.2014.05.012.

Kondoudaki, A., Moira, P. and Mylonopoulos, D. (2017) The application of slow movement to tourism: is slow tourism a new paradigm? *Journal of Tourism and Leisure Studies* 2(2), 1–10. DOI: 10.18848/2470-9336/CGP/v02i02/1-10.

Krumpal, I. (2013) Determinants of social desirability bias in sensitive surveys: a literature review. *Quality & Quantity* 47(4), 2025–2047. DOI: 10.1007/s11135-011-9640-9.

Lin, L.-P., Huang, S.-C. and Ho, Y.-C. (2020) Could virtual reality effectively market slow travel in a heritage destination? *Tourism Management* 78(2020), 104027. DOI: 10.1016/j.tourman.2019.104027.

Lozano-Oyola, M., López-Guzmán, T. and Huertas-Roig, A. (2019) Exploring the slow travel concept: a systematic literature review. *Current Issues in Tourism* 22(11), 1374–1394.

Lukic, T., Mitas, O. and Kovacevic, M. (2018) Slow tourism and regional development: exploring the potential of slow travel for rural areas. In: Timothy, D.J. and Boyd, S.W. (eds) *Tourism and Regional Development: New Pathways*. Springer, Cham, Switzerland, pp. 91–106.

Lumsdon, L.M. and McGrath, P. (2011) Developing a conceptual framework for slow travel: a grounded theory approach. *Journal of Sustainable Tourism* 19(3), 265–279. DOI: 10.1080/09669582.2010.519438.

Manthiou, A., Klaus, P. and Luong, V.H. (2022) Slow tourism: conceptualization and interpretation – a travel vloggers' perspective. *Tourism Management* 93, 104570. DOI: 10.1016/j.tourman.2022.104570.

Marson, D. (2011) From mass tourism to niche tourism. In: Marson, D. (ed.) *Research Themes for Tourism*. CAB International, Wallingford, UK, pp. 1–15. DOI: 10.1079/9781845936846.0000.

McDonald, S., Oates, C.J., Thyne, M., Timmis, A.J. and Carlile, C. (2015) Flying in the face of environmental concern: why green consumers continue to fly. *Journal of Marketing Management* 31(13–14), 1503–1528. DOI: 10.1080/0267257X.2015.1059352.

Meng, B. and Choi, K. (2016) The role of authenticity in forming slow tourists' intentions: developing an extended model of goal-directed behavior. *Tourism Management* 57(2016), 397–410. DOI: 10.1016/j.tourman.2016.07.003.

Milfont, T.L. and Duckitt, J. (2010) The environmental attitudes inventory: a valid and reliable measure to assess the structure of environmental attitudes. *Journal of Environmental Psychology* 30(1), 80–94. DOI: 10.1016/j.jenvp.2009.09.001.

Nederhof, A.J. (1985) Methods of coping with social desirability bias: a review. *European Journal of Social Psychology* 15(3), 263–280. DOI: 10.1002/ejsp.2420150303.

Netemeyer, R.G., Krishnan, B., Pullig, C., Wang, G., Yagci, M. *et al*. (2004) Developing and validating measures of facets of customer-based brand equity. *Journal of Business Research* 57(2), 209–224. DOI: 10.1016/S0148-2963(01)00303-4.

Nikolić, T.M., Pantić, S.P., Paunović, I. and Filipović, S. (2021) Sustainable travel decision-making of Europeans: insights from a household survey. *Sustainability* 13(4), 1960. DOI: 10.3390/su13041960.

Novelli, M., Cheer, J.M., Dolezal, C., Jones, A. and Milano, C. (2022) Introduction to niche tourism: contemporary trends and development. In: Novelli, M., Cheer, J.M., Dolezal, C., Jones, A. and Milano, C. (eds) *Handbook of Niche Tourism*. Edward Elgar Publishing, Cheltenham, UK, pp. xxiii–xxxii. DOI: 10.4337/9781839100185.

Oh, H., Assaf, A.G. and Baloglu, S. (2016) Motivations and goals of slow tourism. *Journal of Travel Research* 55(2), 205–219. DOI: 10.1177/0047287514546228.

Oliveira, C.S. (2020) My trip in my words: subjectivities, time(s) and mobilities in slow travel blogs. *Time & Society* 29(1), 223–255. DOI: 10.1177/0961463X18820740.

Piispa, M. (2021) Travelling by ground – young climate activists and the everyday utopia of sustainable travel. *Applied Mobilities* 8(1), 1–16. DOI: 10.1080/23800127.2021.2015822.

Pulido-Fernández, J.I. and López-Sánchez, Y. (2016) Are tourists really willing to pay more for sustainable destinations? An analysis of tourists' willingness to pay for sustainable tourism certification. *Sustainability* 8(12), 1240.

Rein, H. and Strasdas, W. (2017) *Nachhaltiger Tourismus* [Sustainable tourism], 2nd edn. UVK, Munchen, Germany. DOI: 10.1080/10837450.2017.1395886.

Roblek, V., Drpić, D., Meško, M. and Milojica, V. (2021) Evolution of sustainable tourism concepts. *Sustainability* 13(22), 12829. DOI: 10.3390/su132212829.

Schmied, M. (2004) Umwelt und tourismus: daten, fakten, perspektiven [Environment and tourism: data, facts, perspectives]. In: Hell, W. (ed.) *Erlebniswelten und Tourismus [Experiential Worlds and Tourism]*. Springer, pp. 27–42. DOI: 10.1007/978-3-540-76794-7.

Schmied, M., Götz, K., Kreilkamp, M., Buchert, M., Hellwig, T, T. *et al*. (2009) *Traumziel Nachhaltigkeit: Innovative Vermarktungskonzepte nachhaltiger Tourismusangebote für den Massenmarkt* [Dream destination sustainability: innovative marketing concepts of sustainable tourism offers for the mass market.]. Springer.

Serdane, Z. (2020) Slow philosophy in tourism development in Latvia: The Supply Side Perspective. *Tourism Planning & Development* 17(3), 295–312. DOI: 10.1080/21568316.2019.1650103.

Simpson, M.C., Gössling, S., Scott, D. and Hall, C.M. (2008) *Tourism and Water: Interactions, Impacts and Challenges*. Channel View Publications, Clevedon, UK.

The World Counts (2022) Negative environmental impacts of tourism. Available at: www.theworldcounts.com/challenges/consumption/transport-and-tourism/negative-environmental-impacts-of-tourism (accessed 18 October 2023).

UNEP & (UNWTO) (2005) *Making Tourism More Sustainable – A Guide for Policy Makers*. World Tourism Organization Publications. DOI: 10.18111/9789284408214.

UNWTO (1997) *Agenda 21 for the Travel and Tourism Industry*. World Tourism Organization Publications. DOI: 10.18111/9789284403714.

UNWTO (2017) Sustainable development. Available at: https://www.unwto.org/sustainable-development (accessed 18 October 2023).

UNWTO (2019) *International Tourism Highlights, 2019 Edition*. World Tourism Organization Publications. DOI: 10.18111/9789284421152.

Weidner, I. (2015) *Das Ende der Entschleunigung [The end of deceleration]*. Zeit Online. Available at: https://www.zeit.de/karriere/beruf/2015-09/stress-achtsamkeit-trend-entschleunigung (accessed 18 October 2023).

WTTC (2020) *Travel & tourism – global economic impact & trends 2020*. Available at: https://wttc.org/Portals/0/Documents/Reports/2020/Global%20Economic%20Impact%20Trends%202020.pdf?ver=2021-02-25-183118-360 (accessed 18 October 2023).

WWF (2014) WWF Deutschland Blog. Available at: https://wwfdeutschland-blog.tumblr.com/post/104569524425/alle-wollen-zur%C3%BCck-zur-natur-aber-keiner-zu-fu%C3%9F (accessed 18 October 2023).

13 Tourism Regenesis: Citizen Science and Wildfire Recovery in the Time of Climate Change

Abhijeet Shirsat[1]* and Erik Luvaas[2]
[1]Department of Recreation, Parks and Tourism Administration, California State University, Sacramento, California, USA; [2]Center on Disabilities and Human Development, Department of Movement Sciences, University of Idaho, USA

Abstract
Climate change-fuelled natural disasters, such as wildfires, are becoming a commonplace phenomenon and call for innovative actions to address recovery through sustainable tourism practices. The purpose of this case study is to present an experiential vacation that will have a twofold purpose of regenerating tourism and supporting tourist hedonic and eudaimonic wellbeing, where citizen scientists serve as the catalyst. The case proposes a new perspective of regenerative tourism, coined 'rescue tourism', and an example of how to implement rescue tourism, with prospective locations for implementation. The rescue tourism example is called Engaging Citizen Scientists in Rescue Tourism (ECSRT), which offers activities leveraging tourist motivations for rebuilding wildfire-impacted areas. ECSRT programmes not only offer economic recovery opportunities but also increased awareness about climate change impacts. The case demonstrates that in return for the efforts of these citizen scientists, they may receive personal wellbeing benefits while experiencing a unique vacation.

Introduction

In recent years, there has been a significant change in the way people travel. Due to climate change, environmental conservation is at the forefront of the minds of travellers (Hofman *et al.*, 2022). The purpose of this case study is to provide an example, through the Engaging Citizen Scientists in Rescue Tourism (ECSRT) programme, of how tourism can address wildfire recovery in a time of climate change with the help of citizen scientists.

The proposed ECSRT programmes will address both hedonic and eudaimonic wellbeing of tourists and help improve the wellbeing of the local communities and natural areas of selected sites that have endured natural disasters. Disaster tourism addresses the human desire for hedonic satisfaction and/or self-promotion via observing the ruins of natural disasters. Disaster tourist sites are those that 'market disaster, loss, and tragedy' (Zhang, 2021, p. 1), and they have been promoted as mechanisms to rebuild community resilience, restore economic stability and even address local socio-emotional distress (Prayag *et al.*, 2021). However, this form of tourism still remains largely consumptive, where visitors seek to experience local leisure attractions and tour disaster sites (Biran *et al.*, 2014). Volunteer

*Corresponding author: shirsat@csus.edu

DOI: 10.1079/9781800621411.0013

tourism involves organized volunteering on holiday and addresses the human desire for both hedonic and eudaimonic satisfaction via helping activities at a tourism site (Wearing, 2001). Merging these two umbrellas of tourist motivations creates an opportunity for recovery at disaster sites impacted by wildfire. This recovery may be achieved by engaging tourists, local communities, government agencies and policy-makers as not only visitors but as citizen scientists. As citizen scientists, visitors will be engaged in the research, data collection, analysis process and rebuilding of the wildfire-struck areas (EPA, n.d.).

Tourism and Wildfire

In recent years, the tourism economy, especially in the summer and fall months, can be entirely reactionary to wildfire incidents. As climate-change-driven weather patterns bring hotter, longer warm-weather periods to much of the western USA, the wildfire season will likely continue to extend nearly year-round in some parts of the region and with increasingly severe fire events (IPCC, 2014).

Beyond the localized devastation of buildings, forests and communities, the impact of wildfire smoke has regional, state-wide and sometimes nationwide implications for tourist behaviour. A study of tourist visitation at the five national parks in Utah between 1993 and 2015 found that, as measures of monthly area burned increased, statistically significant decreases in park visitation were observed in four of five national parks (Kim and Jakus, 2019). The researchers estimated a resulting regional economic loss of $2.7–4.5 million in the tourism economy with a disproportionate negative impact on rural tourism-dependent communities. Similarly, in California major wildfires can have lasting negative impacts on local tourism economies and community members. In a mixed-methods study on the 13 largest fires that ravaged Trinity County, California, and its large recreation and tourism economy in 2008, Davis *et al.* (2014) reported that, although employment statistics remained stable immediately following the fires, interview data revealed substantial negative impacts to recreational businesses. Interviewees described how 'loss of one summer's income or depressed future turnout in this industry can cause businesses to close' (p. 991). Concern for long-term recovery of major California tourism regions after wildfires will continue as climate change drives more numerous and damaging wildfires.

The tourism industry is more vulnerable to the negative impacts of wildfires than other economic sectors, and losses are likely to be felt for years after wildfire events (Feo *et al.*, 2020). The state of California is no exception, with its robust outdoor recreation tourism economy often localized around rural communities near protected natural areas. For example, the 2008 Trinity County wildfires burned approximately 241,000 acres and led to a reduction in recreation and tourism spending in the years following the fire (Davis *et al.*, 2014). Similarly, estimates derived from survey data from users of the 33,000 acre San Jacinto Wilderness of the San Bernardino National Forest indicated a $3.7 million welfare loss if trails closed due to a significant wildfire (Sánchez *et al.*, 2016). Since these economic impact studies were conducted, California has seen much larger and more devastating fires that will have long-term impacts on the tourism economy yet to be observed. There is a clear need for innovative models of tourism and pathways to tourism recovery for areas impacted by climate change-driven natural disasters such as wildfire.

Reconceptualizing Tourism for Community and Ecosystem Recovery

The tourism industry could use the ECSRT programme as an opportunity to educate people about ethical and sustainable travelling. In the past, several researchers, tourism professionals and governmental agencies have been involved in educating people on the need for responsible tourism, a new form of experiential vacation that will have a twofold purpose of regenerating tourism and supporting tourist hedonic and eudaimonic wellbeing. This case study suggests the term 'rescue tourism' for this recovery effort.

Regenerative tourism

Regenerative tourism involves activities for the tourists to heal, while counter-balancing social, economic and environmental impacts of tourism (Zaman *et al.*, 2022).

Rescue tourism

Rescue tourism involves responsible citizen scientists who use their vacation time to help with regenerative tourism activities like data collection, analysis and redevelopment activities for areas that are struck with environmental, man-made and medical disasters/emergencies, for the twofold purpose of regenerating tourism and for personal hedonic and eudaimonic wellbeing.

Citizen scientist

The US government uses the following definition for citizen scientist which we have adapted for this case study:

> In citizen science, the public participates voluntarily in the scientific process, addressing real-world problems in ways that may include formulating research questions, conducting scientific experiments, collecting and analyzing data, interpreting results, making new discoveries, developing technologies and applications, and solving complex problems.
>
> (CitizenScience.gov, n.d, para. 3)

Rescue Tourism: Merging Tourism Sectors

The rescue tourism case study addresses both hedonic and eudaimonic wellbeing of tourists and helps improve the wellbeing of the local communities and natural areas of selected sites that have endured natural disasters such as wildfires. In most cases, disaster or dark tourism addresses the human desire for hedonic satisfaction (positive emotion) and/or self-promotion via touring the ruins of natural disasters and/or sharing the experience on social platforms (Zhang, 2021). Dark tourism sites have been promoted as financial mechanisms to rebuild community resilience, restore economic stability, and even address local socio-emotional distress (Zhang, 2021; Prayag *et al.*, 2021). However, dark tourism remains largely a consumptive form of tourism where visitors seek to experience local leisure attractions and tour dark sites (Biran *et al.*, 2014).

Volunteer tourism involves spending vacation time in service to underserved groups or restoration of the environment (Wearing, 2001). Volunteer tourism addresses the human desire for eudaimonic satisfaction via helping activities to address a social need at a tourism site, but it often lacks significant satisfaction of tourist hedonic wellbeing due to the unpleasant nature of the work/service. Combining the umbrellas of disaster/dark and volunteer tourist motivations can create a new subsect of regenerative tourism that addresses both hedonic and eudaimonic wellbeing while benefitting the tourism community and natural area. With the rise of climate change impacts, rescue tourism could be one unique tourism industry solution to address the climate change-driven impacts of natural disasters, such as wildfires, in a more responsible manner.

Tourist Wellbeing

The convergence of volunteer and disaster/dark tourism provides an opportunity to consider tourist wellbeing more comprehensively. Rescue tourism may address both hedonic and eudaimonic wellbeing. This case considers tourist wellbeing as an important aspect of why the tourist would be motivated to participate in the process of wildfire recovery through citizen science activities.

Hedonic wellbeing

In today's fields of psychology and positive psychology, the three major components of hedonic wellbeing are life satisfaction and positive and negative affect (i.e. emotion) (Ryff *et al.*, 2021). Certainly, tourist behaviour is driven, in part, by the desire to experience more positive emotion, less negative emotion, and more life satisfaction via tourist activities such as vacation. Even disaster tourism exemplifies opportunities for hedonic satisfaction. However, the use of positive psychology in tourism studies has been critiqued as having too much focus on contexts where hedonic wellbeing is an outcome (Nawijn, 2016). Other scholars have argued that eudaimonic tourist experiences exist in which the presence of challenge and hardship allow for negative emotions to be experienced via various forms of work, but ultimately leading to the tourist achieving happiness via the creation of meaning, achievement and fulfilment from the activity (Filep, 2016).

Eudaimonic wellbeing

As opposed to hedonic wellbeing, which describes subjective happiness, eudaimonic wellbeing describes how much meaning one derives out of their pursuits and life activities and if one has achieved their human potential. Contemporary models of human flourishing from the field of positive psychology include both hedonic and eudaimonic wellbeing, as in the case of Seligman (2002) authentic happiness theory. Seligman (2011) expands this theory to include a framework identifying components of authentic happiness: PERMA (positive emotions, engagement, relationships, meaning and achievement). Scholars have debated the potential for eudaimonic tourist activities (e.g. volunteer tourism), which include strong components of engagement, meaning and achievement, to produce positive emotion due to the displeasing nature of service work in environments in need of assistance (disaster-struck areas, impoverished communities, etc.). However, some research has found tourists do experience positive emotion in these settings in the form of awe, compassion, gratitude and interest (Nawijn *et al.*, 2015). Volunteer tourism, as previously discussed, is one such example of a form of tourism allowing for some negative emotions via service work but yielding greater meaning and achievement in the process. With the potential for tourist activities to engage both types of wellbeing, hedonic and eudaimonic, we propose that the merging of disaster tourism and volunteer tourism may provide that express possibility through the rescue tourism model of regenerative tourism.

Application of Rescue Tourism on Impacts of Climate Change

Climate change has significantly impacted the tourism industry, especially in California where nature-based tourism accounts for a large portion of the tourism economy. For example, the California Department of Parks and Recreation reports that California state park visitors spend an average of $4.32 billion per annum in visitation expenses (State of California, 2022). Furthermore, attracting non-local tourists is a significant driver of the tourism economy in the state with 11.95% of visitors (or $1.66 billion in tourism revenue) being non-residents of California. The negative impacts of climate change on California tourism can already be observed in the $1.4 billion winter sports tourism sector where the difference between high and low snowfall years is estimated to see a decrease of 4.7% of skiers in the state. That equates to a potential $38–117 million reduced economic value. Along with significant revenue generation, the California winter sports tourism economy employs an estimated 24,000 people, making it the state with the second-highest winter sports economic activity in the USA. With some of the largest and most destructive wildfires in California's history impacting nature-based tourism areas within the past decade, the negative impacts on the tourism industry are only recently unfolding, but they will be significant to communities, tourism operators and the environment. Rescue tourism may provide a solution where the support of the public, tourists (local and international), non-profit agencies, governments and tourism hobbyists can protect and restore the ecological balance that will have lasting impacts on future generations.

This case study merges the two umbrellas of disaster/dark tourism and volunteer tourism and motivates people to act for hedonic and eudaimonic wellbeing benefits, which will also support the disaster sites impacted by wildfire. The mission of this rescue tourism may be achieved by engaging tourists, local communities, government agencies and policymakers as not only visitors but as citizen scientists. As citizen scientists, visitors will be engaged in the research, data collection, analysis process and rebuilding of the wildfire-struck areas (EPA, n.d.).

Engaging Citizen Scientists in Rescue Tourism (ECSRT)

In an effort to regenerate tourism in wildfire-impacted areas, rescue tourism will support the cause. Tourism sustainability does not lie in the hands of just the government or local policymakers; it lies in the hands of the people who are part of the entire society. Tourists are also part of this society since they get involved in visiting these places for a variety of reasons. The devastating effects of climate change and natural disasters like wildfires require immediate attention and a democratic way of enhancing the chances of rebuilding the lost ecosystem. Tourism plays a key role in the revitalization of the economy and the ecosystems. Hence, tourists can be key contributors to tourism revitalization by serving as citizen scientists.

Creating citizens interested in tourism who could participate in responsible tourism activities is one of the goals of tourism education. Tourism hobbyists embody this goal through their life-long leisure interest and engagement in regenerative tourist activities. In this chapter, we motivate these responsible travellers to join in activities that will convert their hobby into data collection and analysis in an effort to rebuild wildfire-struck areas.

The proposed reconceptualization of regenerative tourism transforms tourist experiences through activities that improve the wellbeing of local communities and natural areas. This is accomplished by utilizing a blending of disaster and volunteer tourism and with the inclusion of a key stakeholder, the tourist as citizen scientist.

ECSRT programmes will address both the hedonic and eudaimonic wellbeing of the tourists and help improve the wellbeing of the local communities and natural areas of selected sites that have endured natural disasters. ECSRT programmes are designed around the shared vision of the tourist and disaster-struck communities following the 6Ds guiding activities:

Dream: Tourists will envision the destination before wildfire destruction juxtaposed to the current state and then what the destination could be after regenesis.
Activity: Researchers will recruit people who are curious, explorers and motivated to travel to wildfire-struck areas. These could be previous travellers to the area, new travellers or someone who is visiting to serve the community.

Draw: In conjunction with tour organizers, tourists will plan regenesis activities to mitigate fire damage to the natural landscape and human inhabitants.
Activity: A tour itinerary will be created for these tourists (citizen scientists) which will include activities like planting trees and helping clean up the wildfire areas.

Dare: Tourists will process the gravity of the destruction and prepare emotionally, mentally and physically to consciously engage in regenesis activities with mindful awareness of the impacts of their actions.
Activity: Citizen scientists will be shown real images of the area so that they are aware of the situation and are ready to take on the responsibilities.

Do: Tourists will engage in the service work of citizen science through a variety of activities (i.e. flora and fauna destruction and recovery observation, habitat restoration).
Activity: Citizen scientists will work on the assigned projects while they also enjoy their leisure activities in the wildfire areas.

Distribute: Tourists will bring influx of economic benefit to affected communities and spread awareness of the impact and potential recovery of fire-damaged areas from their service work.
Activity: The financial resource, labour and data collection of these citizen scientists will help benefit local areas to rebuild and make these places more sustainable for future travellers.

Deserve: Global citizens, tourists and local community members can philosophically expect execution of similar natural disaster and climate change mitigation activities from fellow global citizens.
Activity: In return for the support in these regenerative tourism activities, these travellers will be given opportunities to travel to other places where ECSRT programmes have been implemented and where the wildfire-impacted area is under recovery.

Suggested locations for implementing this programme

Based on the need of ECSRT programmes and the recent wildfire-struck areas, we recommend implementing preliminary programmes at the sites of three major fires that struck northern California between 2018 and 2021. These were selected due to the fires' significant impact on the natural topography and ecosystem of the sites and the regional tourism economy and local communities.

California wildfires

Northern California's nature-based tourism identity is a major contributor to the nearly $460 billion US outdoor recreation economy. California contributes more than any other state with nearly $44.5 million and boasts nine national parks, dozens of state parks, and countless other public lands tourist sites (Bureau of Economic Analysis, 2022).

The 2021 Caldor Fire was one of the largest and most destructive fires in California's history, burning 347 square miles near South Lake Tahoe, and destroying 1003 structures (CalFire, 2022). The 2021 CZU lightning complex fire was the largest wildfire seen in the area of Big Basin Redwood State Park near San Francisco in over 100 years, destroying 1490 structures including more than 200 historic buildings, and claiming one life (CalFire, 2022). The Camp Fire was the costliest disaster worldwide in 2018 and the deadliest in California's history with 153,336 acres burned, 18,804 structures destroyed in and around the city of Paradise, and 85 deaths (CalFire, 2022).

Implications, Discussion and Conclusion

Symbiotically, the damaged ecosystems, from which rescue tourists may receive some of these health and wellbeing benefits, may also be the recipients of long-term benefits from those same tourists. Research on nature connectedness, the affective or emotional relationship individuals may develop from spending considerable time in nature, is shown to be positively associated with pro-environmental attitudes and behaviour (DeVille *et al.*, 2021). Recent meta-analyses have also shown that individuals with greater nature connectedness tend to have greater hedonic wellbeing, including subscales of positive affect, vitality and life satisfaction (Capaldi *et al.*, 2014), and greater eudaimonic wellbeing including the subscale for self-reported personal growth (Pritchard *et al.*, 2020).

The global tourism industry has already seen negative impacts of climate change as weather patterns change. For instance, popular skiing destinations are seeing a shortening of the ski season due to less snow; rising sea levels are threatening coastal areas and hurricane damage and wildfire destruction are some of the issues that the tourism industry is facing. Creative solutions to both support the tourism economy and help natural ecosystems to recover from wildfires are necessary in the face of the ongoing effects of climate change. Tourism activities such as those in the ECSRT programmes offer an opportunity to leverage tourist motivations and natural disasters to mitigate wildfire impacts to natural landscapes, tourism-dependent communities and the wellbeing of global citizens. ECSRT programmes not only offer economic recovery, community recovery, tourism recovery and traveller wellbeing, but they also increase awareness that we all need to be citizen scientists and stewards of our natural world.

References

Biran, A., Liu, W., Li, G. and Eichhorn, V. (2014) Consuming post-disaster destinations: the case of Sichuan, China. *Annals of Tourism Research* 47, 1–17. DOI: 10.1016/j.annals.2014.03.004.

Bureau of Economic Analysis (2022) Outdoor recreation satellite account, U.S. and states, 2020. Bureau of Economic Analysis Data. Available at: https://www.bea.gov/data/special-topics/outdoor-recreation (accessed 19 October 2023).

CalFire (2022) Top 20 most destructive California wildfires (1/13/22). [Stats and Events]. State of California. Available at: http://large.stanford.edu/courses/2022/ph240/chunduru1/docs/calfire-24oct22.pdf (accessed 20 October 2023).

Capaldi, C.A., Dopko, R.L. and Zelenski, J.M. (2014) The relationship between nature connectedness and happiness: a meta-analysis. *Frontiers in Psychology* 5, 976. DOI: 10.3389/fpsyg.2014.00976.

CitizenScience.gov (n.d.) About CitizenScience.gov. Available at: https://www.citizenscience.gov/about (accessed 19 October 2023).

Davis, E.J., Moseley, C., Nielsen-Pincus, M. and Jakes, P.J. (2014) The Community economic Impacts of large wildfires: a case study from Trinity county, California. *Society & Natural Resources* 27(9), 983–993. DOI: 10.1080/08941920.2014.905812.

DeVille, N.V., Tomasso, L.P., Stoddard, O.P., Wilt, G.E., Horton, T.H. *et al.* (2021) Time spent in nature is associated with increased pro-environmental attitudes and behaviors. *International Journal of Environmental Research and Public Health* 18(14), 7498. DOI: 10.3390/ijerph18147498.

EPA (n.d.) Basic Information about citizen science. Environmental Protection Agency. Available at: https://19january2021snapshot.epa.gov/citizen-science/basic-information-about-citizen-science-0_.html (accessed 20 October 2023).

Feo, T.J., Mace, A.J., Brady, S.E. and Linsey, B. (2020) The costs of wildfire in California. California Council on Science and Technology. Available at: https://ccst.us/reports/the-costs-of-wildfire-in-california/ (accessed 19 October 2023).

Filep, S. (2016) Tourism and positive psychology critique: too emotional? *Annals of Tourism Research* 59, 113–115. DOI: 10.1016/j.annals.2016.04.004.

Hofman, K., Walters, G. and Hughes, K. (2022) The effectiveness of virtual vs real-life marine tourism experiences in encouraging conservation behaviour. *Journal of Sustainable Tourism* 30(4), 742–766. DOI: 10.1080/09669582.2021.1884690.

IPCC (2014) Contribution of working groups I, II and III to the fifth assessment report of the intergovernmental panel on climate change. Core writing team. In: Pachauri, R.K. and Meyer, L.A. (eds) *Climate Change 2014: Synthesis Report*. IPCC, Geneva, Switzerland.

Kim, M.K. and Jakus, P.M. (2019) Wildfire, national park visitation, and changes in regional economic activity. *Journal of Outdoor Recreation and Tourism* 26, 34–42. DOI: 10.1016/j.jort.2019.03.007.

Nawijn, J. (2016) Positive psychology in tourism: a critique. *Annals of Tourism Research* 56, 151–153. DOI: 10.1016/j.annals.2015.11.004.

Nawijn, J, Isaac, R.K., Gridnevskiy, K. and van Liempt, A. (2015) Holocaust concentration camp memorial sites: an exploratory study into expected emotional response. *Current Issues in Tourism* 21(2), 175–190. DOI: 10.1080/13683500.2015.1058343.

Prayag, G., Buda, D. and Jordan, E.J. (2021) Mortality salience and meaning in life for residents visiting dark tourism sites. *Journal of Sustainable Tourism* 29(9), 1508–1528. DOI: 10.1080/09669582.2020.1823398.

Pritchard, A., Richardson, M., Sheffield, D. and McEwan, K. (2020) The relationship between nature connectedness and eudaimonic well-being: a meta-analysis. *Journal of Happiness Studies* 21(3), 1145–1167. DOI: 10.1007/s10902-019-00118-6.

Ryff, C.D., Boylan, J.M. and Kirsch, J.A. (2021) Eudaimonic and hedonic well-being: an integrative perspective with linkages to sociodemographic factors and health. In: Lee, M.T., Kubzansky, L.D. and VanderWeele, T.J. (eds) *Measuring Well-Being: Interdisciplinary Perspectives from the Social Sciences and the Humanities*. Oxford University Press, Oxford, UK, pp. 92–135. DOI: 10.1093/oso/9780197512531.001.0001.

Sánchez, J.J., Baerenklau, K. and González-Cabán, A. (2016) Valuing hypothetical wildfire impacts with a Kuhn–Tucker model of recreation demand. *Forest Policy and Economics* 71, 63–70. DOI: 10.1016/j.forpol.2015.08.001.

Seligman, M. (2002) *Authentic Happiness: Using the New Positive Psychology to Realize Your Potential for Lasting Fulfillment*. Free Press, New York.

Seligman, M. (2011) *Flourish*. Random House, Sydney, Australia.

State of California (2022) Sustainable tourism in California state parks. California Department of Parks and Recreation. Available at: http://parks.ca.gov/?page_id=26055 (accessed 19 October 2023).

Wearing, S. (2001) *Volunteer Tourism: Experiences That Make a Difference*. CAB International, UK. DOI: 10.1079/9780851995335.0000.

Zaman, U., Aktan, M., Agrusa, J. and Khwaja, M.G. (2022) Linking regenerative travel and residents' support for tourism development in Kaua'i Island (Hawaii): moderating-mediating effects of travel-shaming and foreign tourist attractiveness. *Journal of Travel Research* 64(4), 782–801. DOI: 10.1177/00472875221098934.

Zhang, Y. (2021) Unpacking visitors' experiences at dark tourism sites of natural disasters. *Tourism Management Perspectives* 40, 100880. DOI: 10.1016/j.tmp.2021.100880.

14 Recreation in Forests: Implications for Tourism Product Development Based on the Analysis of Terpenes in Two Forest Types

Monika Bachinger[1]*, Angelina Mattivi[1], Philipp Hilmer[1] and Ahmed A. Karim[2]
[1]University of Applied Forest Sciences, Rottenburg, Germany;
[2]University Clinic of Psychiatry and Psychotherapy, University of Tübingen, Germany and Department of Health Psychology and Neurorehabilitation, SRH Mobile University, Riedlingen, Germany

Abstract

Forests have been found to positively impact human health. Forests, therefore, may help mitigate health crises like the COVID-19 pandemic. However, it is still unclear which forest structures and what forest types contribute to human health and to what extent. This chapter looks at the effects of forest terpenes on human recreation. This is done by measuring the concentration and composition of terpenes together with the subjective judgement of relaxation using psychometric questionnaires in four different experimental conditions. These conditions are two forest types (a mixed forest and a coniferous forest) and two interior settings. The results show that the two forests perform better with regard to relaxation than the interior conditions. However, it is difficult to solely ascribe the relaxation effects to terpenes. The restorative quality of forests seems to profit from other factors like forest soundscape or visual appearance. From this, implications for tourism product design are deduced, which range from seasonal and weather-related recommendations to recommendations on activities and on favourable forest structures.

Introduction

Demographic change, sedentary lifestyles and mental strain in family or job have led to an increasing number of individuals suffering from physical or mental disease (OECD, 2018). This situation was aggravated during the COVID-19 pandemic. Common leisure activities, like going to a gym, were restricted. This had remarkable consequences on large parts of the population, some of them reporting increased mental strain and decreased general life satisfaction (Buecker *et al.*, 2020; Ravens-Sieberer *et al.*, 2021). One of the few recreational spaces that could still be visited was the outdoors, especially green spaces like forests. The importance of forests for human health, therefore, increased during the pandemic (Derks *et al.*, 2020; Grima *et al.*, 2020; Bamwesigye *et al.*, 2021).

As a result, public awareness of the restorative qualities of forests increased (UN, 2020). More and more, forests will be used as restorative environments in tourism, too (Ohe *et al.*, 2017). In addition to medicinal forests, products are marketed specially in the field of forest therapy or forest bathing. The first medicinal forest in Germany,

*Corresponding author: bachinger@hs-rottenburg.de

DOI: 10.1079/9781800621411.0014

for example, was created in 2015. Before that, a new-use category was introduced into forest law which allowed for an explicit dedication of forests to medicinal purposes (BVMV, 2022). Similarly, a 'Power Forest' had been established in Finland a few years earlier (Vattulainen *et al.*, 2011; Korpela *et al.*, 2017). Overall, however, practical implementation of forest-based health tourism throughout Europe is still rare.

The missing link for using the restorative qualities of forests in tourism product development is knowledge of how exactly forests contribute to human health. Although there is an abundance of evidence on health effects of forests (Kotte *et al.*, 2019; Wen *et al.*, 2019; Rajoo *et al.*, 2020), very few studies investigate the mechanisms of how forests impact on human health (Kuo, 2015). This is underpinned by the fact that the majority of studies do not distinguish various forest types (Karim *et al.*, 2020). Instead, 'forest' as a natural environment is treated as a generic surrounding, not taking into account that forests in their structures, tree composition and visual appearance may differ widely (Grilli and Sacchelli, 2020). Some exceptions to this exist, for example in Lee *et al.* (2011) and Cho *et al.* (2017). However, these studies were undertaken in Asian forests.

The aim of this chapter, therefore, is to investigate how forests exactly contribute to human health, with a focus on psychophysiological effects of forests in terms of stress reduction and relaxation. This is done with a focus on different forest types, which are characterized by their terpene concentration and composition as measurable features of forests. Secondly, based on the findings, the chapter formulates recommendations for tourism product development. This is done from the perspective that forests are crucial social infrastructures, especially in times of health-related crises like COVID-19 (Derks *et al.*, 2020). In the following sections, the conceptual basics of recreation in forests, the health effects of forests and their use in tourism are explained. After that, the empirical study is introduced, which sheds light on the question of how forest terpenes impact on human health, in particular on human restoration. The analysis of the terpenes' effects on human recreation is based on measurements in two forest types and two interior settings. In the last part of the chapter, the results are discussed and recommendations for tourism product development are offered.

Literature Review

Recreation in forests

Forests are special environments from multiple perspectives. In terms of their natural set-up, forests display a multi-layered structure consisting of herb and shrub layers, trunks and canopy. Forests, therefore, exhibit a pronounced, visual three-dimensionality (Cervinka *et al.*, 2014). Furthermore, forests have a characteristic micro-climate. The forest climate contains fewer dust particles, less noise, and different light spectra as compared to open land or urban environments. Forests are characterized by lower temperatures and a higher humidity, especially in summer, and, most importantly, forest air contains essential oils – the terpenes. It is against this background that forests are said to have positive physical, mental and social effects on human health (Harting *et al.*, 2011; Cervinka *et al.*, 2014).

From an environmental psychology perspective, forests provide a great variety of stimuli (Gibson, 1979). Forest visitors may perceive and respond to forest environments differently, according to their expectations or previous experiences (Tsaur *et al.*, 2014). In contrast to urban environments, forests do not prescribe certain courses of action (i.e. like traffic lights in cities); forests, rather, are open to subjective interpretation and reaction. They therefore provide an environment which stimulates creativity and self-expression (Yu and Hsieh, 2020). They contribute to the development of identity and self-efficacy (Hong *et al.*, 2021). This is especially the case, because forests provide for ample opportunity to engage all the five senses (Antonelli *et al.*, 2020). The interplay between light and shadow, various forms and colours of leaves, forest scents and the sounds address human senses intensively (Clifford, 2018; Li, 2018; Kotte *et al.*, 2019).

From a touristic point of view, forests allow for a great variety of recreational activities. Actually, outdoor recreation in forests has not only seen an increase in numbers, but also in types of activities (Lupp *et al.*, 2016). Typical forest activities are hiking, jogging, cycling, but also horse riding, having a picnic or geocaching (Hegetschweiler *et al.*, 2022). The motives for these activities are manifold. They cover motives like taking in 'good fresh air', 'being in nature' and 'exercising' as well as enjoying 'quietness' (Ensinger *et al.*, 2012). Many forest visitors describe being in forests as experiencing a counter-world to everyday life. Forests appear as a 'refuge from everyday strain' and 'a break-away from modern civilization' (Ensinger *et al.*, 2012, p. 79).

The importance of forests for human health became very clear during the COVD-19 pandemic (Immich and Schuh, 2021). During that time, forest visits increased through many parts of the world. In a forest near Bonn in Germany, for example, more than twice as many visitors were counted during the first lockdown in 2020 than had been recorded previously. Not only numbers increased but also visitation patterns. While there were clear visitation peaks previously, the pandemic period saw visitors distributed evenly throughout the day, and weekdays hardly differed from weekends (Derks *et al.*, 2020). A study from a forest in Burlington, Vermont, USA, showed similar patterns, with around 69% of 400 respondents visiting the forest more frequently during the pandemic (Grima *et al.*, 2020).

Effects of forests on human health

In general, human health is conceived of as consisting of mental, physical and social aspects (WHO, 1948). Forests may contribute to these three dimensions of human health. As for the physical aspect, forests provide space for various sporting activities like walking or cycling. These activities promote physical fitness, for example, by positively impacting on the cardiovascular system (Kaczynski and Henderson, 2007). As regards mental aspects, forests provide a counter-world to urban life suitable to calm visitors down and relieve their tension (Ohe *et al.*, 2017; Bielinis *et al.*, 2020). Due to their natural structure, forests may contain a great number of people before feeling overcrowded. They therefore provide privacy and space for retreat and inner contemplation (Kim and Shin, 2021). Compared to other leisure facilities, forests are not expensive (O'Brien and Morris, 2014). Access is mainly unrestricted, thus blurring social differences. As for social health, forests provide the opportunity to meet people, to socialize and to engage in conversation. Forests therefore fulfil an important social and integrative function (Jay and Schraml, 2009).

Lately, and especially during COVID-19, research focus is on the psychophysiological effects of forests in terms of stress reduction and relaxation (Karim *et al.*, 2020; Lee *et al.*, 2022). It has been shown that being in forests reduces sympathetic nerve activity and increases parasympathetic activity (Lee *et al.*, 2011; Ochiai *et al.*, 2015). In addition, several studies show that mindful practices in forests lead to a significant reduction in the stress hormone cortisol (Park *et al.*, 2007, 2010). Forests thus contribute to stress reduction, higher concentration, healthier sleep, inner peace and increased mental resilience (Keniger *et al.*, 2013; Takayama *et al.*, 2014).

In addition, health effects of forests are examined from a biochemical perspective. The focus is on volatile organic compounds (VOCs). A large proportion of VOCs belong to the class of terpenes, which are formed by plants as an important component of their defence against pathogens. Terpenes, in turn, may impact positively on human health (Abe *et al.*, 2008; Putra *et al.*, 2018). For example, Lee *et al.* (2011) found that terpenes support human recovery processes. The study shows that forest terpenes promote physiological and psychological relaxation in humans. However, the importance of concentration and composition of terpenes remains unclear.

Akakabe (2012) addresses this problem by showing that a certain composition of terpenes from the bark of *Lindera umbellata* (fever shrub, laurel family) increases the parasympathetic activity. Kawai and Miyachi (2016) follow an almost identical approach, investigating the effects of juniper oil on the autonomic nervous system. Li *et al.* (2007), in turn, investigate the influence of terpenes on the human immune system and show that staying in forests increases the activity of natural killer cells (Li *et al.*, 2007).

Lastly, Cho *et al.* (2017) examine the effect of terpenes on a wide range of health aspects, such as anti-inflammatory effects, effects on tumours or neuroprotective effects. With regard to further fields of research, they come to the conclusion that, 'although data elucidating the possible action mechanisms of these compounds are increasing, numerous studies still present only preliminary screening results' (p. 102). This applies, in particular, to Europe, as most of the studies were done in Asian countries (especially Japan and South Korea) (Stier-Jarmer *et al.*, 2021).

Health-related tourism products based on forests

There are at least two forms of forest-based health tourism. Whereas the first category aims at restoring physical or mental health and therefore targets individuals with a clinical indication or who already suffer from an ailment, the other category is leisure-oriented. This latter form takes a rather broad approach to restoration, including good accommodation, food and entertainment. It therefore may be better referred to as health-related or health-oriented tourism, instead of health tourism in a narrow sense (Hofer *et al.*, 2012). In the following example, two categories of forest-based health tourism are introduced. These are cure and healing forests, and forest therapy and forest bathing.

Cure and healing forests

In Germany, medicinal forests were first designated in Mecklenburg-West Pomerania. There, a new-use-category was introduced in the federal state's forest law. This allowed forest administrations to declare medicinal forests, in addition to the recreational forests that had existed before. These forests are characterized by a special regime of usage rights, but also usage limitations (ICOHF, 2022; BVMV, 2022). There are two categories of medicinal forests. Whereas the so-called healing forests address individuals with a clinical indication that is a diagnosed disease, the cure forests address a broad group of visitors that visit the forests for leisure and recreation. Whereas healing forests are located near clinics and serve therapeutic purposes, cure forests are used for health promotion or recreation (BVMV, 2022).

For both categories there are quality requirements. Cure forests, for example, need to have patches of dense tree stands to ensure noise insulation, privacy and wind protection. Besides this, they need clearings or areas with a light stock of trees that invite visitors to rest and lie down. The forest paths should be unsealed, with a natural surface. There should be clean and fresh air. Noise pollution from traffic should be avoided. Overall, the infrastructure should enable a safe and comfortable visit. There should be safe and sufficient parking. All roads and infrastructures should be signposted. An information board at the entrance of the forest should contain information on trails but also on rescue points. Some parts of the cure forest need to be accessible for visitors in wheelchairs or with sensory impairments, etc. The requirements for healing forests are higher. For example, air quality and absence of noise pollution have to be proven by an expert (ICOHF, 2022; BVMV, 2022).

Forest therapy and forest bathing

Forest bathing or forest therapy both relate to the Japanese term of *shinrin-yoku*, which was coined by the Ministry of Agriculture, Forestry and Fisheries of Japan in 1982 and which has been defined as 'taking in the forest atmosphere' (Tsunetsugu *et al.*, 2010, p. 27). Since both terms come from the same root, they are often used interchangeably. However, forest therapy is aimed at treating sick people, while forest bathing appeals to a broad target group for recreational purposes (Immich and Schuh, 2021).

The contents of *shinrin-yoku* are undefined and comprise a lot of different activities from walking in the forests or just spending a quiet time there and 'resetting one's heart and body', to therapeutic work and approved rehabilitation methods (Clifford, 2018; Li, 2018). Basic practices are sensory exercises for sight, hearing, taste, smell and touch to increase sensual awareness (Wen *et al.*, 2019). Exercises are, for example, walks, meditation, yoga or breathing exercises as well as handicraft activities, such as laying mandalas out of natural materials such as pine cones (Kotera *et al.*, 2022; Schuh and Immich, 2022).

Forest therapy can be practised both with instruction by a trainer and individually on one's own. It may be practised alone or in groups. Group sizes may vary. In Germany, forest therapy normally includes around ten participants.

However, there are also individual tours or tours with much more than ten people. As for duration, forest therapy can be practised within 15 min or take several days, for example as part of a therapeutic treatment. Besides instructing participants in forest bathing practices, trainers usually point out special features of forests. They, thus, aim at increasing knowledge on nature and nature-connectedness as precursors for nature protection (Kotera *et al.*, 2022; Schuh and Immich, 2022).

Forest therapy has been practised in Japan for several years, partly in a therapeutic context in so-called forest therapy centres (Li, 2018). In Germany, forest therapy has increasingly gained attention throughout the last ten years. Only recently, the Bundesverband Waldbaden e.V. in Germany has been founded, an association that combines nature protection with promoting the health effects of forests (see www.bundesverband-waldbaden.de [accessed 19/10/23]). In addition, interested individuals may take part in numerous training and educational courses in the field of forest therapy. In the aftermath of COVID-19, forest therapy has been discussed as an adjunct therapy for Long-COVID patients (McEwan *et al.*, 2022).

Methodology

The aim of this chapter is to investigate the effects of terpenes on human relaxation. To this end, an interdisciplinary research design was developed, which covered biochemical analysis as well as analysis of human restoration. For biochemical analysis, the environments' trace gases were recorded with a focus on terpenes. Recovery in humans was analysed by subjective assessment of the participants. Terpenes and human recovery were recorded in parallel to identify causal relationships.

The study consisted of four test conditions: (i) a mixed forest with coniferous and broad-leaf trees (MF = mixed forest); (ii) a coniferous forest (CF = coniferous forest); (iii) a terpene-scented interior (IS = interior scented); and (iv) an interior without terpene scent (IU = interior unscented, neutral condition) (Fig. 14.1). Through this set-up, the olfactory effects of terpenes could be studied separately from visual or auditory effects of the forest. Participants were randomly allocated to the four test conditions.

The interior conditions were found in the cultural centre of the village Bad Peterstal-Griesbach (Baden-Württemberg, Germany). The room had an air volume of approx. 360 m^3. To scent the interior, 20 branches of fir and 12 branches of spruce, each 60–100 cm long, were manually worked on regularly – starting 1 h before the sessions, and every 10 min during the sessions. Additionally, the air around the participants was swirled with a fan. After the end of the sessions, all doors and windows were opened to ventilate the room.

The requirements for the two forest conditions were of a qualitative nature. The areas should differ in terms of their tree population or the composition of tree species, and they should be easily accessible. Two patches of forest close to the village of Bad Peterstal-Griesbach were selected. One patch was a mixed forest, the other one a coniferous forest.

The MF area is about 710 m above sea level. The area faces north-east. Its maximum slope gradient is 12%. The nutrient balance of the soil is oligotrophic, which leads to a rather poor and uniform herb layer. The ratio of coniferous and broad-leafed trees is almost balanced. In terms of tree species, beech (*Fagus sylvatica*) dominates, with 53% mixed with 31% silver fir (*Abies alba*). The trees are around 80 years old. Participants were placed in the midst of light forest structures.

The CF area is about 620 m above sea level. The patch faces south and exhibits a maximum slope gradient of 35%. The herbaceous layer is largely dominated by blueberries (*Vaccinium myrtillus*). The CF area is characterized by a high percentage of coniferous trees (96%). The tree species are silver fir (*Abies alba*) to 73% and red spruce (*Picea abies*) to 18%. The trees are around 100 years old. The participants were placed in the midst of a rather densely stocked part of the forest.

Data were collected between 8 July and 15 July 2019. In all four experimental conditions, the subjects sat on chairs or on tree stumps. Each session lasted about 90 min. There was no movement allowed during the session. The subjects were asked not to talk to each other. The collection of data comprised the following steps: (i) before the session: survey on subjective

Fig. 14.1. Visual impression of the four test conditions; IU = interior unscented; IS = interior scented; CF = coniferous forest; and MF = mixed forest.

wellbeing, measurement of cortisol and oxygen saturation of the blood; (ii) during the session: collection of psychophysiological data – heart rate variability and terpene data; (iii) after the session: survey on subjective wellbeing, on perceived restoration outcome, and on perceived restorative quality of the test condition, plus measurement of cortisol level and oxygen saturation of blood. Before participants came to the sessions, they filled out a questionnaire on socio-demographic data, their basic wellbeing, mindfulness in everyday life and their nature-connectedness. During the sessions, two interventions with mindful practices took place. In this chapter, only parts of the data are reported. These are the data on recorded terpenes, on restoration outcome and perceived restorative quality of the test conditions. In addition, the socio-demographic data and the participants' nature-connectedness are included.

Detection of terpene concentration and composition

To determine the composition and quantity of terpenes, a defined volume of air (10–12 litres during each session) was sucked into air collection tubes using air-sampling pumps (210–1003 MTX, SKC, Dorset, UK). These tubes were covered with special adsorbents (Tenax TA) and located in the immediate vicinity of each participant. In the laboratory, the accumulated trace gases were desorbed from the tubes in a thermal desorption system (TDU-CIS, Gerstel, Germany), frozen out in a cold trap at −70°C and, after a heat pulse, transferred to a gas chromatograph (GC model 7890B, Agilent Technologies, Böblingen, Germany). After separating the individual components, detection was carried out using a mass spectrometer. With the help of external standards, the trace gases in the air samples were identified and their concentrations quantified.

Analysis of human restoration

To determine the recreational effects of terpenes, a survey was carried out using established psychometric questionnaires. This was for nature-connectedness the Inclusion of Nature in Self Scale (Schultz, 2002). The Restoration Outcome Scale (ROS; Korpela *et al.*, 2008) was used for measuring the participants' recovery, the Perceived Restorativeness Scale (PRS; Pasini *et al.*, 2014) to record the perceived restorative quality of the test conditions. The difference between ROS and PRS is that while ROS measures the outcome of a recovery process, PRS describes how participants subjectively perceive the characteristics of the test conditions and how these conditions contribute to their recovery.

Both instruments were adapted to the study's requirements. As regards ROS, the scale was shortened by one item and contained the following five items: (i) I feel calmer after being here; (ii) I feel restored and relaxed; (iii) I have received energy for my everyday tasks; (iv) Visiting here was clearing and clarifying my thoughts; (v) I forgot everyday worries here. The items could be answered on a five-point scale (1 = does not apply at all; 5 = fully applies). The PRS was measured using four items instead of eight. These items were: (i) Places like this are fascinating; (ii) To get away from things that usually demand my attention, I like to go to places like this; (iii) In places like this everything seems to have its proper place; (iv) In places like this my attention is drawn to many interesting things. These items could also be answered using a five-point scale (1 = does not apply at all; 5 = fully applies).

A confirmatory factor analysis was performed for both scales. Both ROS and PRS showed a satisfactory one-factorial structure in each test condition. With regard to PRS, the internal consistency was at a Cronbach's α of 0.84–0.93. The factor loadings were at 0.78–0.96. Against this background, the items of PRS were aggregated into a single variable for each test condition. The procedure for ROS was identical. Cronbach's α was at 0.83–0.92, the factor loadings at 0.71–0.93. One of the test conditions showed a two-dimensional instead of a one-dimensional factor structure. However, due to the fact that the variable was of sufficient quality in all other test conditions and due to the small sample size, the items were aggregated into one variable in this case.

Results

Sample description

The sample consisted of 24 participants. Most of them (29.2%) were between 50 and 59 years old, and another large group between 60 and 69 years old (25%). In terms of gender, the sample was almost equally distributed between men and women. A large proportion of the sample (45.8%) was retired, most of the rest (37.5%) worked full-time.

As regards forest visits, the majority answered to visit forests 'a few times a week' (58.3%). More frequent visits (every day or more than once a day) were made by 16.6%. About 12.5% of the sample visited forests rather rarely (once every 2–3 months or 1–2 times a month). Asked to judge their nature-connectedness, 56.5% of the respondents considered they were very close to nature; 39.1% showed a medium degree of nature-connectedness. There was hardly anyone who did not consider his- or herself as being close to nature.

Results of terpene detection

A total of 51 different trace gases were detected in the four test conditions, of which 30 components could be identified as terpenes. The terpene concentrations for mixed and coniferous forests were 0.89 ± 0.53 and 1.58 ± 0.9 ng l^{-1}, which is within the usual range for forest ecosystems (Janson, 1992). As expected, the terpene concentration in IU was lower than in the forest conditions (0.49 ± 0.15 ng l^{-1}). However, after scenting the interior room with fir and spruce branches, terpene concentrations increased considerably, so that they were significantly higher than in the forest conditions, even after ventilating the room (155 ± 56 ng l^{-1}). The terpene concentrations in IU, thus, were partly about twice as high as in the forest locations.

Nevertheless, the differences in terpene concentrations in the various test conditions were mostly highly significant. This applies to differences between the unscented and the scented interior (IU vs IS), as well as to the differences between the two forest conditions (CF vs MF). Only the difference between the unscented interior condition after ventilation and the coniferous forest was not significant.

Important monoterpenes in the samples were camphene, limonene, tricyclene, terpinolene and δ−3-carene. The dominant oxygenated monoterpene was the compound bornyl acetate, which made up more than 90% of this class of compounds. Bisabolene, which is found in fir trees, dominated the group of highly reactive sesquiterpenes. In the scented interior conditions (IS), the sesquiterpene caryophyllene played a prominent role. This compound is commonly emitted during plant stress, which seems plausible because fir and spruce branches were worked on manually during the sessions.

The analysed terpenes were subjected to multivariate data analysis to test whether characteristic terpene patterns were present in each test condition. In fact, most of the conditions showed clearly definable characteristics. Particularly noticeable was that the samples from the scented interior conditions (IS) differed largely in their terpene patterns from the other conditions. The samples from the unscented interior condition (IU) also formed a distinguishable cluster with a characteristic terpene composition. Somewhat surprisingly, the two forest conditions formed one cluster, indicating only slight differences in their terpene patterns.

Results on human restoration

To examine the effects of terpenes on human restoration, the terpene concentrations measured in the test conditions were correlated with the participants' recreation outcome (ROS) and with the perceived restorative quality of the places (PRS). In addition, both variables (ROS and PRS) were examined for differences in the test conditions.

Analysis of differences between test conditions

Descriptively, the restorative quality (PRS) of the two forest conditions (MF and CF) was rated higher than the two interior conditions (IS and IU). The mean value of the two forest conditions is at 4.54 and 4.50. For the two interior conditions, the mean is at 2.29 and 2.05, which is remarkably lower. The standard deviation across all four conditions is between 0.56 and 1.02. Similar results were found for the restoration outcome (ROS). Also there, participants rated recovery higher in the two forest conditions (4.08 and 4.25) than in the interior conditions (3.23 and 3.58). The standard deviation is rather large across the test conditions.

Analysing the differences of the conditions, results show that values of restoration outcome (ROS) differ significantly between all four test conditions (Friedman test: $Chi^2 = 19.452$, $p=0.0002$, panel X). Values increase, starting from IU (3.23) through IS (3.58) to CF (4.08) and MF (4.25) (Fig. 14.2). Thus, according to the survey, participants attributed MF the highest recreational effect.

The analysis on perceived restorative quality (PRS) showed that the two interior conditions differ significantly from the two forest conditions (Friedman's test: $F = 60.38$, $p<0.05$). Participants thus judged the interior conditions less relaxing than the forest places. However, no differences were found between IU (2.05) and IS (2.29) or between CF (4.54) and MF (4.50) (Fig. 14.3).

Relation between terpene concentration and restoration

To test whether the terpene concentration had an influence on the perceived recreational effect, a regression was calculated with ROS and PRS as dependent variables and the terpene concentration as independent variable. Results show that there is no significant relationship between ROS and the terpene concentration. However, there is a significant impact of the terpene concentration on PRS (Fig. 14.4). The regression coefficient is negative; that is, more terpenes led to less PRS. Since data from all four test conditions were used, other influencers (like sounds or visual

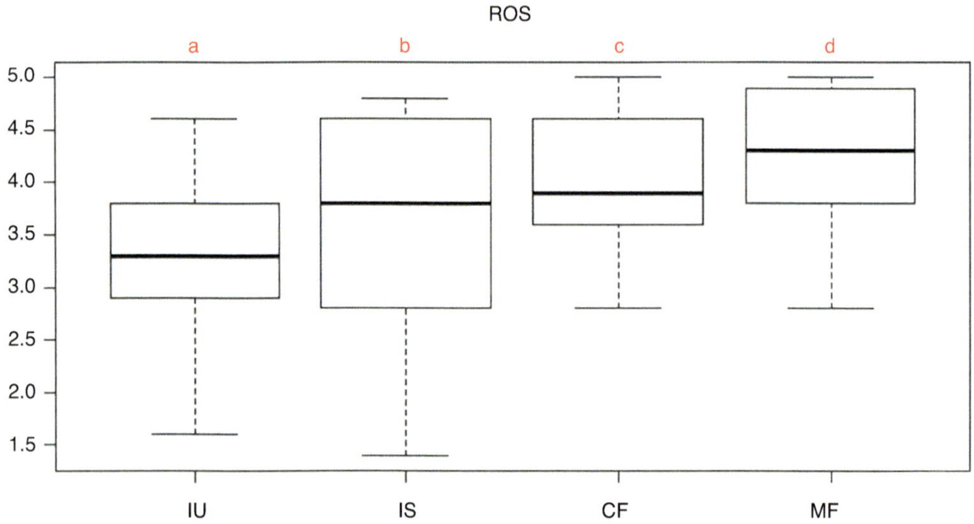

Fig. 14.2. Boxplot with significant results (a, b, c, d) for the ROS values within the four conditions (IU = interior unscented, IS = interior scented, CF = coniferous forest, MF = mixed forest).

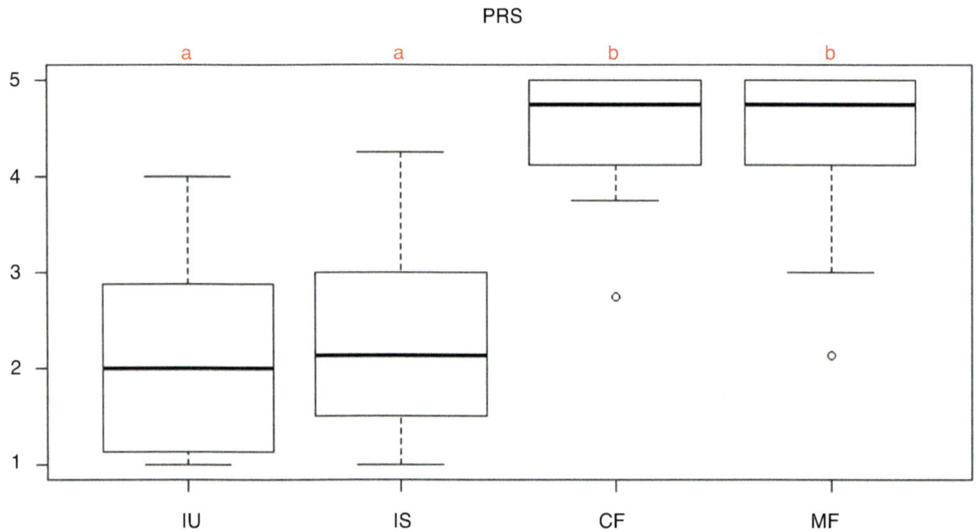

Fig. 14.3. Box plot with significant results (a, b) for the PRS values within the four conditions (IU = interior unscented, IS = interior scented, CF = coniferous forest, MF = mixed forest).

appearance of the environments) could not be excluded from having an impact on this result. Therefore, IU and IS were analysed separately from the two forest conditions, to isolate the impact of terpenes on ROS and PRS. For PRS, no meaningful correlation could be found. For ROS, results showed a positive correlation, although not significant.

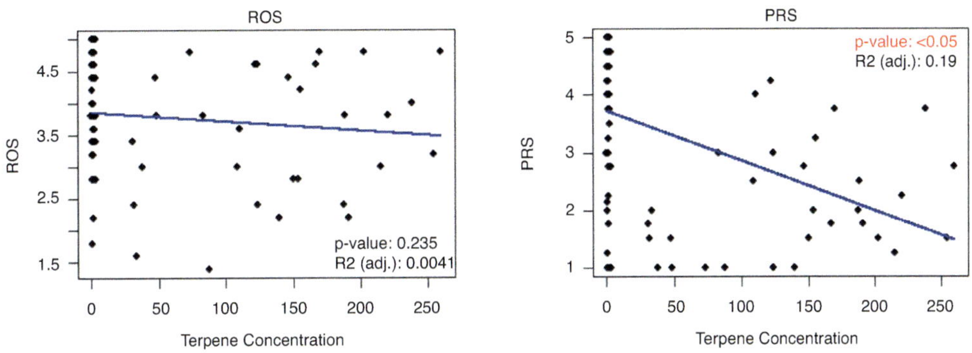

Fig. 14.4. Results of GLM regression for ROS and PRS.

Discussion

Forests have been proven to have positive impacts on human health (Park *et al.*, 2007, 2010; Ochiai *et al.*, 2015; Kotte *et al.*, 2019). However, it is still unclear which forest structures and what forest types contribute to human health and to what extent. This study examined the effects of forest terpenes, as a measurable forest component, on human recreation. This was done by measuring the concentration of terpenes in parallel to a subjective judgement of relaxation in four test conditions.

Results show that the two forest conditions perform better with regard to relaxation than the two interior conditions. This applies to ROS as well as to PRS. With regard to ROS, participants rated MF as the most relaxing environment. But also, the recreational effect of CF was significantly higher than of the interior settings. Similarly, participants judged the restorative quality of the forest environments as significantly higher than the one of the interior conditions (see above – 'Analysis of differences between test conditions'). These findings are consistent with other work, such as by Tyrväinen *et al.* (2014), according to which PRS in forests is significantly higher than in cities.

It was revealed that terpene concentrations were higher in CF and MF than in IU, but lower than in IS. A regression analysis was done to analyse whether the positive recreational effect of the two forest conditions was induced by the terpene concentrations. The results showed that the concentration of terpenes impacted negatively on perceived restorative quality (PRS). That is, environments with more terpenes were judged less restorative. The results for ROS were not significant; however, they also showed a negative tendency. These results somewhat contradict findings from other works that found positive effects of forest terpenes on human health (Li *et al.*, 2007; Lee *et al.*, 2011; Akakabe, 2012; Kawai and Miyachi, 2016). However, they might be explained by the large concentration of terpenes in IS.

As a result of this, the analysis was taken one step further. According to Wen *et al.* (2019), forests support human relaxation in multiple ways by stimulating different human senses. The next step was to isolate the effects of terpenes from other influencing variables. For this, the terpenes of the two interior conditions were assessed separately from the forest conditions. The difference between the two conditions was the terpene concentration, since IS was scented with fir and spruce branches. All other parameters (like light and temperature) were kept constant. A correlation analysis was done to evaluate the relationship between ROS and terpenes. Results show a positive relationship between ROS and terpene concentration, albeit on an insignificant level. In contrast to this, the restorative quality (PRS) of IS and IU was neither positively nor negatively influenced by the terpene concentration.

From this, it can be deduced that human restoration is positively influenced by forest terpenes. First and foremost, this applies to restoration outcome (ROS). The observed higher restoration outcome in the two forest environments (MF and CF) may be attributed to the terpenes in these two conditions, as deduced from the correlation between terpenes and ROS. However, the higher PRS values in the two forest conditions may not be explained by terpenes alone, since higher terpene concentrations induced a decreasing PRS value. PRS, therefore, needs to be influenced by other factors, such as the forest's soundscape or visual appearance. The restorative quality of a place, therefore, needs more than a pleasant forest scent to be judged as attractive. Natural spaces, as recreation spaces, therefore benefit from a meaningful combination of audio-visual, tactile and olfactory components. In general, it has to be assumed that although terpenes are important, they are not the sole factor in human recreation. Instead, a complex interplay of diverse forest components is at work. A better understanding of this web of factors is necessary, which calls for further, interdisciplinary research.

Lastly, results showed that participants did not judge MF and CF differently in terms of the restorative quality of the place (PRS). Both place qualities were judged equally restorative. With regard to ROS, however, there were differences: relaxation after visiting MF was judged higher than after being in CF. This suggests that different types of forests exhibit different impacts on human recreation. However, the perceived restorative quality of a forest (and thus its visual or audial appearance) does not predict its recreational impact to the full. Instead, other indicators, such as psychophysiological and neuroendocrine indicators, need to be taken into account.

These results also apply in a post-COVID situation. During the pandemic, not only were recreational activities restricted, but also stay-home orders were issued (ECDC, 2023). As a result, one can assume that the difference in the perceived restorative quality of the two forest conditions and the two interior conditions would have been judged even higher if the study had been performed during COVID-19. Secondly, many people visited forests more frequently during the pandemic, because forests were one of the few recreational places left to visit. Given this, one may assume that frequent visitors might have developed nature appreciation for forests. Both assumptions imply that the study's results might have become even more relevant. This is because the study's sample represents an even larger part of the population than before COVID-19. The sample was characterized by frequent forest visitors with a high degree of nature-connectedness.

Implications for Tourism Product Development

The present study shows that different forest types differ in their terpene concentration. Since the concentration of terpene impacts positively on the perceived restorativeness of a forest visit, different forest types may foster human recreation differently. Forest types that produce higher concentrations of terpenes may be regarded as more restorative than forests with a poor terpene emission. Forest-based health products, that is medicinal forests as well as forest therapy, therefore should take place in forests which are rich in terpenes. Oak forests, for example, are poor terpene emitters. Under the criterion of terpene concentration, therefore, they would not qualify as a place for forest-based health tourism.

In addition, seasonal and weather conditions should be taken into account. Forest-based health tourism should take place in the summer months, during dry and warm weather, and not on sunless or windy days, because forests produce terpenes particularly in warm and dry weather conditions. This is not to say that forest bathing may not take place in winter and generate positive effects on human health (Bielinis *et al.*, 2018). However, in this case, effects depend most probably on other factors such as visual or acoustic factors and not on terpene concentration. If the effects of terpenes are to be used on rainy and cooler days, indoor products can be designed. As seen in the study, interior settings scented with fir and spruce branches successfully contributed to the participants' relaxation. That is, a man-made forest environment indoors may be an alternative for people who do not want or cannot visit forests or for interventions that take place on a rainy day.

In the study, the two forest conditions were perceived as more restorative than the two interior conditions, although the terpene concentration indoors was higher. This means that terpenes are not the sole factor determining the restorative qualities of an environment. There are other influencers like sounds or visual appearances. Forests, therefore, are a complex web of factors that altogether make up an attractive and restorative environment. For tourism product design, this means taking into account landscape quality. Attractive forest-based health products need attractive forest environments. In our study, relaxation effects in the mixed forest was highest. This forest was characterized by fir and beech trees, and a thin herbaceous layer. Participants were placed in the midst of light tree stands. Light forests with different tree species therefore may be recommended for forest-based health tourism products. However, it is important to keep in mind that visual qualities will not predict restoration outcome alone.

Furthermore, the participants of this study were surveyed while sitting in the four test conditions. They were not to move or become physically active. Even under this condition, recreational effects of terpenes and other forest components could be confirmed. It is against this background that forest-based health tourism products do not necessarily need active parts, like walking or hiking. Instead, forest-based health tourism profits from a slow-down of participants. This means that simply being in the forest, or doing mindful practices in forests, could be central parts of tourism offers. In this, they would differentiate against other offers which already exist and which rely on sporting activities to a larger extent. Simultaneously, this kind of product would fit the visitation motives of 'being in nature' and finding 'quietness' (Ensinger *et al.*, 2012).

During COVID-19, forests were one of the few places people could go to for relaxation and leisure. Forests, therefore, have been called crucial infrastructures during times of health crises (Derks *et al.*, 2020). Due to the sheer number of visitors, however, usage conflicts and overcrowding occurred (Hansen *et al.*, 2022). Besides terpenes, factors like soundscape or the forest's visual appearance impact on relaxation effects. Forest tourism providers, therefore, should make sure that natural sounds prevail in medicinal forests. The same applies to the forest's visual appearance, because the restorative quality of forests may erode with too many people visiting them. Natural surroundings and natural sounds, therefore, have been defined as quality requirements for cure and healing forests in Germany (BVMV, 2022). With an increasing number of forest visitors, due to COVID-19, maintaining these quality standards has become even more important.

Lastly, it needs to be added that the study's results are preliminary. It is therefore difficult to formulate recommendations especially for tourism products which aim at treating existing clinical indications or ailments. Consistent with other studies, for example Rajoo *et al.* (2020), this study's recommendations rather target preventive tourism products like forest bathing and cure forests than healing forests and forest therapy.

Limitations

The study compared a specific mixed forest with a specific coniferous forest. The results cannot be generalized to all mixed or coniferous forests. The data, therefore, provide first evidence that different kinds of forests may have differential effects on human health. However, the researched forests were not representative. They consisted of a specific mixture of fir, spruce, beech and other tree species. Future studies should include representative forests. Furthermore, they should try to keep other forest components than terpenes constant or to include them into the study's design (such as light conditions, temperature and wind).

The study took place in early summer during a period with a relatively cold temperature. Higher temperatures would have led to significantly higher terpene emissions from trees and from ground litter, which would likely have resulted in a higher terpene concentration in the two forest conditions. Furthermore, the cool temperature also impacted on the participants' wellbeing. Higher temperatures would most probably have been more pleasant with possible repercussions on the participants' subjective judgement on restoration outcome and the restorative quality of the place (ROS, PRS). Furthermore, it was difficult to ventilate the scented interior in such a way that the terpenes completely disappear from the room. Therefore, the neutral control condition (IU) partly was not empty of terpenes, but contained varying concentrations of terpenes.

The sample examined was very small in terms of participants ($N = 24$). It furthermore consisted mainly of older subjects between 50 and 69 years of age. For future studies, it would be desirable to include a significantly larger number of participants, with different classes of age and with different cultural or social backgrounds.

Acknowledgements

This chapter is based on selected results of a study funded by the Ministry of Justice and for Europe Baden-Württemberg (Germany), and the Association for Research in Balneology and Climatic Physiology Baden-Württemberg (Förderkreis für Balneologische und Klimaphysiologische Forschung Baden-Württemberg e.V., Germany). The study was carried out on behalf of Marketing Association of Spas and Health Resorts in Baden-Württemberg (HKM-Heilbäder und Kurorte Marketing GmbH Baden-Württemberg, Germany). We would like to thank the Municipality of Bad Peterstal-Griesbach (Germany), and MEDICLIN Clinic Schlüsselbad (MEDICLIN Schlüsselbadklinik Bad Peterstal, Germany) for their support. Professor Christiane Werner, Dr Jürgen Kreuzwieser and Michel Grün, Chair of Ecosystem Physiology at the University Freiburg, Germany, contributed to this study by analysing the terpene concentration and composition in the four test conditions. Silke Muck, SRH Mobile University Riedlingen, Germany, supported data collection.

References

Abe, T., Hisama, M., Tanimoto, S., Shibayama, H., Mihara, Y. *et al.* (2008) Antioxidant effects and antimicrobial activites of phytoncide. *Biocontrol Science* 13(1), 23–27. DOI: 10.4265/bio.13.23.

Akakabe, Y. (2012) The relaxation effect of woody aroma in forest bath (shinrinyoku) on human. *Aroma Research* 13(3), 229–234.

Antonelli, M., Donelli, D., Barbieri, G., Valussi, M., Maggini, V. *et al.* (2020) Forest volatile organic compounds and their effects on human health: a state-of-the-art review. *International Journal of Environmental Research and Public Health* 17(18), 6506. DOI: 10.3390/ijerph17186506.

Bamwesigye, D., Fialová, J., Kupec, P., Łukaszkiewicz, J. and Fortuna-Antoszkiewicz, B. (2021) Forest recreational services in the face of COVID-19 pandemic stress. *Land* 10(12), 1347. DOI: 10.3390/land10121347.

Bielinis, E., Jaroszewska, A., Łukowski, A. and Takayama, N. (2020) The effects of a forest therapy programme on mental hospital patients with affective and psychotic disorders. *International Journal of Environmental Research and Public Health* 17(1), 118. DOI: 10.3390/ijerph17010118.

Bielinis, E., Takayama, N., Boiko, S., Omelan, A. and Bielinis, L. (2018) The effect of winter forest bathing on psychological relaxation of young Polish adults. *Urban Forestry & Urban Greening* 29, 276–283. DOI: 10.1016/j.ufug.2017.12.006.

Buecker, S., Horstmann, K.T., Krasko, J., Kritzler, S., Terwiel, S. *et al.* (2020) Changes in daily loneliness for German residents during the first four weeks of the COVID-19 pandemic. *Social Science & Medicine* 265, 113541. DOI: 10.1016/j.socscimed.2020.113541.

BVMV (Baederverband Mecklenburg-Vorpommern) (2022) Gesundheitspotenzial Kur- und Heilwald [Healing capacities of cure and healing forests]. Available at: https://www.mv-baederverband.de/de (accessed 20 October 2023).

Cervinka, R., Höltge, J., Pirgie, L., Schwab, M., Sudkamp, J. *et al.* (2014) Zur Gesundheitswirkung von Waldlandschaften [The healing effects of forest landscapes]. *BfW-Berichte* 147 Vienna.

Cho, K.S., Lim, Y.-R., Lee, K., Lee, J., Lee, J.H. *et al.* (2017) Terpenes from forests and human health. *Toxicological Research* 33(2), 97–106. DOI: 10.5487/TR.2017.33.2.097.

Clifford, A. (2018) *Your Guide to Forest Bathing.* Conari Press, Newburyport, Massachusetts.

Derks, J., Giessen, L. and Winkel, G. (2020) COVID-19-induced visitor boom reveals the importance of forests as critical infrastructure. *Forest Policy and Economics* 118, 102253. DOI: 10.1016/j.forpol.2020.102253.

ECDC (European Centre for Disease Prevention and Control) (2023) Data on country response measures to COVID-19. Available at: https://www.ecdc.europa.eu/en/publications-data/download-data-response-measures-covid-19 (accessed 20 October 2023).

Ensinger, K., Wurster, M., Selter, A., Jenne, M., Bethmann, S. *et al.* (2012) Eintauchen in eine andere Welt – Untersuchung über Erholungskonzepte und Erholungsprozesse im Wald [Immersion into another world – recovery concepts and recovery processes in forests]. *AFZ der Wald* 184(3/4), 70–83.

Gibson, J. (1979) *The Theory of Affordances: The Ecological Approach to Visual Perception*. Houghton Mifflin, Boston, Massachusetts.

Grilli, G. and Sacchelli, S. (2020) Health benefits derived from forest: a review. *International Journal of Environmental Research and Public Health* 17, 6125. DOI: 10.3390/ijerph17176125.

Grima, N., Corcoran, W., Hill-James, C., Langton, B., Sommer, H. *et al.* (2020) The importance of urban natural areas and urban ecosystem services during the COVID-19 pandemic. *PLoS One* 15(12), e0243344. DOI: 10.1371/journal.pone.0243344.

Hansen, A.S., Beery, T., Fredman, P. and Wolf-Watz, D. (2022) Outdoor recreation in Sweden during and after the COVID-19 pandemic – management and policy implications. *Journal of Environmental Planning and Management* 66(7), 1472–1493. DOI: 10.1080/09640568.2022.2029736.

Harting, T., van den Berg, A.E., Hagerhall, C.M., Tomalak, M., Bauer, N. *et al.* (2011) Health benefits of nature experience: psychological, social and cultural processes. In: Nilsson, K., Sangster, M., Gallis, C., Hartig, T. and de Vries, S. *et al* (eds) *Forests, Trees and Human Health*. Springer, New York, pp. 127–168. DOI: 10.1007/978-90-481-9806-1.

Hegetschweiler, K.T., Stride, C.B., Fischer, C., Ginzler, C. and Hunziker, M. (2022) Integrating recreation into national forest inventories – results from a forest visitor survey in winter and summer. *Journal of Outdoor Recreation and Tourism* 39, 100489. DOI: 10.1016/j.jort.2022.100489.

Hofer, S., Honegger, F. and Hubeli, J. (2012) Health tourism: definition focused on the Swiss market and conceptualisation of health(i)ness. *Journal of Health Organization and Management* 26(1), 60–80. DOI: 10.1108/14777261211211098.

Hong, J., Park, S. and An, M. (2021) Are forest healing programs useful in promoting children's emotional welfare? The Interpersonal relationships of children in foster care. *Urban Forestry & Urban Greening* 59, 127034. DOI: 10.1016/j.ufug.2021.127034.

ICOHF (International Certification Office of Healing Forests) (2022) High quality standards. Audited by independent experts. Available at: https://www.healing-forest-certification.org/ (accessed 20 October 2023).

Immich, G. and Schuh, A. (2021) A conceptual framework of forest therapy as an innovative health approach combined with local health resort medicine in Alpine regions to increase mental health and well-being. *Physical Medicine and Rehabilitation International* 8(2), id1179. DOI: 10.26420/physmedrehabilint.2021.1179.

Janson, R. (1992) Monoterpene concentrations in and above a forest of scots pine. *Journal of Atmospheric Chemistry* 14(1–4), 385–394. DOI: 10.1007/BF00115246.

Jay, M. and Schraml, U. (2009) Understanding the role of urban forests for migrants – uses, perception and integrative potential. *Urban Forestry & Urban Greening* 8(4), 283–294. DOI: 10.1016/j.ufug.2009.07.003.

Kaczynski, A.T. and Henderson, K.A. (2007) Environmental correlates of physical activity: a review of evidence about parks and recreation. *Leisure Sciences* 29(4), 315–354. DOI: 10.1080/01490400701394865.

Karim, A.A., Khalil, R. and Schmitt, M. (2020) Wald reloaded – Die Neuentdeckung des Waldes aus gesundheitspsychologischer Sicht [Forests reloaded – rediscovery of forests from a health psychological perspective]. *Zeitschrift für Komplementärmedizin* 12(2), 24–30. DOI: 10.1055/a-1140-1148.

Kawai, F. and Miyachi, E. (2016) Heart rate variability analysis on effects of the odor of juniper's essential oil on the autonomic nervous system. *Aroma Research* 17(1), 58–63.

Keniger, L.E., Gaston, K.J., Irvine, K.N. and Fuller, R.A. (2013) What are the benefits of interacting with nature? *International Journal of Environmental Research and Public Health* 10(3), 913–935. DOI: 10.3390/ijerph10030913.

Kim, J.-G. and Shin, W.-S. (2021) Forest therapy alone or with a guide: is there a difference between self-guided forest therapy and guided forest therapy programs? *International Journal of Environmental Research and Public Health* 18(13), 6957. DOI: 10.3390/ijerph18136957.

Korpela, K., Savonen, E.-M., Anttila, S., Pasanen, T. and Ratcliffe, E. (2017) Enhancing wellbeing with psychological tasks along forest trails. *Urban Forestry & Urban Greening* 26, 25–30. DOI: 10.1016/j.ufug.2017.06.004.

Korpela, K.M., Ylén, M., Tyrväinen, L. and Silvennoinen, H. (2008) Determinants of restorative experiences in everyday favorite places. *Health & Place* 14(4), 636–652. DOI: 10.1016/j.healthplace.2007.10.008.

Kotera, Y., Richardson, M. and Sheffield, D. (2022) Effects of shinrin-yoku (forest bathing) and nature therapy on mental health: a systematic review and meta-analysis. *International Journal of Mental Health and Addiction* 20(1), 337–361. DOI: 10.1007/s11469-020-00363-4.

Kotte, D., Li, Q., Shin, W.S. and Michalsen, A. (2019) *International Handbook of Forest Therapy*. Cambridge Scholars Publishing.

Kuo, M. (2015) How might contact with nature promote human health? Promising mechanisms and a possible central pathway. *Frontiers in Psychology* 6, 1093. DOI: 10.3389/fpsyg.2015.01093.

Lee, D.-G., Kim, J.-G., Park, B.-J. and Shin, W.S. (2022) Effect of forest users' stress on perceived restorativeness, forest recreation motivation, and mental well-being during COVID-19 pandemic. *International Journal of Environmental Research and Public Health* 19(11), 6675. DOI: 10.3390/ijerph19116675.

Lee, J., Park, B.J., Tsunetsugu, Y., Ohira, T., Kagawa, T. *et al.* (2011) Effect of forest bathing on physiological and psychological responses in young Japanese male subjects. *Public Health* 125(2), 93–100. DOI: 10.1016/j.puhe.2010.09.005.

Li, Q., Morimoto, K., Nakadai, A., Inagaki, H., Katsumata, M. *et al.* (2007) Forest bathing enhances human natural killer activity and expression of anti-cancer proteins. *International Journal of Immunopathology and Pharmacology* 20(2 Suppl 2), 3–8. DOI: 10.1177/03946320070200S202.

Li, Q. (2018) *Shinrin-Yoku: The Art and Science of Forest Bathing*. Penguin, London.

Lupp, G., Melber, M., Hirschbeck, T., Ritter, A., Brockard, M. *et al.* (2016) Outdoor recreation in urban forests: user patterns and impacts. Examples from southern Germany. In: Francis, R., Millington, J. and Chadwick, M. (eds) *Urban Landscape Ecology: Science, Policy and Practice*. Routledge, London and New York, pp. 209–228.

McEwan, K., Collett, H., Nairn, J., Bird, J., Faghy, M.A. *et al.* (2022) The feasibility and impact of practising online forest bathing to improve anxiety, rumination, social connection and Long-COVID symptoms: a pilot study. *International Journal of Environmental Research and Public Health* 19(22), 14905. DOI: 10.3390/ijerph192214905.

O'Brien, L. and Morris, J. (2014) Well-being for all? The social distribution of benefits gained from woodlands and forests in Britain. *Local Environment* 19(4), 356–383. DOI: 10.1080/13549839.2013.790354.

Ochiai, H., Ikei, H., Song, C., Kobayashi, M., Miura, T. *et al.* (2015) Physiological and psychological effects of a forest therapy program on middle-aged females. *International Journal of Environmental Research and Public Health* 12(12), 15222–15232. DOI: 10.3390/ijerph121214984.

OECD (2018) *Health at a Glance: Europe 2018: State of Health in the EU Cycle*. OECD Publishing, Paris and Brussels. DOI: 10.1787/health_glance_eur-2018-en.

Ohe, Y., Ikei, H., Song, C. and Miyazaki, Y. (2017) Evaluating the relaxation effects of emerging forest-therapy tourism: a multidisciplinary approach. *Tourism Management* 62, 322–334. DOI: 10.1016/j.tourman.2017.04.010.

Park, B.-J., Tsunetsugu, Y., Kasetani, T., Hirano, H., Kagawa, T. *et al.* (2007) Physiological effects of shinrin-yoku (taking in the atmosphere of the forest) – using salivary cortisol and cerebral activity as indicators. *Journal of Physiology and Anthropology* 26(2), 123–128. DOI: 10.2114/jpa2.26.123.

Park, B.J., Tsunetsugu, Y., Kasetani, T., Kagawa, T. and Miyazaki, Y. (2010) The physiological effects of shinrin-yoku (taking in the forest atmosphere or forest bathing): evidence from field experiments in 24 forests across Japan. *Environmental Health and Preventive Medicine* 15(1), 18–26. DOI: 10.1007/s12199-009-0086-9.

Pasini, M., Berto, R., Brondino, M., Hall, R. and Ortner, C. (2014) How to measure the restorative quality of environments: the PRS-11. *Procedia – Social and Behavioral Sciences* 159, 293–297. DOI: 10.1016/j.sbspro.2014.12.375.

Putra, R.R., Veridianti, D.D., Nathalia, E., Brilliant, D., Rosellinny, G. *et al.* (2018) Immunostimulant effect from phytoncide of forest bathing to prevent the development of cancer. *Advanced Science Letters* 24(9), 6653–6659. DOI: 10.1166/asl.2018.12804.

Rajoo, K.S., Karam, D.S. and Abdullah, M.Z. (2020) The physiological and psychosocial effects of forest therapy: a systematic review. *Urban Forestry & Urban Greening* 54, 126744. DOI: 10.1016/j.ufug.2020.126744.

Ravens-Sieberer, U., Kaman, A., Erhart, M., Devine, J., Schlack, R. *et al.* (2021) Impact of the COVID-19 pandemic on quality of life and mental health in children and adolescents in Germany. *European Child & Adolescent Psychiatry* 31(6), 879–889. DOI: 10.1007/s00787-021-01726-5.

Schuh, A. and Immich, G. (2022) *Forest Therapy – The Potential of the Forest for Your Health*. Springer, Berlin, Heidelberg. DOI: 10.1007/978-3-662-64280-1.

Schultz, P.W. (2002) Inclusion with nature: the psychology of human-nature relations. In: Schmuck, P. and Schultz, W.P. (eds) *Psychology of Sustainable Development*. Springer, Boston, Massachusetts, pp. 61–78. DOI: 10.1007/978-1-4615-0995-0.

Stier-Jarmer, M., Throner, V., Kirschneck, M., Immich, G., Frisch, D. *et al.* (2021) The psychological and physical effects of forests on human health: a systematic review of systematic reviews and meta-analyses. *International Journal of Environmental Research and Public Health* 18(4), 1770. DOI: 10.3390/ijerph18041770.

Takayama, N., Korpela, K., Lee, J., Morikawa, T., Tsunetsugu, Y. *et al.* (2014) Emotional, restorative and vitalizing effects of forest and urban environments at four sites in Japan. *International Journal of Environmental Research and Public Health* 11(7), 7207–7230. DOI: 10.3390/ijerph110707207.

Tsaur, S.H., Liang, Y.W. and Weng, S.C. (2014) Recreationist-environment fit and place attachment. *Journal of Environmental Psychology* 40, 421–429. DOI: 10.1016/j.jenvp.2014.10.003.

Tsunetsugu, Y., Park, BJ. and Miyazaki, Y. (2010) Trends in research related to ‘Shinrin-yoku’ (taking in the forest atmosphere or forest bathing) in Japan. *Environmental Health and Preventive Medicine* 15(1), 27–37. DOI: 10.1007/s12199-009-0091-z.

Tyrväinen, L., Ojala, A., Korpela, K., Lanki, T., Tsunetsugu, Y. *et al.* (2014) The influence of urban green environments on stress relief measures: a field experiment. *Journal of Environmental Psychology* 38, 1–9. DOI: 10.1016/j.jenvp.2013.12.005.

UN (United Nations Department of Economic and Social Affairs) (2020) Forests – at the heart of a green recovery from the COVID-19 pandemic. Policy Brief No. 80. Available at: https://www.un.org/development/desa/dpad/wp-content/uploads/sites/45/publication/PB_80.pdf (accessed 20 October 2023).

Vattulainen, K., Sarjala, T., Savonen, E.M. and Korpela, K. (2011) Elpymiskokemuksia metsästä, Voimapolun käyttäjäkysely [Restorative experiences from the forest user survey of the power trail]. Working papers of the Finnish Forest Research Institute. Available at: https://jukuri.luke.fi/handle/10024/536112 (accessed 20 October 2023).

Wen, Y., Yan, Q., Pan, Y., Gu, X. and Liu, Y. (2019) Medical empirical research on forest bathing (Shinrin-yoku): a systematic review. *Environmental Health and Preventive Medicine* 24(1), 70. DOI: 10.1186/s12199-019-0822-8.

WHO (1948) *Constitution of the World Health Organization*. World Health Organization, Geneva, Switzerland.

Yu, C.-P. and Hsieh, H. (2020) Beyond restorative benefits: evaluating the effect of forest therapy on creativity. *Urban Forestry & Urban Greening* 51, 126670. DOI: 10.1016/j.ufug.2020.126670.

15 Sustainability and Post-COVID-19 Tourism Strategies in Natural Protected Areas: Spain as a Case Study

Margarita Capdepón Frías*
Departamento de Análisis Geográfico Regional y Geografía Física, University of Alicante, Spain

Abstract

Beyond the undeniable negative impacts on tourism, the COVID-19 outbreak in 2020 also represented an opportunity to transform the sector and produced significant changes in demand. A certain part of that demand is increasingly interested in outdoor activities, in contact with nature and away from crowded places. Within this context, special attention should be paid to the natural environment, and particularly to natural protected areas (NPAs) as rising tourism destinations.

With the aim of giving a precise perspective, this chapter is focused on Spain as a case study. The central purpose of this work is to analyse the role played by the Spanish NPAs in the recent tourism transition processes and in the post-pandemic sector recovery. The work provides an insight into the main national policies for the alleviation of the impacts on tourism both during the first months of the pandemic and within current recovery processes.

Introduction

Protecting the biodiversity and developing nature-based solutions can generate multiple benefits. Today, these solutions constitute some of the most relevant responses to several current global challenges, including reducing the risk of disasters such as the COVID-19 pandemic or improving the physical and mental wellbeing of human beings (Hockings *et al.*, 2020). According to Hockings *et al.* (2020, p. 15), a healthy natural environment, managed as NPAs, forms 'the backbone of a recovering planet'. To do this, it is necessary to manage various sources of finance, including sustainable tourism.

The natural environment and protected spaces constitute one of the bases of tourism activity, particularly in the light of the recent changes caused by the pandemic and the growing motivational diversification of the demand, which increasingly seeks experiences in contact with nature. Therefore, it is necessary to develop a well-planned tourism strategy with capability not only to help finance the protection and management of these territories, but also to raise awareness about environmental values among visitors and increase their economic importance (OECD, 2021).

This chapter focuses on Spain as a case study. It is one of the world's leading tourism destinations and a European country with one

*margarita.capdepon@ua.es

 Nature-based Tourism and Wellbeing: Impacts and Future Outlook (eds F. Niccolini *et al.*)
DOI: 10.1079/9781800621411.0015

of the largest surfaces under protection. The main purpose of this work is to analyse the role played by the Spanish NPAs in the tourism policies implemented by the central government during the pandemic. The secondary objective is to identify and explain some of the main measures developed to boost nature-based tourism with sustainability context. Finally, the text considers the probable impacts caused by the increase in the number of visitors and discusses some of the challenges for the management of the natural areas in the future.

The first part of the text reviews the immediate consequences of the pandemic on natural territories around the world. The second part focuses on the Spanish case and the relevance of its NPAs as rising tourism destinations. The work provides an insight into the national policies for the alleviation of the impacts on tourism both during the first months of the pandemic and during recent recovery processes.

The Repercussions of COVID-19 for Natural Protected Areas

The beginning of the health crisis and the resulting reduction in the inflow of visitors to protected areas during the first few months generated different effects on their conservation. Although the impacts of COVID-19 varied depending on distinct factors, the destinations most affected were those most dependent on international markets (OECD, 2020). The severe restrictions to travel and the closure of parks during the first semester of 2020 led to a drastic reduction in the number of visitors to NPAs and the loss of economic income, as happened in Brazil, in many African countries (Spenceley, 2021a) and in European ones such as Austria (OECD, 2021), the UK, Italy or Spain (McGinlay *et al.*, 2020).

These impacts included the deterioration of wildlife protection due to the reduced capacity, funding and efficiency of the actions, difficulties in applying the management plans and the development of new ones, cuts in conservation programmes and the suspension of voluntary actions, among others. Examples of this have been collected by authors such as Spenceley (2021a) (South Africa, Nepal, Bolivia) and Hockings *et al.* (2020) (Australia, Cape Verde, Ecuador, Costa Rica). In places such as South Africa (Spenceley, 2021a), Kenya, Uganda or Ecuador (Hockings *et al.*, 2020), many of the communities living around NPAs were harmed by the reduction in the wages of the employees of the tourism sector and of protected spaces. In the same way, declining incomes affected small businesses within the tourism value chain.

On the contrary, we can refer to more positive implications for the conservation of the natural environment. The wildlife in NPAs in countries such as India, Thailand, Italy, UK, Israel or USA (Spenceley, 2021a) prospered in the absence of visitors. The suspension of the tourism activity helped to slow down, at least temporarily, overcrowding and over-tourism, giving rise to lower levels of pollution and a lower habitat loss (Spenceley, 2021a). As reported by some authors, in Peru, Bolivia or Australia (Spenceley, 2021a) and in several European NPAs (McGinlay *et al.*, 2020; OAPN, 2020; OECD, 2021) the closure of the protected areas provided opportunities for their managers to assess the state of the biodiversity and adequately plan it for when the tourism activity resumed. In other words, they were granted a moment to focus on observing the biodiversity and establish a much more resistant and sustainable nature-based tourism sector.

In addition, the pandemic gave rise to major changes in the motivations of the tourism demand and the emergence of new attitudes. They would define the development of the activity after the end of the lockdowns and the lifting of restrictions. As indicated by Marques Santos *et al.* (2020), the change of mentality reinforced fundamental trends for post-COVID-19 tourism such as less overcrowding and activities in nature, outdoors and internal tourism. As could be expected, tourists prioritized health, hygiene, social distancing and being away from the crowds. They also expressed a greater interest in sustainable tourism products, the tourism experience in open-air spaces and in contact with nature, and the search for activities with a greater participation of the local economies and the protection of the natural environment (Hockings *et al.*, 2020; WTTC, 2020; Böhme *et al.*, 2021).

Similarly, given the restrictions on international mobility, domestic tourism was given a considerable boost once tourism had been reactivated. Tourists sought shorter trips and destinations closer to home (UNWTO, 2020; Böhme *et al.*, 2021). Less-crowded rural areas, national parks and other natural areas became 'the winners of leisure and travel during the pandemic' (Böhme *et al.*, 2021, p. 138), especially for national visitors. Indeed, such processes occurred in NPAs of China, Indonesia, USA (Spenceley, 2021a), Germany (OECD, 2020), Denmark (Böhme *et al.*, 2021), Croatia, Serbia (Dinarides Park, WWF, 2021) and Sweden (McGinlay *et al.*, 2020), to mention a few examples. There was a boom in proximity tourism in which NPAs played a fundamental role (Romagosa, 2020) after a long period of confinement or isolation (Hockings *et al.*, 2020). Hence, authors such as Spenceley (2021b) predicted that nature tourism would be one of the first to recover.

Methodology

As mentioned above, the main purpose of this work is to analyse the role played by the Spanish NPAs in the tourism policies implemented by the central government during the pandemic. To that end, an exploratory-descriptive method has been carried out, gaining better sense of the research gap, providing further knowledge regarding the topic, as well as presenting some of the recent and current trends.

Firstly, the work is based on a narrative review of different secondary sources of information such as academic references (Hockings *et al.*, 2020; McGinlay *et al.*, 2020; Böhme *et al.*, 2021; Spenceley, 2021a; among others) and institutional reports (from the OECD, the UNWTO and the WTTC). They are some of the premier reliable publications that provided qualitative data on the research topics covered in the first part of the work.

Secondly, a documentary review has been performed for the Spanish case study. For this purpose, the main official tourism planning documents developed by the Spanish central administration have been consulted: the Plan to Boost the Tourism Sector: Towards Safe and Sustainable Tourism (June 2020), the Plan for the Modernisation and Competitiveness of the Tourism Sector (June 2021) and the Sustainability Strategy in Tourism Destinations (revised 2023 version). In addition, other governmental sources have been added, especially some reports from the Autonomous Authority for National Parks (hereafter OAPN, its Spanish abbreviation). It is a public body responsible for the protection and co-ordination of the National Park Network and Spanish Network of Biosphere Reserves.

Spain as a Case Study

According to the latest *Anuario del estado de las áreas protegidas en España* (EUROPARC-España, 2021), 35.2% of Spain's land territory and 12.3% of its marine area is protected. It is the European nation that most contributes to the Natura 2000 network. It has 170 areas protected by different international bodies and stands out as the territory with the most biosphere reserves in the world, with a total of 53. On a state level, there are 16 national parks and 1825 NPAs declared under a wide range of national and regional categories.

It is estimated that before the pandemic, the volume of visitors to the Spanish NPAs exceeded 31 million (EUROPARC-España, 2021). This number was higher, given that the available information varies depending on the protection body and territory. In any case, it is necessary to emphasize the role of the national parks network which received about 14.8 million visitors in 2019 (OAPN, 2021), just before the pandemic. In 2020, the number of visitors fell to 9.3 million, before recovering significantly to 12 million in 2021. It is worth checking whether the trends continue in the coming years to confirm a return to normality. Also, it should be noted that there are significant differences among the parks, so the distribution of visits is highly heterogeneous.

The Spanish natural protected areas as a tourist attraction during the pandemic

The Spanish government declared a state of emergency in March 2020, announcing measures such as the confinement of the population, limitations to mobility and the closure of borders. Within this context, NPAs, in general, and the national parks in particular, remained closed until mid-June. Based on OAPN data (OAPN, 2020), between January and June 2020, the 16 Spanish national parks recorded drastic reductions in the number of visitors with respect to the same period in 2019; decreases of up to 80 and 90% in some cases. The closure was used by the employees of many parks to advance in certain management tasks, prepare future revitalization activities, review action plans and implement training programmes.

After the end of the lockdown and the reopening of these spaces from July 2020, there were significant increases in visitors in several national parks compared to the previous year (Cabañeros – 31.4%, Sierra de Guadarrama – 28.1%, Monfragüe – 11.9%, Picos de Europa – 8.4%, Ordesa y Monte Perdido – 7%, and Sierra Nevada – 6.3%). Similarly, research has found (Ibarra and Rabanaque, 2020; Izcara and Cañada, 2021; Margalef, 2021) that there was greater tourist frequency in other natural areas of Spain. In many cases, this meant additional pressure for the NPAs, particularly during the first months of the pandemic. As a result, these areas suffered from overcrowding problems, parking difficulties, more traffic, rubbish and waste dumping and the development of illegal activities, among other effects. Furthermore, it is worth highlighting that the new visitors were less responsible, with no knowledge of the regulations and rules and unaware of the environment with less respect for nature (OAPN, 2020; McGinlay *et al.*, 2020). On the contrary, there were also parks that continued to experience significant decreases in the number of visits, particularly those located in the Canary Islands (Timanfaya – 66.3%, Garajonay – 3.8%, Caldera de Taburiente – 43.6%) and Balearic Islands (Cabrera – 44.1%) (OAPN, 2020).

As indicated by EUROPARC-España (2021), the situation experienced with COVID-19 not only revealed the connections between human wellbeing and access to natural spaces, but also a change in the dynamics of the affluence of visitors (saturation in some places or an absence of visitors in others). In other words, it meant new challenges for the management of NPAs.

Principal national measures for boosting nature-based tourism

The urgent response: the plan to boost the tourism sector: towards safe and sustainable tourism

In response to the health crisis, the Spanish government was obliged to react urgently and passed the Plan to Boost the Tourism Sector: Towards Safe and Sustainable Tourism in June 2020. This instrument sought to facilitate the rapid recovery of the Spanish tourism sector, to mitigate the adverse effects of the pandemic and to ensure the continuity and competitiveness of the sector. The plan was articulated around five large pillars and 28 measures.

One of the most remarkable actions was the Programme of Tourism Sustainability Plans in Destinations aimed at invigorating and organizing tourism destinations. Two types of plans were contemplated: (i) tourism sustainability in pioneering destinations (for example, the coastal ones); and (ii) tourism sustainability plans in rural destinations that included, precisely, the NPAs as one of the main categories. They were focused on the creation of sustainable tourism products based on natural resources and the creation of facilities capable of differentiating the destination (Gobierno de España, 2020). Initially, it was a short-term measure that has been continued in the following years as part of the recovery policies.

Another strategy included in the plan was the elaboration of the Guides to Reduce the Risk of the Transmission of SARS-CoV-2 in the Tourism Sector. This was a unique health protocol for COVID-19, co-ordinated by the Secretariat of State of Tourism, together with other regional and local bodies. Its main purpose was to elaborate hygiene-sanitary good practice to protect tourism workers in an attempt to facilitate the reopening of the tourism sector after the relaxation of the confinement measures. Twenty-two regulations were designed, aimed at specific tourism subsectors. The one corresponding to the natural areas established recommendations to be applied by the managing entities of the NPAs, mainly those recognized by the Natura 2000 network and the areas protected by international instruments (the UNESCO Biosphere Reserves and Geoparks), aiming to minimize the risk of infection with the SARS-CoV-2 virus in facilities, services and activities of public use.

In addition, the government contemplated information and communication actions carried out by the OAPN, which opened a channel of communication through social networks regarding aspects with a naturalistic, operational or managerial interest related to the parks (OAPN, 2020). Furthermore, a guide was published with recommendations for visitors and protection measures to adopt for the management of the national parks in the de-escalation process. This was a period when the recovery of the activity in the national parks had to combine the return to the activities of using and enjoying the parks with the protection of the visitors against COVID-19. In this respect, the Spanish central government insisted on the fundamental role of NPAs in the re-emerging tourism sector.

Along the same lines, actions were developed with respect to environmental education and the transfer of knowledge. The OAPN also created several virtual classrooms adapted to diverse groups (national park employees, tourist guides, teachers, etc.) in which specific thematic training courses were given in each case (OAPN, 2020). Moreover, Turespaña (the public body responsible for the marketing of Spain as a travel destination in the world) and the Ecotourism Club in Spain organized different educational activities such as webinars. The objective was publicizing 'an opportunity to gain a better understanding of Spain's ecotourism proposition and learn about the diversity of the natural and cultural heritage that make for fantastic and safe holiday experiences in a Covid-19 context' (Ecotourism Club in Spain, 2021).

The recovery measures: the tourism sustainability plans in destinations

In June 2021, the Spanish government launched the Recovery, Transformation and Resilience Plan to use the 'next-generation EU' funds, aimed at repairing the damage caused by the pandemic. As part of this major economic recovery strategy, the Plan for the Modernisation and Competitiveness of the Tourism Sector is being developed at present. It includes the Tourism Sustainability Plans in Destinations (hereafter PSTDs by its Spanish abbreviation) that take over from the measures implemented in 2020. They have an initial budget of €1858 million, the highest of the whole tourism strategy (Gobierno de España, 2021). In other words, they became a key element for the recovery of the Spanish tourism sector.

According to the Spanish government (Gobierno de España, 2023), the plans are instruments for co-operative action between the three levels of government: central government, the Autonomous Communities (regional level) and local authorities, as well as the private sector. Their objective is to undertake transformative actions in the tourism sector in areas such as green transition and the improvement of tourism competitiveness. The plans are also aimed at tackling the demographic challenge and combating depopulation in rural areas. The management and implementation of the funds are the responsibility of the local authorities in charge of tourism destinations. The plans have a period of completion of 3 years and are expected to be completed by 2026.

Again, natural areas (protected or not) are considered as one of the main PSTDs categories. In this case, the objectives to achieve include: boosting the tourism offer based on natural heritage; designing public use models in NPAs to make their conservation compatible with sustainable tourism; creating tourism products linked to the ecosystem services of protected areas (e.g. ecotourism and nature-based tourism); and involving all stakeholders in the tourism system in the protection of resources, among others. The PSTDs will have a special impact on both the protection of natural resources and the boost of ecotourism.

Since the approval of the first projects in September 2020, more than €112 million euros have been allocated in nearly 40 Spanish destinations aiming to develop a sustainability plan in their NPAs (usually National Parks, UNESCO Biosphere Reserves, Geoparks and Natura 2000 sites). It is worth mentioning some remarkable places such as the Doñana, Picos de Europa and Sierra Nevada national parks, as well as the Menorca and La Palma biosphere reserves, among others.

Conclusions

Among its multiple effects on the tourism industry, COVID-19 has turned NPAs into places of special interest. Firstly, as areas that were increasingly visited during the initial stages of the pandemic, then as key tourism destinations for economic recovery within the boost of typologies that have been gaining great relevance in recent years, such as nature-based tourism. Therefore, it is necessary to think about some future implications.

On the one hand, we should consider whether the changes in the demand and the latest trends that marked the development of the activity during the pandemic are temporary or will become definitive changes. In any case, although some of the actions taken by the Spanish government to promote tourism in NPAs were short-term emergency measures, most of them remain at the present time, especially the PSTDs as a vital support to the transformation processes of Spanish tourism destinations, particularly NPAs. In fact, it has become one of the main lines of action for the current tourism recovery processes.

On the other hand, there has been an increase in the number of visitors to many of Spain's national parks during the pandemic, with the resulting generation of impacts derived from overcrowding and negative demand attitudes. In many cases, this greater tourist frequentation is forecast to prevail in the long term and will remain a persistent problem. In response to this scenario, it is decisive to assess whether the protected areas in Spain are sufficiently prepared to take on the new challenges in terms of funding, management tools, conservation and human capital, among other aspects. Within this context, there are other issues to consider. For example, the possibility of limiting the number of visitors in overcrowded NPAs or, even, an eventual entrance fee to use and enjoy this type of space. Both practices are uncommon in Spain and could lead to different difficulties due to the impossibility for the territories to establish limits to access. Also, these kind of measures could generate exclusive and elitist spaces that are only accessible to certain sectors of society.

Finally, it should be noted that this work has focused on the national scale and the role of central government. However, environmental and conservation responsibilities, as well as tourism development around NPAs, lie primarily with regional and local governments. Given this limitation, future research should be carried out to understand the policies developed at different scales. Another limitation is determined by the time horizon of this study, which runs until the end of 2022. Hence, it is difficult to know the real impact of several measures explained in the text, especially the PSTDs passed in the tourism destinations with NPAs. Most of them have started recently so there are not clear and definitive outcomes yet. Therefore, the study of specific cases based on qualitative and quantitative data would be a subject of debate in the coming years.

References

Böhme, K., Haarich, S., Toptsidou, M., Besana, F., Corbineau, C. *et al.* (2021) *Regional Impacts of the Covid-19 Crisis on the Tourist Sector. Final Report*. European Commission. Available at: https://ec.europa.eu/regional_policy/en/information/publications/studies/2021/regional-impacts-of-the-covid-19-crisis-on-the-tourist-sector (accessed 20 October 2023).

Dinarides Park, W.W.F. (2021) Impact of COVID-19 pandemic on protected areas in the Dinarides region.

Ecotourism Club in Spain (2021) Webinar ecotourism in Spain in a covid-19 context. Available at: https://ecotouristinspain.com/webinar-may-2021 (accessed 20 October 2023).

EUROPARC-España (2021) *Anuario 2020 del estado de las áreas protegidas en España*. Ed. Fundación Fernando González Bernáldez, Madrid. Available at: https://redeuroparc.org/wp-content/uploads/2022/01/anuario2020finalweb.pdf (accessed 20 October 2023).

Gobierno de España (2020) *Plan de impulso turístico para el sector turístico*. Available at: https://www.lamoncloa.gob.es/serviciosdeprensa/notasprensa/industria/Documents/2020/20062020_PlanTurismo.pdf (accessed 20 October 2023).

Gobierno de España (2021) *Plan de modernización y competitividad del sector turístico*. Available at: https://www.lamoncloa.gob.es/temas/fondos-recuperacion/Documents/16062021-Componente14.pdf (accessed 20 October 2023).

Gobierno de España (2023) Estrategia de Sostenibilidad Turística en Destinos. Available at: https://www.mintur.gob.es/es-es/recuperacion-transformacion-resiliencia/Documents/2023_Estrat_Sosten_Tur_Destino.pdf (accessed 18 January 2024).

Hockings, M., Dudley, N., Elliott, W., Napolitano Ferreira, M. and MacKinnon, K. (2020) Editorial essay: COVID-19 and protected and conserved areas. *PARKS* 26(1), 7–24. DOI: 10.2305/IUCN.CH.2020.PARKS-26-1MH.en.

Ibarra, P. and Rabanaque, I. (2020) La marca de calidad territorial y el paisaje como estrategia para el turismo interior de la comarca del Matarraña (Teruel). In: Pons, G.X., Blanco, A., Navalón, R., Troitiño, L. and Blázquez, M. (eds) *Sostenibilidad Turística: overtourism vs ondertourism*. pp. 69–81. Available at: https://dspace.uib.es/xmlui/bitstream/handle/11201/154055/557909.pdf?sequence= (accessed 20 October 2023).

Izcara, C. and Cañada, E. (2021) *Espacios naturales saturados: colisión de derechos*. Available at: https://www.albasud.org/noticia/es/1369/espacios-naturales-saturados-colision-de-derechos

Margalef, C.S. (2021) Masificación turística en espacios naturales: el caso del Parque Natural del Delta del Ebro en tiempos de COVID-19. *Barataria* 30, 68–85. DOI: 10.20932/barataria.v0i30.607.

Marques Santos, A., Madrid, C., Haegeman, K. and Rainoldi, A. (2020) *Behavioural changes in tourism in times of Covid-19: employment scenarios and policy options*. Publications Office of the European Union. DOI: 10.2760/00411.

McGinlay, J., Gkoumas, V., Holtvoeth, J., Armas, R.F., Bazhenova, E. *et al*. (2020) The Impact of COVID-19 on the management of European protected areas and policy implications. *Forests* 11. DOI: 10.3390/f11111214.

OAPN (Organismo Autónomo de Parques Nacionales) (2020) *Los parques nacionales tras la alarma sanitaria*. Available at: https://www.miteco.gob.es/es/ceneam/grupos-de-trabajo-y-seminarios/red-parques-nacionales/los-parques-nacionales-tras-alarma-sanitaria_tcm30-516749.pdf (accessed 20 October 2023).

OAPN (Organismo Autónomo Parques Nacionales) & Ministerio para la Transición Ecológica y el Reto Demográfico (2021) *Memoria de la Red de Parques Nacionales 2019*. Available at: https://www.miteco.gob.es/es/red-parques-nacionales/divulgacion/memoria-red-2019_tcm30-525158.pdf (accessed 20 October 2023).

OECD (Organisation for Economic Co-operation and Development) (2020) Tourism policy responses to the coronavirus (COVID-19). Available at: https://www.oecd.org/coronavirus/policy-responses/tourism-policy-responses-to-the-coronavirus-covid-19-6466aa20/ (accessed 20 October 2023).

OECD (Organisation for Economic Co-operation and Development (OECD) (2021) Managing tourism development for sustainable and inclusive recovery. OECD Tourism Papers, 2021/01. OECD Publishing, Paris. Available at: https://www.oecd.org/publications/managing-tourism-development-for-sustainable-and-inclusive-recovery-b062f603-en.htm (accessed 20 October 2023).

Romagosa, F. (2020) The COVID-19 crisis: opportunities for sustainable and proximity tourism. *Tourism Geographies* 22(3), 690–694. DOI: 10.1080/14616688.2020.1763447.

Spenceley, A. (2021a) The future of nature-based tourism: impacts of COVID-19 and paths to sustainability. Luc Hoffmann Institute. Available at: chrome-extension://oemmndcbldboiebfnladdacbdfmadadm/https://www.humanrights-in-tourism.net/sites/default/files/media/file/2021/rc136the-future-nature-based-tourism-1654.pdf (accessed 20 October 2023).

Spenceley, A. (2021b) Tourism and visitation in protected areas amid COVID-19: guidance for protected area authorities and managers. European Commission. Available at: https://op.europa.eu/en/publication-detail/-/publication/fe764777-c990-11eb-84ce-01aa75ed71a1 (accessed 20 October 2023).

UNWTO (2020) Understanding domestic tourism and seizing its opportunities. UNWTO briefing note – Tourism and COVID-19. Issue 3. UNWTO, Madrid. DOI: 10.18111/9789284422111.

WTTC (World Travel and Tourism Council) (2020) To recovery & beyond. The future of travel & tourism in the wake of COVID-19. Available at: https://wttc.org/Initiatives/To-Recovery-Beyond (accessed 20 October 2023).

16 The Tourism Transition of Green Destinations in the Post-pandemic

Rita Salvatore*
Department of Political Science, University of Teramo, Teramo, Italy

Abstract
This chapter explores the notion of tourism as a driver of change within green destinations and highlights pros and cons related to the enhancement of environmental resources. If nature-based tourism has always been considered a favourable context to orient tourist development towards sustainability, its recent growing trend has posed new critical issues which require a reassessment of the topics related to both environmental and social impacts. In some cases, these changes have happened at an unpredicted speed, eventually turning tourism into an undesired source of development. The risk of over-tourism is highlighted as a phenomenon associated not only with large numbers of tourists but also referring to small numbers of unexpected visitors who may cause concern for small and unprepared local communities if their arrival is not adequately managed. Some indications related to the methodology of transition management are examined as a strategy to address these criticalities.

Nature-Based Tourism, Ecotourism and Green Destinations

As widely acknowledged, nature protection and conservation were the framework where sustainability in tourism was first discussed and questioned. In 1993, the European Federation for Parks and Protected Areas (EUROPARC) set up a task force to reflect on the role of tourism within the context of environmental conservation, aiming at addressing the growing interest from tourists towards nature and protected places. The working group edited a report titled *Loving them to death? Sustainable tourism in Europe's nature and national Parks* (EUROPARC, 1993), in which the first guidelines for the development of sustainable nature-based tourism were presented. Two years later, two important documents were issued: (i) the so-called Lanzarote Charter, containing the main principles about sustainable tourism; and (ii) the European Charter for Sustainable Tourism in protected Areas (ECST), an important management tool and a process of certification aimed at supporting protected areas and local stakeholders in pursuing a sustainable tourism strategy to be developed through specific guiding principles and a participatory action plan.

Since then, sustainable nature-based tourism has always been thought of as a viable tool for favouring and supporting environmental protection also through tourist visits. Hence, protected areas have been thought to have the potential to act as catalysts of environmental, social and economic sustainability and as agents of change in lifestyles at local, regional and national levels.

Against this framework, we can consider nature-based tourism as the wider phenomenon at the base of ecotourism, with this last referring

*rsalvatore@unite.it

DOI: 10.1079/9781800621411.0016

to the principles and activities tourists may hold and practise regarding nature preservation. In regard to the different acceptations of ecotourism given in literature (Donohoe and Needham, 2006), here we refer to the definition and principles suggested by TIES (The International Ecotourism Society), a non-profit network of ecotourism professionals, scholars and travellers, founded in 1990 and committed to the development of responsible nature-based travel around the world. On its website, we can read that 'ecotourism is defined as responsible travel to natural areas that conserves the environment, sustains the wellbeing of the local people, and involves interpretation and education' (TIES, 2015). Even if the eco prefix clearly recalls the 'green' character of the phenomenon, ecotourism is not only focused on nature but also refers to the categories of culture and social impact. These aspects are even more explicit and stressed in the Principles of Ecotourism that TIES addresses to anyone who implements or participates in marketing ecotourism activities. These principles include:

- Minimize physical, social, behavioural and psychological impacts.
- Build environmental and cultural awareness and respect.
- Provide positive experiences for both visitors and hosts.
- Provide direct financial benefits for conservation.
- Generate financial benefits for both local people and private industry.
- Deliver memorable interpretative experiences to visitors that help raise sensitivity to host countries' political, environmental and social climates.
- Design, construct and operate low-impact facilities.
- Recognize the rights and spiritual beliefs of the indigenous people in your community and work in partnership with them to create empowerment.

Compared to the first definition of ecotourism proposed by Ceballos-Lascuráin (1996), which was mainly focused on what tourists do, TIES's outline integrates the so-called demand-side variables (tourists' needs and wants) with aspects related to the supply side (community awareness and social empowerment). The care and sense of responsibility towards nature resources is matched with socio-cultural aspects, related on the one side to the involvement of local communities (Jones, 2005) and on the other to the minimization of the impact from tourists on the environment (Perkins and Brown, 2012). Due to these peculiarities, ecotourism has been considered as a means to equally rebalance nature and wildlife protection with social development and economic growth.

The need to keep these aspects in balance has been one of the hardest challenges EUROPARC has been addressing through the implementation of the European Charter of Sustainable Tourism so far (Balandina *et al.*, 2012; Salvatore and Chiodo, 2017; Piscè, 2021). Tourism can certainly represent a way to enhance the value of natural assets in an economic perspective and trigger processes of development for local communities. On the other hand, though, an unmanaged growth of tourist flows may have an impact that could cause the depletion of natural resources and even the degradation of the environment, eventually impacting on the attractiveness of the location. As well known in literature and as the Tourist Area Life Cycle model (TALC) (Butler, 1980, 2004) has clearly highlighted, each destination is exposed to the risk of losing its quality due to excessive tourist fruition and, as in a 'fairly obvious and often-repeated pattern' (Butler, 2004, p. 162), each one of them, sooner or later, might reach the limit of carrying capacity. Tourism is always a driver for change and very few destinations probably remain unchanged without losing their attractiveness. Notwithstanding in an environmentally protected context, these processes are definitively more fragile and need special care, principally in relation to the intrinsic meaning and value of conservation (Butler and Boyd, 2000).

For the specific purpose of keeping these aspects under control and in order to manage and monitor the impact of tourism, the European Union has set up a specific indicator system, the so-called European Tourism Indicator System (ETIS), which has envisioned a toolkit to support destinations in carrying out their plans towards long-term sustainability. Before describing the whole dataset in detail, ETIS underlines what the meaning of destination is. Even if every single corner of the world could be entitled to host visitors, only those that can provide beds and other hospitality services can be considered tourist destinations. Besides, ETIS underlines that a destination is not only a geographic area that is attractive to tourists but also a place where a range of facilities, products and services for tourists are available on the basis of a visitor management process that includes 'a range of public and private-sector stakeholders together with the host community' (European Union, 2016, p. 10).

Based on this definition and framework, we can define a sustainable green destination as a place where tourism in general and ecotourism in particular have reached such a level of management and organization that the balance between nature conservation and economic development can be constantly observed and monitored, thanks to the commitment from institutions, private actors, citizens and tourists and through the use of a specific system of indicators.

From this standpoint, we can understand how the idea of sustainable tourism has evolved during the past 30 years. From representing the objective of an ideally perfect process of development or the outcome on a linear path, it has turned out to be a far more complex praxis where all the interests need to be balanced, considering the constant interaction among the different subsystems (socio-ecological, socio-technical, socio-political) within the whole system of a destination (Scuttari *et al.*, 2016). Sustainability itself becomes an ongoing process, a step-by-step transition governance practice (Kemp *et al.*, 2007; Gössling *et al.*, 2012), which involves the commitment from a set of various stakeholders (public institutions and professionals, technicians and private entrepreneurs, local communities and tourists) aiming at combining the economic capital with the safeguard of the natural, social and cultural ones.

Green Destinations During the Post-Covid: From Non-Popular to Highly Attractive Places

During the past 30 years, tourism has gone through considerable changes and both the choice of destinations and itineraries have ceased to be attributable to unique experiences or to a unified model of leisure-time fulfilment as they used to be in the previous decades (Cohen, 1979). A growing number of tourists (the so-called 'new tourists') (Poon, 1994) have started to look for new destinations apart from the masses, and from highly standardized supply services towards more 'authentic' experiences and places (MacCannell, 1973). Moreover, the search for emotional experiences and for territorial peculiarity (Jepson and Sharpley, 2015), as well as the possibilities offered by the green economy (Reddy and Wilkes, 2015), have further broadened the range of tourist choices, eventually bringing the possibility of even less-popular places and destinations to address the challenge of a tourism transition (Salvatore *et al.*, 2018). As a consequence of this, tourism has ceased to move its flows only towards its own dedicated places, that is towards popular destinations whose local economies were highly seasonal and mainly based on tourist revenues. This has led to the consideration that today each place in the world is itself a potential tourist destination and can turn into an attractive place for tourists if local economic actors get involved in the management and the territorial organization of tourist services and experiences (Savelli, 2012).

Nature-based tourism has been particularly impacted by these changes as new tourists' needs have been increasingly oriented in looking for 'non-traditional tourist destinations' (European Commission, 2003) as places particularly suitable to match with the idea of wellbeing, reconnection with nature and culture, ethical consumption, and environmental awareness and care (Winter *et al.*, 2019). Responding to this trend, motivations for travelling towards green destinations and travellers' pro-environmental behaviour had already gained attention within both tourism studies and policies (Kim and Koo, 2020) before the COVID-19 pandemic blast.

Therefore, if green destinations had been partly implemented as favourable contexts to orient tourist development in a sustainable direction, in the post-COVID time they have certainly performed a further growing trend. As highlighted by Seraphin and Dosquet (2020), the outbreak of the pandemic has made the 'Janus-faced character' of tourism definitively more evident than before. If, at first, the travel limitations related to the total lockdown triggered one of the most critical crises tourism has ever experienced due to the collapse of demand (Vu *et al.*, 2022) – causing during the first semester of 2020 a fall of 850 million to 1.1 billion international tourists and a loss of $910 billion to $1.2 trillion in export revenues from tourism, and putting 100–120 million direct-tourism jobs at risk (UNWTO, 2020) – in some countries, such as France, Italy and the USA, a high percentage of citizens escaped from big cities towards mountain second homes and remote places, leading to a meaningful growth in domestic tourism (OECD, 2020; Seraphin and Dosquet, 2020; Cocco *et al.*, 2020). Most of these people must have considered the possibility of living closer to nature as a good chance for their physical and social health and as a way to keep safe. Something similar was also happening in relation to amenity migration, a phenomenon developed on the idea that mountainous contexts can offer the possibility to live in a sort of 'idyllic elsewhere' (Bourdeau, 2008). If this choice may appear as something different from simply being a tourist – since it assumes a longer movement than the short time of a vacation – nevertheless, it has disclosed a renewed value of wider green spaces that also reinforced their tourist attractiveness. Furthermore, the possibilities related to smart-/homeworking have widened choices about residence as people can also live quite far away from their offices/workplaces without any particular impact. These situations have certainly blurred the border between either being a resident or a tourist and suggested new categories like those of 'temporary citizenship' (Belleri and Ratti, 2022) or 'workation' (Voll *et al.*, 2022).

An online survey conducted by the OUT (Osservatorio Universitario sul Turismo – University of Naples Tourism Research Observatory) to examine Italian tourists' future intentions and propensities to travel confirmed the increase in the attractiveness of green destinations in the pandemic aftermath (Corbisiero and Monaco, 2021). The great majority of their interviewees (82.62%), for example, declared that they would have preferred to travel towards places with large open spaces such as mountain areas, countryside and natural parks. At the same time, most of them appeared to be particularly oriented to spend their vacation in more proximal places. These findings have led the authors to hypothesize that holidays will increasingly focus on subjective motivations rather than on the choice of the place and that probably the 'outdoors will reign in the short-term with specific attention to health, safety and social, and environmental sustainability' (Corbisiero and Monaco, 2021, p. 412).

Therefore, the initial need to respect social distance seems to have somehow revealed the value of closer and greener spaces as more suitable contexts in which to live sustainable and high-quality tourist experiences. UNWTO experts believe that open-air activities and nature-based and rural tourism are among the major travel trends that will continue shaping tourism (UNWTO, 2022).

For these reasons, both scholars and policy-makers have agreed that nature-based tourism is likely to perform a further growing trend in the following years than in pre-COVID, particularly from domestic tourists (OECD, 2020; Wendt *et al.*, 2022; Williams and Ólafsdóttir, 2022). In line with these expectations, the World Travel and Tourism Council states that spending in domestic tourism out of the whole travel and tourism spending increased from almost 51% in 2019 to almost 63% in 2020 (World Economic Forum, 2022).

This desire to travel towards nearby green destinations has certainly made it possible for some left-behind natural places, like, for example, some mountain areas (Romeo *et al.*, 2021), to experiment with new paths of local development based on green hospitality. On the other hand, it has resurfaced the issue of overcrowding and overwhelming, once again, and with an even greater concern in some of those places that had already begun to witness it before COVID (Honey and Frenkiel, 2021).

During the last summer seasons, some natural parks around the world were literally mobbed with tourists. On 19 August 2020, the Italian newspaper *Il Dolomiti* headlined 'Val di Fassa is Stormed with Tourists' [trans.] (Andreazza, 2020). More or less the same kind of phenomenon happened in France where several natural sites, such as the cliffs of Étretat in Normandy or the Chartreuse Natural Park in the Alps, had to face too many tourists and, after finding this unsustainable, decided to limit visitor numbers (Makooi, 2022). Less popular and almost unknown places faced more or less the same situation, too. Talking about the unusual presence of tourists in the Italian Val Tramontina (Natural Park of Dolomiti Friuliane), the journalist Padovan (2022) wrote: 'Pozze Smeraldine from hidden wonder to mass tourism destination: how to destroy a natural show'. Also, some parks in the USA had to address the same issue, with *The New York Times* writing that 'Americans are flocking to national parks in record numbers' (*New York Times*, 2022).

Such cases seem to suggest that maybe these changes have happened at an unpredicted speed, eventually turning tourism into a diminished experience for visitors, into an undesired novelty for local community, and causing new problems and concerns in relation to nature safeguarding. In such situations, the relationship between tourism, local livelihood and environmental conservation becomes really challenging and may trigger a vicious loop, with different conflicts arising because of the particular fragilities of these areas.

The Need to Manage the Change

The way green destinations deal with these new opportunities, concerns and threats is a central issue for a sociological theory of tourism more and more oriented to analysing development and wider societal change. The lesson learned along these pandemic years is that, in order to be affordable in the longer term, tourism transition of natural places cannot be left to each case nor thought of as merely spontaneous.

For example, the above-mentioned cases suggest that when the number of tourists becomes excessive or grows at an unpredicted speed tourism might even become an undesired option of change for local communities. The risks and implications related to overtourism (Séraphin *et al.*, 2020) in terms of local conflict and antagonistic encounters between residents and tourists have been recently highlighted (Dodds and Butler, 2019; Seraphin and Dosquet, 2020) in urban contexts such as Venice, Madrid and other tourist cities as well as in natural areas. However, this phenomenon is not just the consequence of large numbers of tourists, and it can also be referred to in its relative terms. In some local contexts – for example rural areas or small municipalities that are not used to experiencing hospitality at all – small numbers of visitors may cause concerns if their arrivals are unexpected (Butler, 2020).

Paradoxically, what at first appears as an important driver of change and as a way to requalify locally based economies in a qualitatively appreciable way may turn eventually to a source of social conflict. The increase in the number of visitors, as well as the growing diversification of motivations from different types of tourists, may bring a rapid spread of commercial aspects and to the replacement of some elements related to social identity in favour of what the customer requires. All these may hit the identity of a place, causing some undesirable changes in terms of socio-cultural values in relation to local heritage. Social psychology has described this phenomenon as loss of perceived control over place identity and the gradual loss of sense of place (Gössling *et al.*, 2020). As a consequence, this loss might weaken both attachment to place (as residents' sense of belonging) (Wang, 2016) and place experience (as tourists' emotional satisfaction) (Wang and Chen, 2015).

Is there any way to control and prevent these drifts of change? In order to look for some suggestions in literature, we have to take a step back, looking at the social dynamics that nature-based tourism may trigger. Environmental landscape, natural resources, local culture and wildlife can all become important assets for tourism development as far as they are elements of distinctiveness representing the value upon which tourist experience gains its deep sense (Jepson and Sharpley, 2015). Therefore, in sociological terms, these material and immaterial capitals can be moved and 'touristically translated' into the new processes characterizing the change related to tourism.

When a local community finds itself able to take advantage from becoming a tourist destination, it opens itself to new and different social forces. Some of these are already within the system. They may be active, but they may also be latent (i.e. locally based resources and local enterprises). Some others are completely external and may become part of the environment (i.e. tour operators, guides, private investors).

In any case, the territory becomes a 'transition social arena' (Kemp *et al.*, 2007) where each different influence plays an equally important role. The way the whole system metabolizes these forces is a central issue for a sociological theory of change in tourism, because tourism positions itself as a condition for local societies to re-elaborate their own identities, also in relation to their social sense of place and belonging (Alleyne, 2002; Cocco *et al.*, 2020).

As already stated, especially in green destinations, all these subsystems (cultural, socio-economic and environmental), as well as their interaction, can be very volatile and fragile. For this reason, all issues related to hospitality cannot be left to chance or to spontaneity. On the contrary, a focused co-ordination and management is needed (Guevara Manzo and Constantin, 2017; Honey and Frenkiel, 2021), along with a co-ordinated monitoring system, for example ETIS.

If adequately planned, changes related to tourism may be considered as a possibility to redirect local development in a sustainable way, with sustainability meant as an ongoing process rather than a clear end point. These processes are to be interpreted as co-evolutionary dynamics, including different subsystems shaping each other in interaction and involving changes in institutions, values, culture and practices (Kemp *et al.*, 2007). From this standpoint, the social change may be managed through an accurate 'transition management', which is a multi-level model of governance helping societies 'to transform themselves in a gradual, reflexive way' (p. 78) and to determine to what extent their development objectives are desirable in terms of environmental, socio-economic and cultural change. Therefore, it includes the interaction among all relevant social actors on different levels and is aimed at bridging the gap between pure top-down strategies and bottom-up ones. It does so by promoting an innovative idea of a bottom-up model which includes a wider co-ordination from the higher levels of governance, while fostering self-organization through cycles of mutual learning and action. Managers' actions are led by a scenario about where they wish the community to be at a certain time and try to move towards those goals on the basis of a flexible and mutual building of the experience (Kemp *et al.*, 2007).

Gössling *et al.* (2012) experimented with the application of transition management to the field of sustainable tourism as an attempt to reduce the gap between the rapid growth of tourism on a global scale and the reduced number of initiatives based on pro-environment tourist actions. The aspect they refer to as being most interesting is that the time this methodology takes to manage change is usually no longer than a couple of decades. It is usually based on a common vision that a group of individuals and private and public organizations can share, starting from a situation of crisis. Actually, working as a meta-policy, it can be considered as a perspective which guides the choice of the most appropriate policy instruments towards the reaching of a common vision.

Some further food for thought closely related to the methodology of 'transition management' comes from an action-research that Moriggi (2020) conducted in Finland within the framework of Green Care practices. As highlighted previously, the potential of place-based entrepreneurship committed with sustainability is not only to provide services and products but also to contribute to socio-economic change by shifting from systems of knowing to acting upon specific challenges. A lot of strategies and methods can be applied to these aims but the participatory approaches remain paramount to ensure co-evolution and interactive reflexive learning since relations and contexts are crucial to success. 'A relational approach highlights the intricate web of connections and processes that enmesh people and places, actors and their context: entities do not exist on their own, but are co-constructed and co-evolving' (Moriggi, 2020, p. 439).

Moving to COVID recovery strategies, most European funds related to the next generation will be addressed in supporting projects aimed at rebalancing territorial disparities, also through tourism. This is the case with the European Regional Development Fund (ERDF) and Cohesion Fund whose specific objectives are dedicated to exploiting the potential of culture and tourism for an economic recovery able to couple both social inclusion and environmental and economic sustainability. The Cohesion Fund, in particular, can support tourism-related investments in the environment with a particular focus on strengthening economic, territorial and social cohesion in the European Union by reducing development imbalances among regions.

If this scenario is a desirable one for the resilience of tourism (Vu *et al.*, 2022) as well as for the fight against socio-economic marginalization of most European mountain areas, some policies and management tools have to be promoted in order to address future challenges and to limit an uncontrolled and unregulated fruition of fragile environments, as green destinations are. Some valid technical tools useful to support a self-assessing system are already available at European level, such as the already-mentioned ECST, aimed at practising sustainable tourism within a protected area. This Charter has clearly stated the need to manage the change through the application of specific principles which somehow recall the same perspective as the transition management's one (Piscè, 2021). They are principles that each protected area has to commit to formally, once it applies to the Charter. In particular, the third principle clearly refers to the need to protect and enhance the area's natural and cultural heritage from excessive tourism development by:

- monitoring impact on flora and fauna and controlling tourism in sensitive locations;
- encouraging activities, including tourism uses, which support the maintenance of historic heritage, culture and traditions;
- controlling and reducing activities, including tourism impacts, which adversely affect the quality of landscapes, air and water; use non-renewable energy; and create unnecessary waste and noise; and
- encouraging visitors and the tourism industry to contribute to conservation.

(EUROPARC, 2010)

The way the Charter aims to reach these objectives is tightly intertwined with its first principle, which is aimed at involving all those implicated by tourism in and around the protected area in its development and management. However, even if the Charter and its principles sound like excellent tools to manage the change, there are still limits and challenges related to the lack of funding aimed at supporting their concrete application in a broad way.

On a wider international level, more recently the Global Sustainable Tourism Council has launched an initiative aimed at self-monitoring the sustainability at destination level. It is the 'destination stewardship', which is particularly aimed at providing practical information and visions for those actors committed to improving the quality and sustainability supply within destinations post-COVID (Destination Stewardship Center, 2021).

Moreover, the support from both international and European institutions responsible for preserving the environment, and maintaining and balancing ecosystems, including through sustainable development such as the IUCN (International Union for the Conservation of Nature), has never ceased the issue of specific guidelines, for example the Marseille Manifesto (IUCN, 2021), aimed at addressing new emergencies and challenges.

Conclusions

When it comes to tourism, the COVID-19 pandemic has definitely turned out to be double-edged. The restrictions related to the repeated blasts in the two-year period 2020–2021 caused one of the worst world crises in tourism ever, with dramatic loss for the whole sector around the globe (UNWTO, 2020). At the same time, it has shown that a territorial cohesion and a more widespread sustainability in terms of enhancement of domestic and proximity tourism and movements of the tourist flows towards non-traditional tourist places is possible. Therefore, we can really see that the aftermath is offering a precious opportunity to rethink and reset tourism towards a better path for the future (Higgins-Desbiolles, 2020). American parks for example have reassessed the need to experiment with strategies and tactics aimed at managing tourist flows, including restraining tourist numbers at more fragile sites or at certain times, establishing specific reservation systems in order to prevent overcrowding, and favouring both conservation and sustainable local economies (Honey and Frenkiel, 2021).

In the Italian peripheral protected areas, the post-COVID tourism recovery could also represent a turning point and a means of testing local communities' capacity to adapt their needs to the opportunities offered by a new political and cultural context, within the framework of tourism transition (Salvatore *et al.*, 2018). For the last 2 years, remote areas have become more attractive both for holidays and for an alternative lifestyle that might be more suited to post-pandemic times (Cocco *et al.*, 2020). This situation has shown that, from a public health as well as from a development perspective, it could be worth considering the possibility to reinvest in the strategic role of greener and protected areas all over the country.

However, looking at the future of green destinations, the key turning point of their transition from non-traditional tourist places to sustainably attractive ones is played at a micro-level. In this sense, parks, institutions and local communities should act as paramount partners for a sustainable regeneration of tourism flows, and the hope is that an important part of the funds will be somehow redirected in developing new professionals working for this purpose. Future research paths that aim at monitoring the impact of over-tourism mitigation tactics and strategies in different international natural areas are not only desirable but also necessary, and should feature in future funding plans and tourism policies.

In short, a successful and sustainable way out of this crisis can only be centred on the vision and planning local communities can develop about the goals they want to reach in terms of balance between the value of the natural heritage they have and the amount/quality of visitors they can host. Such a perspective requires a community-based tourism framework and adequate planning that redefines and reorients choices related to the management of tourist flows to safeguard natural resources as well as preserving the rights and interests of local societies. Said differently, by quoting Higgins-Desbiolles (2020), it represents a way to recentre the tourism industry on the public good, and this is now essential for a sustainable tourism that is aimed at being accountable to the social and ecological limits of the planet.

Acknowledgements

This work has been supported by the European Union – FSE REACT-EU, PON research and innovation 2014–2020.

References

Alleyne, B. (2002) An idea of community and its discontents: towards a more reflexive sense of belonging in multicultural Britain. *Ethnic and Racial Studies* 25(4), 607–627. DOI: 10.1080/01419870220136655.

Andreazza, L. (2020) Code lunghissime, distanziamento azzerato e mascherine (quando ci sono) sotto il mento, la val di Fassa presa d'assalto. *Il Dolomiti*. Available at: www.ildolomiti.it/cronaca/2020/code-lunghissime-distanziamento-azzerato-e-mascherine-quando-ci-sono-sotto-il-mento-la-val-di-fassa-presa-dassalto (accessed 11 May 2022).

Balandina, A., Lovén, L., Ostermann, O. and Partington, R. (2012) European Charter Parks—a growing network for sustainable tourism development in protected areas. *Parks* 18(2), 133–144. DOI: 10.2305/IUCN.CH.2012.PARKS-18-2.AB.en.

Belleri, D. and Ratti, C. (2022) Reconnecting civitas and Urbs: making tourists and remote workers as temporary urban citizens. *Journal of E-Learning and Knowledge Society* 18(3), 45–51. DOI: 10.20368/1971-8829/1135820.

Bourdeau, P. (2008) Amenity migration and post-tourism: a geo-cultural approach to the alpine case. In: Moss, L.A.G., Glorioso, R.S. and Krause, A. (eds) *Understanding and Managing Amenity-Led Migration in Mountain Regions*. Banff Centre, Alta, Canada.

Butler, R. (1980) The concept of a tourist area cycle of evolution: implications for management of resources. *Canadian Geographies / Géographies Canadiennes* 24(1), 5–12. DOI: 10.1111/j.1541-0064.1980.tb00970.x.

Butler, R (2004) The tourism area life cycle in the twenty-first century. In: Lew, A.A., Hall, C.M. and Williams, A.M. (eds) *A Companion to Tourism*. John Wiley and Sons, Hoboken, New Jersey, pp. 159–169. DOI: 10.1002/9780470752272.

Butler, R. (2020) Overtourism in rural areas. In: Séraphin, H., Gladkikh, T. and Vo Thanh, T. (eds) *Overtourism: Causes, Implications and Solutions*. Palgrave-Macmillan, London. DOI: 10.1007/978-3-030-42458-9.

Butler, R. and Boyd, S.W. (eds) (2000) *Tourism and National Parks*. Wiley, Chichester, UK.

Ceballos-Lascuráin, H. (1996) *Tourism, Ecotourism and Protected Areas: The State of Nature-Based Tourism around the World and Guidelines for Its Development*. IUCN Publications, Cambridge, UK. DOI: 10.2305/IUCN.CH.1996.7.en.

Cocco, E., Farrell Mines, A. and Salvatore, R. (2020) Dreaming of remoteness, coping with emptiness in post-lockdown (under)tourism scenarios for the inner areas of Southern Italy: a fieldwork based reflection. *Fuori Luogo. Rivista Di Sociologia Del Territorio, Turismo, Tecnologia* 7(1). DOI: 10.6092/2723-9608/7124.

Cohen, E. (1979) A phenomenology of tourist experiences. *Sociology* 13(2), 179–201. DOI: 10.1177/003803857901300203.

Corbisiero, F. and Monaco, S. (2021) Post-pandemic tourism resilience: changes in Italians' travel behavior and the possible responses of tourist cities. *Worldwide Hospitality and Tourism Themes* 13(3), 401–417. DOI: 10.1108/WHATT-01-2021-0011.

Destination Stewardship Center (2021) *Destination Stewardship Yearbook 2020–2021*. Available at: https://www.gstcouncil.org/wp-content/uploads/Destination-Stewardship-Yearbook-2020-2021-online-ver.pdf (accessed 18 May 2022).

Dodds, R. and Butler, R. (2019) The phenomena of overtourism: a review. *International Journal of Tourism Cities* 5(4), 519–528. DOI: 10.1108/IJTC-06-2019-0090.

Donohoe, H.M. and Needham, R.D. (2006) Ecotourism: the evolving contemporary definition. *Journal of Ecotourism* 5(3), 192–210. DOI: 10.2167/joe152.0.

EUROPARC (1993) *Loving Them to Death? Sustainable Tourism in Europe's Nature and National Parks*. EUROPARC, Grafenau, Germany.

EUROPARC (2010) European Charter for Sustainable Tourism in Protected Areas. Available at: https://www.europarc.org/wp-content/uploads/2015/05/2010-European-Charter-for-Sustainable-Tourism-in-Protected-Areas.pdf (accessed 11 May 2022).

European, Commission (2003) *Using Natural and Cultural Heritage to Develop Sustainable Tourism in Non-traditional Tourist Destinations*. EU Publications, Brussels.

European Union. (2016) The European system of tourist indicators. ETIS Toolkit for sustainable destination management. DOI: 10.2873/983087.

Gössling, S., Hall, C.M., Ekström, F., Engeset, A.B. and Aall, C. (2012) Transition management: a tool for implementing sustainable tourism scenarios? *Journal of Sustainable Tourism* 20(6), 899–916. DOI: 10.1080/09669582.2012.699062.

Gössling, S., McCabe, S. and Chen, N.C. (2020) A sociopsychological conceptualisation of overtourism. *Annals of Tourism Research* 84, 102976. DOI: 10.1016/j.annals.2020.102976.

Guevara Manzo, G. and Constantin, M. (2017) *Coping with Success: Managing Overcrowding in Tourism Destinations*. McKinsey & Company and WTTC.

Higgins-Desbiolles, F. (2020) Socialising tourism for social and ecological justice after COVID-19. *Tourism Geographies* 22(3), 610–623. DOI: 10.1080/14616688.2020.1757748.

Honey, M. and Frenkiel, K. (2021) *Overtourism: Lessons for a Better Future*. Island Press, Washington.

IUCN (2021) *The Marseille Manifesto*. 10 September 2021. Available at: https://iucn.s3.eu-west-3.amazonaws.com/en/CGR-2021-1.6-2_Marseille_Manifesto_IUCN_World_Conservation_Congress_10_%20September_2021.pdf (accessed 28 July 2022).

Jepson, D. and Sharpley, R. (2015) More than sense of place? Exploring the emotional dimension of rural tourism experiences. *Journal of Sustainable Tourism* 23(8–9), 1157–1178. DOI: 10.1080/09669582.2014.953543.

Jones, S. (2005) Community based ecotourism: the significance of social capital. *Annals of Tourism Research* 32(2), 303–324. DOI: 10.1016/j.annals.2004.06.007.

Kemp, R., Loorbach, D. and Rotmans, J. (2007) Transition management as a model for managing processes of co-evolution towards sustainable development. *The International Journal of Sustainable Development & World Ecology* 14(1), 78–91. DOI: 10.1080/13504500709469709.

Kim, M. and Koo, D.W. (2020) Visitors' pro-environmental behavior and the underlying motivations for natural environment: merging dual concern theory and attachment theory. *Journal of Retailing and Consumer Services* 56, 102147. DOI: 10.1016/j.jretconser.2020.102147.

MacCannell, D. (1973) Staged Authenticity: arrangements of social space in tourist settings. *The American Journal of Sociology* 79(3), 589–603. DOI: 10.1086/225585.

Makooi, B. (2022) Faced with too many tourists, France's natural sites push back. *France 24*. Available at: https://www.france24.com/en/france/20220710-faced-with-too-many-tourists-france-s-natural-sites-push-back (accessed 10 January 2023).

Moriggi, A. (2020) Exploring enabling resources for place-based social entrepreneurship: a participatory study of Green Care practices in Finland. *Sustainability Science* 15(2), 437–453. DOI: 10.1007/s11625-019-00738-0.

New York Times (2022) How crowded are America's national parks? See for yourself. Available at: https://www.nytimes.com/2021/07/08/travel/crowded-national-parks.html (accessed 10 January 2023).

OECD (2020) The future of tourism in natural areas: impact, governance, financing. Available at: https://www.oecd.org/cfe/leed/2020-TourNat-agenda.pdf (accessed 18 May 2022).

Padovan, L. (2022) Pozze Smeraldine da meraviglia nascosta a meta del turismo di massa: così si rovina uno spettacolo. *Il Gazzettino.it*. Available at: www.ilgazzettino.it/nordest/pordenone/pozze_smeraldine_tramonti_di_sopra_turismo_friuli-6875559.html?refresh_ce (accessed 11 May 2022).

Perkins, H.E. and Brown, P.R. (2012) Environmental values and the so-called true ecotourist. *Journal of Travel Research* 51(6), 793–803. DOI: 10.1177/0047287512451133.

Piscè, G.C. (2021) The 'Torre del Cerrano' marine protected area and the European charter for sustainable tourism in protected areas as an Ecotourism management tool. *International Journal of Economic Behavior* 11(1), 79–97. DOI: 10.14276/2285-0430.3091.

Poon, A. (1994) The 'new tourism' revolution. *Tourism Management* 15(2), 91–92. DOI: 10.1016/0261-5177(94)90001-9.

Reddy, M.V. and Wilkes, K. (eds) (2015) *Tourism in the Green Economy*. Routledge, London. DOI: 10.4324/9781315885681.

Romeo, R., Russo, L., Parisi, F., Notarianni, M., Manuelli, S. *et al.* (2021) *Mountain tourism – towards a more sustainable path*. FAO, Rome. DOI: 10.4060/%20cb7884en.

Salvatore, R. and Chiodo, E. (2017) *Non più e non ancora. Le aree fragili tra conservazione ambientale, cambiamento sociale e sviluppo turistico*. FrancoAngeli, Rome.

Salvatore, R., Chiodo, E. and Fantini, A. (2018) Tourism transition in peripheral rural areas: theories, issues and strategies. *Annals of Tourism Research* 68, 41–51. DOI: 10.1016/j.annals.2017.11.003.

Savelli, A. (2012) *Sociologia del turismo*. Hoepli, Milan, Italy.

Scuttari, A., Volgger, M. and Pechlaner, H. (2016) Transition management towards sustainable mobility in Alpine destinations: realities and realpolitik in Italy's South Tyrol region. *Journal of Sustainable Tourism* 24(3), 463–483. DOI: 10.1080/09669582.2015.1136634.

Seraphin, H. and Dosquet, F. (2020) Mountain tourism and second home tourism as post COVID-19 lockdown placebo? *Worldwide Hospitality and Tourism Themes* 12(4), 485–500. DOI: 10.1108/WHATT-05-2020-0027.

Séraphin, H., Gladkikh, T. and Thanh, T.V. (eds) (2020) *Overtourism: Causes, Implications and Solutions*. Palgrave-Macmillan, London. DOI: 10.1007/978-3-030-42458-9.

TIES (2015) What is ecotourism? Available at. Available at: www.ecotourism.org/what-is-ecotourism (accessed July 2022).

UNWTO (2020) International tourism and Covid-19. Available at: https://www.unwto.org/international-tourism-and-covid-19 (accessed 10 July 2022).

UNWTO (2022) Tourism grows 4% in 2021 but remains far below pre-pandemic levels. News release. Available at: https://webunwto.s3.eu-west-1.amazonaws.com/s3fs-public/2022-01/220118-unwto-barometer-en.pdf?VersionId=vUmCK4Vel8HXI.49ZvSI44gCZDrZVzpR (accessed 3 July 2022).

Voll, K., Gauger, F. and Pfnür, A. (2022) Work from anywhere: traditional workation, coworkation and workation retreats: a conceptual review. *World Leisure Journal* 65(2), 150–174. DOI: 10.1080/16078055.2022.2134199.

Vu, H.D., Nguyen, A.T.N., Nguyen, N.T.P. and Tran, D.B. (2022) Impacts and restoration strategy of the tourism industry post-COVID-19 pandemic: evidence from Vietnam. *Journal of Tourism Futures* 8(1), 1–13. DOI: 10.1108/JTF-09-2021-0218.

Wang, S. (2016) Roles of place identity distinctiveness and continuity on resident attitude toward tourism. *European Journal of Tourism Research* 13, 58–68. DOI: 10.54055/ejtr.v13i.231.

Wang, S. and Chen, J.S. (2015) The influence of place identity on perceived tourism impacts. *Annals of Tourism Research* 52, 16–28. DOI: 10.1016/j.annals.2015.02.016.

Wendt, M., Sæþórsdóttir, A.D. and Waage, E.R.H. (2022) A break from overtourism: domestic tourists reclaiming nature during the COVID-19 pandemic. *Tourism and Hospitality* 3(3), 788–802. DOI: 10.3390/tourhosp3030048.

Williams, A. and Ólafsdóttir, R. (2022) Nature-based tourism as therapeutic landscape in a COVID era: autoethnographic learnings from a visitor's experience in Iceland. *GeoJournal* 88(2), 1737–1754. DOI: 10.1007/s10708-022-10713-5.

Winter, P.L., Selin, S., Cerveny, L. and Bricker, K. (2019) Outdoor recreation, nature-based tourism, and sustainability. *Sustainability* 12(1), 81. DOI: 10.3390/su12010081.

World, Economic Forum (2022) *Travel and tourism development index 2021: rebuilding for a sustainable and resilient future*. Insight Report.

17 Natural Heritage as a Tool to Promote Wellbeing and Quality of Life: The Case of I Giganti della Sila, Italy

Sonia Ferrari[1]*, Nicolaia Iaffaldano[2] and Simona Lo Bianco[1]
[1]University of Calabria, Italy; [2]University of Bari, Italy

Abstract

The natural site called *I Giganti della Sila* (The Giants of Sila) is a monumental wood and a living testimony of one of the most ancient forests in Calabria, a southern Italian region. Since 2016, this special place has been managed by Fondo per l'Ambiente Italiano (FAI), an Italian non-profit environmental association, with the objective to protect this area representing the authentic Silan landscape that survived a postwar deforestation.

The aim of this exploratory research is to understand not only the degree of visitors' interest in nature and nature experiences, and how it is changing after COVID-19, but also to know how these experiences influence visitors' wellbeing and satisfaction after visiting the park and the nature reserve.

Introduction

The natural site called *I Giganti della Sila* (The Giants of Sila) is a monumental wood and a living testimony of one of the most ancient forests in Calabria, a southern Italian region. It hosts at least 60 specimens of larch pine trees, an endemic species, whose impressive dimensions have made them famous: their height can reach 147 feet (45 m) with trunks up to 7 feet (2 m) wide. Their estimated average age is 400 years.

In 1987, the 17th-century pine forest, which is located inside the Sila National Park, was declared Biogenetic Protected Reserve by the Ministry of Environment. Since 2016, this special place has been managed by FAI (Fondo per l'Ambiente Italiano), an Italian non-profit environmental association. Thanks to a partnership with Sila National Park, FAI aims to protect this area representing the authentic Silan landscape that survived a postwar deforestation.

FAI deals with the promotion, preservation and enhancement of the landscape in all its material, immaterial, cultural and social facets, often involving institutions and communities. It was founded in 1975 and in more than 40 years it has grown steadily, taking action, on the one hand, to increase citizens' awareness of national heritage protection and, on the other, to carry out initiatives aimed at saving, restoring and opening to the public important works of art, historic buildings and areas of outstanding natural beauty. Thanks to the generous help of many citizens and companies, today FAI has several thousand members and ensures its direct protection to a vast cultural heritage.

In Italy, there are 68 FAI properties. I Giganti della Sila is the only FAI property in

*Corresponding author: sonia.ferrari@unical.it

 Nature-based Tourism and Wellbeing: Impacts and Future Outlook (eds F. Niccolini *et al.*)

DOI: 10.1079/9781800621411.0017

Calabria. Despite its geographical position (an internal and mountain area) and high seasonality (it is open only 6 months a year because of local weather conditions), the number of visitors has steadily grown since its establishment, and since 2020 it has been the most visited FAI property in the summer with 29,561 visitors in 2020, 31,186 in 2021 and more than 32,000 in 2022.

This trend confirms that the COVID-19 pandemic has not negatively impacted on nature-based *tourism*. On the contrary, today, tourists are more interested in natural destinations than in the past and people are looking for a renewed contact with nature (Levinger *et al.*, 2021; Cheng *et al.*, 2022). However, this is only a partial effect of the pandemic. Moreover, people's awareness of the positive influence of the contact with nature on personal wellbeing and physical and psychological health is increasing (White *et al.*, 2019; Dobson *et al.*, 2021). There is an extensive literature on this subject and the offer of services and activities for people visiting natural areas and destinations is constantly increasing; it includes nature experiences such as forest bathing, forest therapy, outdoor sports in natural environments, trekking in natural areas, trails in the parks, barefoot in the forest, silence in nature, relaxation, and learning activities concerning how to use natural resources for wellbeing (Ferrari and Gilli, 2016; Azara *et al.*, 2018; Martin *et al.*, 2020; Farkic *et al.*, 2021; Richardson *et al.*, 2021).

I Giganti della Sila, as well as the other FAI properties, organizes its offer and initiatives according to a 3 year operational and strategic plan that meets its sustainable development goals. The nature reserve staff take care of the relationship with visitors with increasing intensity and plan experiences to fulfil all kinds of desires and expectations. The reserve management is aware of the growing importance of relaxation, various activities and nature experiences to increase visitors' wellbeing and satisfaction. In fact, in addition to their educational programmes they offer visitors guided tours, several kinds of entertainment events, local food and wine tasting, as well as health and wellbeing experiences.

The aim of this exploratory research is to understand not only the degree of visitors' interest in nature, nature experiences and how these are changing after COVID-19, but also to see how these experiences influence their wellbeing and satisfaction after visiting the park and the nature reserve.

Methodology

This research represents the beginning of a subsequent, more in-depth, research project. After a literary review and qualitative interviews with some stakeholders, a quantitative survey was carried out through a structured self-compiled questionnaire available online. The target population of the survey was composed of all the Italians that visited I Giganti della Sila, because the number of foreign visitors has been very low in recent years. The percentage of foreign visitors was 0.08% in 2021 and 0.22% in 2020. However, the percentage of foreigners during 2019 reached 4.75% of the total; in more detail, most visitors were from France (32.9%), Germany (21.4%) and the USA (11.2%). The comparison between the data of 2021 and 2019 shows the effects of COVID-19 on international tourism in Italy but, fortunately, a fast recovery of this segment in future years was forecast as well.

The questionnaire focused on the main research questions. Primarily, it aimed to assess the visitors' motivations and their type of trip (alone, with friends, etc.). The questions were also related to the visitors' interest in natural resources and their opinion about the impact of tourism on these resources. Secondly, something about natural areas and wellbeing was asked, focusing on the type of experiences that visitors considered important from that point of view and on their related levels of satisfaction as well. Finally, a part of the questionnaire was dedicated to the impact of COVID-19 on tourism.

A non-probability sample of 157 respondents took part in the survey through a *convenience sampling*. The questionnaire was promoted on social media by the reserve management and the researchers. The survey was carried out in July and August 2022. The main results are presented as follows.

Findings and Discussion

Regarding the demographic characteristics of the sample, the group of respondents is composed of 88 women (56.1%) and 69 men (43.9%). They mainly live in Calabria (93.6%), the region where the Giganti is located, in Cosenza province, and most of them (63.7%) live in Cosenza.

As for their age distribution, the most important segment is that of '26–40 years old', which is about half of the sample (48.4%). The '41–60 years old' segment is the second (33.8%), followed by the over 60s (9.6%). The category of the youngest visitors (0–25 years old) is only 4.5% of the total. The percentage of those who did not answer is 3.8%.

The majority of the sample (86.6%) is composed of people who regularly visit natural destinations. This result suggests a high level of love for nature and awareness about the issue of natural environment conservation. However, more than half of the respondents do not think that tourism could represent a danger for natural resources (56.1%), while 26.1% are aware of this potential danger, and 17.9% do not have a clear idea about this issue or did not answer this question.

The questionnaire goes on with a question about the possible benefits of nature-based tourism, with many different answers to choose from and with a lot of answering possibilities for each respondent. The results stress that almost all respondents believe that tourism can help the territory economically (86.6%). The number of respondents who believe that tourism can help respect nature and know it better (73.9%) and can promote local development (65.0%) is also very high. More than half of the respondents believe that nature tourism can promote environmental education (63.1%) and increase the psychological wellbeing of visitors (51.6%). Finally, a smaller number of respondents believe that nature-based tourism can help improve the health state of visitors (35.0%). This percentage seems quite low considering the increasing awareness of public opinion about the importance of the experiences in natural environments and their effects on wellbeing and physical and mental health, as numerous studies assert.

Regarding the age distribution of the respondents that chose the item 'tourism based on nature helps the territory economically', 41.4% of them belong to the segment 26–40 years old, while 29.9% are in the 41–60 age segment. These results reflect the age distribution of the sample.

As regards their interest in visiting natural areas, the visitors declared their main motivations (a. sole motivation; b. one out of two; c. one out of three or more): 'To spend a pleasant day in contact with nature' was chosen by 122 respondents (77.7% of the sample) (a. 26 respondents or 16.6% of the sample; b. 29 respondents or 18.5% of the sample; c. 67 respondents or 42.7% of the total). 'To recharge your batteries and relax' was chosen by 72 respondents (45.9%) (a. 8 or 5.1%; b. 16 or 10.2%; and c. 48 or 30.6%). 'To introduce nature to your family and friends' was chosen by 68 respondents (43.3%) (a. 9 or 5.7%; b. 10 or 6.4%; c. 49 or 31.2%). The other motivations suggested in the questionnaire did not reach significative values. As for the age distribution of the results regarding the main interest in the most important motivation ('To spend a pleasant day in contact with nature'), the answers are in line with the age composition of the sample, because the segment 26–40 (48.4% of the sample) in 40.8% of the cases indicated this item. The second-largest segment (41–60; 33.8% of the sample) reached 22.3%.

In general, the respondents prefer to live the mountain experience in groups (56.1%), alone (23.6%), in pairs (12.7%) or with their family (7.6%). As regards the organization of their trip in Sila, they came mainly with friends/family (47.1%); some preferred to go on holiday with a partner/spouse (19.7%), a partner/spouse and children (17.8%), an organized group (17.2%) or alone (10.8%). It seems that in this case they chose to travel above all with family and/or friends. There are not huge sex differences in this field, but men choose more to travel alone and women with friends and family. These results suggest that the experience is mainly a moment of enjoyment and relaxation with family and friends.

Regarding the impact of COVID-19 on tourism and on travellers' preferences, 103 respondents (65.6% of the sample) answered that after the pandemic they became more interested in visiting natural places. Answering an open question about this subject, they explained that the main reasons for this increased interest are: 'Open air guarantees greater safety as it generates little crowding'; 'You feel the need to go out after the lockdowns'; 'Natural environments allow to live carefree without the risk of contagions'; 'You feel the need to re-establish the psychophysical wellbeing given by nature'.

Answering another open question, the respondents stated that their travel preferences have changed after the pandemic; today they prefer natural destinations, villages and rural areas to the Grand Tour destinations, and choose less-crowded places and proximity trips.

The following section of the questionnaire tries to understand the main motivations for the visit to the Sila National Park and to the Giganti. The respondents declared that the most important motivation is 'Enjoying the natural environment (walking, relaxing, picnicking, etc.)', which was ticked 124 times (by 79.0% of the sample). It was indicated as sole motivation by 93 people (59.2%) and as one out of two by 26 (16.6%). Regarding the age of the respondents that chose this item, 43.3% of them belong to the 26–40 segment, while 24.2% are in the 41–60 segment. This distribution of results reflects the age composition of the sample, where the first segment is the greatest and the other one is the second.

The second item, 'Carrying on activities/ sports (birdwatching, trekking, rock climbing, etc.)', which was indicated 27 times (by 17.2% of the sample), is much less important for the respondents (10 or 6.4% of the respondents chose it as sole answer, 14 or 8.9% one out of two). The other items are: 'Studying natural resources', which was indicated by 20 respondents (6 or 3.8% as sole answer, 8 or 5.1% one out of two), and 'Working as a volunteer in the protected area for environmental associations (FAI, WWF, Legambiente, etc.)', which was indicated by 7 people (respectively, 2 or 1.3% and 5 or 3.2%). Some of the respondents suggested other motivations, such as a visit to the Giganti, a press tour, work or knowing the park.

These findings confirm that the visit is mostly a way to take a break from daily routine and experience nature by carrying on slow activities.

Regarding the question of whether they visit the Giganti for the first time or not, 58.6% of the sample declared they had been there before (43.9% several times, 14.6% at least once), while for 41.4% of them it was the first time. Furthermore, 119 visitors (75.8% of the sample) already know the FAI, while 38 (24.2%) don't know it. Only 31 interviewees are enrolled in FAI (19.7%).

Large percentages of the sample are satisfied (30.6%) or very satisfied (53.5%) with the visit. They also show a high level of satisfaction for the welcome they received by tour operators and the local community (satisfied: 27.4%, very satisfied: 48.4%). Answering an open question, they explained that the main reasons for satisfaction are: courtesy and helpfulness in the welcoming, professionalism and preparation of the operators, and attention to detail.

Most of the interviewees, 153 (97.5% of the sample), answered that the visit positively affected their wellbeing. Even if, in general terms, the respondents are not widely aware of the positive impact of nature on people's personal wellbeing (it is possible to see that way of replying in the first part of the questionnaire), this answer shows that a large majority of them actually consider this impact as positive. This confirms and is coherent with the extensive literature on the subject which has been partially cited above.

Answering an open question on this subject, the interviewees indicate that the reasons for this positive impact are: 'Nature releases a positive energy that benefits mind and body, improves mood, reduces stress, gives peace, relaxation, and good vibes'; 'Natural environments stimulate outdoor sports'; 'Nature offers a variety of local products that reflect the Sila culture and history, their gastronomic traditions, their genuineness'; 'Eating well is part of good living'; 'The beauty of the forest and the contact with nature improve personal wellbeing and physical and psychological health'; 'The Giganti are unique secular trees that transmit positive energy and wisdom'.

Respondents chose to experience the Giganti through the following activities: guided tours (85.4%), forest bathing (19.7%), night events (7.0%), or carrying on 'different' activities indicated by those who provided answers such as independent visits, environmental education and mountain-bike excursions (3.8%). Few of them answered yoga (3.2%) or did not indicate a preference (2.5%). The majority is very satisfied (66.2%) or satisfied (22.9%) with these experiences.

Answering an open question, they explained that the main elements of satisfaction are: courtesy and helpfulness in welcoming, professionalism and enthusiasm of the operators, useful signs, illustrative panels, and rest areas.

The high rates of satisfaction with visits and activities and the belief that the visit can help improve wellbeing explain the substantial level of visitors' loyalty.

Conclusions

Visitors who are more sensitive and attentive to environmental issues nowadays often require ad hoc and customized services for their needs and want to play an active role during their holidays. The demand for nature experiences is growing but is also becoming more and more diversified every day. The analysis of the findings of the research show that the respondents are mainly nature lovers, people who usually visit nature destinations and are informed about the problems related to tourism development in these areas. They are aware of the importance of staying in contact with nature, especially after the pandemic, and some of them think that this contact could have a positive influence on their personal wellbeing and psychophysical health. They are interested in different aspects of the visit to the park and the Giganti. Even if most respondents are not inside the younger age segment, they are interested in many different activities, sports and experiences. However, their main motivation is that of enjoying nature by carrying out relaxing activities such as walking or picnicking, with others or even alone. Their levels of satisfaction are very high, thus confirming that the reserve's offer is ideal for this type of visitor.

In the light of this research, it can be stated that certainly one of the reasons for the success of the Giganti is that every year it offers innovative nature experiences to an increasingly demanding public of loyal visitors. The objectives should be the reduction of seasonality and the limitation of daily and proximity tourism, as they could be an important step towards environmental sustainability. These are directives to invest in, so as to promote the reserve, attract a greater number of people from other regions all year long, increase visitors' satisfaction, and also improve visitors' and local communities' quality of life and wellbeing. In this way, the Giganti della Sila could become an instrument to create a favourable relationship with all the stakeholders of the reserve, showing them the socio-cultural sustainable development created as a central element of FAI strategy in Sila.

This is an initial first exploratory research. The limits of the survey are related to the small and non-probability sample of respondents, who are mainly local visitors, and the restricted period in which the questionnaire was administered. In fact, it considers only one season, but the type of visitors to the area actually changes in the different periods of the year. For all these reasons the results of the survey are not generalizable.

The connection between nature and personal wellbeing reached through visits to natural protected areas should be studied in more depth in the future, considering its growing importance for tourists but also for residents. It could be interesting to make comparisons with other natural and cultural FAI proprieties in order to analyse new ways of how the contact with the natural environment, beauty and art can influence people's wellbeing and quality of life.

References

Azara, I., Michopoulou, E., Niccolini, F., Taff, B.D. and Clarke, A. (eds) (2018) *Tourism, Health, Wellbeing and Protected Areas*. CAB International, Wallingford, UK. DOI: 10.1079/9781786391315.0000.

Cheng, Y., Hu, F., Wang, J., Wang, G., Innes, J.L. *et al.* (2022) Visitor satisfaction and behavioral intentions in nature-based tourism during the COVID-19 pandemic: a case study from Zhangjiajie National

Forest Park, China. *International Journal of Geoheritage and Parks* 10(1), 143–159. DOI: 10.1016/j.ijgeop.2022.03.001.

Dobson, J., Birch, J., Brindley, P., Henneberry, J., McEwan, K. *et al.* (2021) The magic of the mundane: the vulnerable web of connections between urban nature and wellbeing. *Cities* 108, 102989. DOI: 10.1016/j.cities.2020.102989.

Farkic, J., Isailovic, G. and Taylor, S. (2021) Forest bathing as a mindful tourism practice. *Annals of Tourism Research Empirical Insights* 2(2), 100028. DOI: 10.1016/j.annale.2021.100028.

Ferrari, S. and Gilli, M. (2016) Protected natural areas as innovative health tourism destinations. In: Smith, M.K. and Puczkó, L. (eds) *The Routledge Handbook of Health Tourism*. Routledge, London, pp. 419–429.

Levinger, P., Cerin, E., Milner, C. and Hill, K.D. (2021) Older people and nature: the benefits of outdoors, parks and nature in light of COVID-19 and beyond- where to from here? *International Journal of Environmental Health Research* 32(6), 1–8. DOI: 10.1080/09603123.2021.1879739.

Martin, L., White, M.P., Hunt, A., Richardson, M., Pahl, S. *et al.* (2020) Nature contact, nature connectedness and associations with health, wellbeing and pro-environmental behaviours. *Journal of Environmental Psychology* 68, 101389. DOI: 10.1016/j.jenvp.2020.101389.

Richardson, M., Passmore, H.A., Lumber, R., Thomas, R. and Hunt, A. (2021) Moments, not minutes: the nature-wellbeing relationship. *International Journal of Wellbeing* 11(1), 8–33. DOI: 10.5502/ijw.v11i1.1267.

White, M.P., Alcock, I., Grellier, J., Wheeler, B.W., Hartig, T. *et al.* (2019) Spending at least 120 minutes a week in nature is associated with good health and wellbeing. *Scientific Reports* 9(1), 1–11. DOI: 10.1038/s41598-019-44097-3.

18 Motivations, Outcomes and Emotional Responses of Visitors Attending Personal Interpretation Programmes: A Case Study of Provincial Parks in Alberta, Canada

Glen Hvenegaard[1]*, Elizabeth Halpenny[2] and Clara-Jane Blye[2]
[1]Department of Science, Augustana Campus, University of Alberta, Camrose, Alberta, Canada; [2]Faculty of Kinesiology, Sport and Recreation, University of Alberta, Edmonton, Alberta, Canada

Abstract
This chapter examines the connections between outcomes and emotions resulting from visitors participating in personal interpretation programmes in provincial parks in Alberta, Canada. Visitor responses were high for outcomes such as programme satisfaction, intentions for some park-friendly behaviours, connections to place, and positive memories, but slightly less so for knowledge increase. The emotional responses from the programmes for pride and inspiration were higher than responses for anger, fear and guilt. There were more positive correlations between the positive emotions (i.e. pride and inspiration) and visitor outcomes than between the negative emotions (i.e. anger, fear and guilt) and visitor outcomes. Interpreters can incorporate the positive emotions of pride and inspiration into the planning and delivery of their programmes to help enhance visitor experiences and to improve resource management by targeting behavioural intentions.

Introduction

Parks and conserved areas are regular attractions for nature-based tourism, since they have mandates to protect natural environments and provide outdoor recreational experiences (Leung *et al.*, 2018). Human experiences in parks bring many health, wellbeing, and awareness benefits to visitors (Lemieux *et al.*, 2016). Heritage interpretation, regularly offered by park agencies, and often participated in by visitors during their park experiences, is 'a purposeful approach to communication that facilitates meaningful, relevant, and inclusive experiences that deepen understanding, broaden perspectives, and inspire engagement with the world around us' (National Association for Interpretation, 2022). Our study focused on personal interpretation, involving visitors interacting directly with interpreters (e.g. guided hike, outdoor theatre, bus tours and pop-up displays) (Hvenegaard and Shultis, 2016).

Interpretation can improve the wellbeing of visitors, the ecological integrity of the park

*Corresponding author: gth@ualberta.ca

DOI: 10.1079/9781800621411.0018

and the park agency (Hvenegaard and Shultis, 2016). Visitor outcomes from interpretation are varied and site-specific, and generally relate to satisfaction (or enjoyment), increased knowledge, improved attitudes, behavioural intentions and actual behaviours regarding park-friendly activities (Cook *et al.*, 2021; Blye *et al.*, 2022). Past studies suggest that satisfaction and knowledge outcomes are more easily achieved for visitors than outcomes related to attitudes and behaviours (Hvenegaard and Shultis, 2016). Researchers know less about the development of connections to place and positive memories (e.g. Blye *et al.*, 2022). Emotions such as anger, guilt and pride have the potential to promote environmentally friendly behaviour, but little is known about the pathways involved (He *et al.*, 2022).

Researchers often focus on behaviours because of the potentially tangible ways to reduce impact on a park environment. The theory of planned behaviour (TPB) (Ajzen, 2011) is commonly used to predict a specific behaviour by way of a respondent's behavioural intention, which, in turn, is predicted by attitudes, subjective norms and perceived behavioural control associated with the behaviour in question (Fishbein and Manfredo, 1992). Researchers also use the value-belief-norm theory to predict behavioural intentions by examining a respondent's worldviews and values (van Riper and Kyle, 2014). The norm activation model (Baumeister, 1998) uses emotions, such as guilt, anger or pride, to predict how respondents might develop a sense of obligation to change behaviours.

The aim of this project was to examine the visitor outcomes and emotions associated with attending personal interpretation programmes in provincial parks in Alberta. We hypothesized that visitor outcomes for satisfaction and knowledge gain would exceed those of other outcomes. We also hypothesized that there will be a positive correlation between visitor outcomes and emotional responses. Because very few studies have examined the role of emotions in interpretive outcomes (e.g. Bamberg and Möser, 2007), this study responds to a call for more research on factors, such as emotions, affecting visitor outcomes from interpretation (Stern and Powell, 2021). A deeper understanding of how emotions affect interpretation outcomes will aid in programme development and delivery.

Methods

Study site

Alberta Parks, the provincial agency for protected areas, operates 473 sites, attracting 8.7 million visits per year, of which 2 million are overnight camping visits (Alberta Parks, 2016). The agency seeks to 'inspire people to discover, value, protect, and enjoy the natural world and the benefits it provides for current and future generations' (Government of Alberta, 2009, p. 3). Several parks offer personal interpretation programmes. Public interpretation programmes, plus environmental education programmes for schoolchildren, reached about 450,000 people per annum in 2016 (Alberta Parks, 2016).

Data collection

During June to September in 2018 and 2019, we visited 11 parks (1–2 times each on weekdays and weekends) offering personal interpretation to collect data from camping visitors (only those ≥18 years of age) using on-site, self-administered surveys. We surveyed campers at high-use locations, such as outdoor theatres, visitor information centres and campsites. We used a stratified, convenience-based sampling strategy, randomized by interviewing the next available camper. We asked questions about demographics, trip information, along with outcomes and emotions resulting from their interpretation experience. Regarding interpretive programme outcomes, we used 1–7 scales to ask about frequency of attendance while camping ('never' to 'every time'), satisfaction ('very dissatisfied' to 'very satisfied'), knowledge increase ('no increase' to 'significant increase'), attitudes about the programme ('boring' to 'interesting', 'unpleasant' to 'pleasant'), behavioural intentions ('definitely not' to 'definitely'), connections to nature ('very disconnected' to 'very connected') and positive memories ('not positive' to 'very positive'). For emotions, we asked about reactions to the interpretive programme (1 = not angry, fearful, guilty, proud or inspired, to 7 = very angry, fearful, guilty, proud or inspired).

For each year of sampling, we followed up with respondents 1 year later (in 2019 or 2020) via email or telephone to measure changes in selected variables. Respondents rated only two of the outcomes above, satisfaction and knowledge increase.

Data analysis

Working with SPSS 28 to analyse relationships among variables, we used independent samples t-tests (reporting t-values and Cohen's *d*) and Pearson correlation tests (reporting r-values). We used a threshold of $p<0.05$.

Results

We received responses from 763 attendees of interpretive programmes (98% response rate), of which 38% were male, 59% were female and 3% were undisclosed. The median age was 44 years. The median household income was 100,000–149,999 CAD. The median highest level of education was a bachelor's degree. On average, there were 2.9 adults and 2.5 children in the group attending a programme. The mean park trip length was 6.4 days. Respondents attended interpretive programmes frequently (5.2) during their visits to provincial parks.

Visitor outcomes

Respondents rated interpretive programmes as satisfying (6.6 out of 7), decreasing slightly to 6.5 1 year later (Fig. 18.1). Regarding increased knowledge, respondents' mean score was 5.6, remaining static 1 year later. For attitudes, respondents thought that the programmes were pleasant (6.6) and interesting (6.5). Concerning behavioural intentions resulting from the programme, respondents were very likely to refrain from feeding wild animals (6.5), refrain from building large campfires (6.4) and visit a different park within a year (6.3). Respondents were moderately likely to pick up litter (5.8), visit the

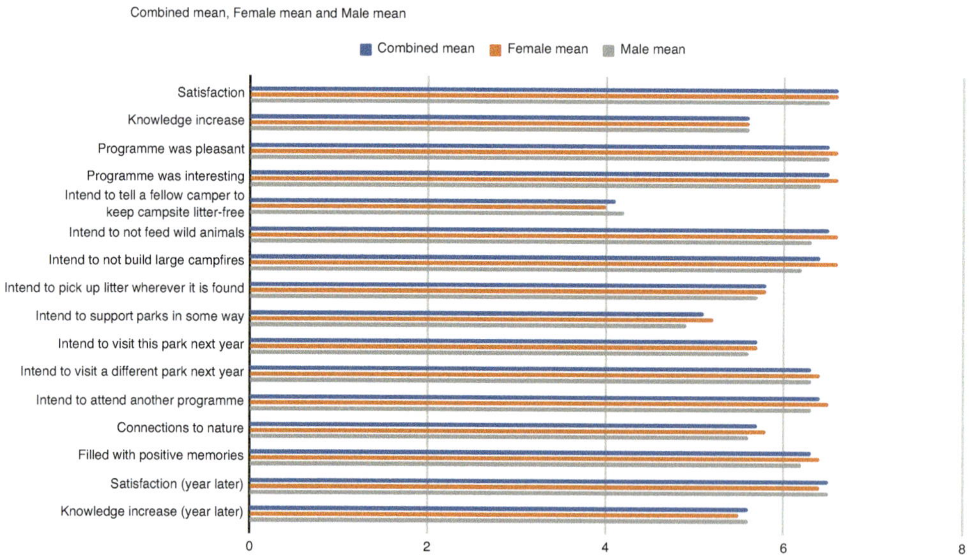

Fig. 18.1. Visitor outcomes from attending an interpretive programme in Alberta's provincial parks by gender (2018–2019).

same park within a year (5.7), attend another interpretive programme (5.7) and support parks in some way (5.0). Respondents were less likely to commit to other behaviours, such as telling a fellow camper to keep their campsite free of litter (3.9). Respondents gained some connections to nature (5.7) and positive memories (6.3).

A few visitor outcomes differed between genders, but the effect sizes were small (< 0.5). Thus, the magnitude of these relationships may have limited practical implications (Fig. 18.1). Females were more satisfied, said the programmes were more interesting, were less likely to feed wild animals, were less likely to build large campfires, were more likely to go to another programme, gained more connections to nature, and were more filled with positive memories than males. Visitor educational levels were correlated with satisfaction a year later ($r=0.124$, $n=356$, $p=0.020$), but not for other outcomes. Household income was positively correlated with knowledge increase ($r=0.135$, $n=598$, $p<0.001$), but negatively correlated with the intention to pick up litter ($r=-0.166$, $n=243$, $p=0.009$). The number of children attending a programme was negatively correlated with feeding wild animals ($r=-0.140$, $n=687$, $p<0.001$). The number of days on a trip was positively correlated with the intention to tell a fellow camper to keep their campsite litter-free ($r=0.103$, $n=692$, $p=0.006$).

Emotional responses

Respondents were modestly 'inspired to do more to protect the environment' (5.3), 'proud of what people are doing to protect the planet' (4.8), 'angry about how humans treat the environment' (4.1) and 'fearful about my future in the environment' (4.0). Respondents were neutral about feeling 'guilty about what people are doing to protect the planet' (3.2). Of these emotional responses, females felt prouder of what people are doing to protect the planet (4.9 vs 4.7; $t=1.82$, $df=688$, $p=0.034$, $d=0.14$) and more inspired to do more to protect the environment (5.5 vs 5.1; $t=3.50$, $df=691$, $p<0.001$, $d=0.27$) than males. There was a negative correlation between respondent age and feeling guilty ($r=-0.109$, $n=698$, $p=0.004$).

Relationships between visitor outcomes and emotional responses

Anger was positively correlated with programme satisfaction, knowledge gain, programme pleasantness, programme was interesting, intention to tell others to keep their campsite litter-free, intention to not build large campfires, intention to pick up litter, intention to support parks, intention to visit this park again next year, connections to place and positive memories (Table 18.1). Fear was positively correlated with programme satisfaction, knowledge increase, pleasantness, interesting programme, intention to tell a fellow camper to keep their campsite litter-free, intention to pick up litter, intention to attend another programme, connections to nature and positive memories. Guilt was positively correlated with the intention to tell a fellow camper to keep their campsite litter-free, intention to pick up litter, intention to support parks and connections to nature. Pride was positively correlated with all of the outcomes, except for intending to not feed wild animals, intending to not build large campfires and knowledge increase a year later. Lastly, inspiration was positively correlated with all of the outcomes, except for not feeding wild animals.

Discussion

This chapter sought to determine the visitor outcomes and emotions associated with attending personal interpretation programmes in provincial parks in Alberta. As expected, satisfaction from programmes was high, as were attitudes towards the programmes, intentions for some park-friendly behaviours, connections to place and positive memories. Unexpectedly, respondents rated knowledge increase slightly less positively than other outcomes. Normally, satisfaction and knowledge gain are easily accomplished goals from park interpretive programmes, with attitude change and behaviour change harder to accomplish (Skibins *et al.*, 2012). Roberts *et al.* (2014) also found that knowledge gain ranked lower than satisfaction, attitude change and intent to change behaviour for interpretive programmes in Kruger National Park, South Africa. In our study, satisfaction and

Table 18.1. Relationships between visitor outcomes and emotional responses from attending interpretive programmes in Alberta's provincial parks (2018–2019).

	Emotional response from programme				
Outcome	Angry	Fearful	Guilty	Proud	Inspired
Satisfaction	0.108**	0.131**	0.056	0.108**	0.222**
Knowledge increase	0.157**	0.101**	0.033	0.133**	0.268**
Programme was pleasant	0.133**	0.146**	0.049	0.094*	0.212**
Programme was interesting	0.122**	0.095*	0.159	0.129**	0.269**
Intend to tell a fellow camper to keep campsite litter-free	0.292**	0.295**	0.185**	0.154**	0.332**
Intend to not feed wild animals	0.042	0.010	–0.007	–0.038	0.023
Intend to not build large campfires	0.082*	0.055	–0.071	0.034	0.093*
Intend to pick up litter wherever it is found	0.239**	0.199**	0.205**	0.161**	0.352**
Intend to support parks in some way	0.297**	0.211**	0.135**	0.124*	0.371**
Intend to visit this park next year	0.128**	0.078*	0.025	0.130**	0.166**
Intend to visit a different park next year	0.008	–0.002	0.041	0.089*	0.102**
Intend to attend another programme	0.063	0.079*	0.027	0.171**	0.220**
Connections to nature	0.264**	0.234**	0.129**	0.286*	0.486**
Filled with positive memories	0.179**	0.157**	0.031	0.230*	0.391**
Satisfaction (year later)	0.076	0.083	0.039	0.148**	0.158**
Knowledge increase (year later)	0.064	0.066	–0.033	0.076	0.130*
# of significant relationships (**+*)	11 = 10+1	11 = 8+3	4 = 4+0	13 = 8+5	15 = 13+2

Note: *r*-values reported; statistically significant at ** = 0.01 level or * = 0.05 level knowledge gain remained at similar levels a year after the programmes, suggesting that these visitor outcomes were relatively stable over time. There have been few longer-term studies with which to compare how these outcomes last over time (Zeppel, 2008).

As expected, the emotions of inspiration, anger, fear and pride were associated with many desired visitor outcomes. However, guilt had far fewer associations with desired visitor outcomes. Both positive and negative emotions can influence behaviour change (Pelletier and Sharp, 2008), but positive emotions are good predictors of pro-environmental behaviour (Harth *et al.*, 2013; Coelho *et al.*, 2017). In our study, the positive emotions of inspiration and pride showed more and longer-lasting visitor outcomes than the negative emotions of anger, fear and guilt. In a recent meta-analysis, Shipley and van Riper (2022) document pride's superior performance in comparison to guilt in studies of pro-environmental behaviour prediction. Strong persuasive messages that target affective attitudes and emotions have been successfully employed in a few environmental education and interpretation studies to inform environmental threats such as litter and bear safety (Brown *et al.*, 2010), influence whale tourists' intentions for pro-environmental behaviour (Jacobs and Harms, 2014) and complement arts-based climate change outcomes (Barrett and Mowen, 2014). In summarizing studies on interpretive best practices, Skibins *et al.* (2012) report that employing affective (emotional) practices is associated with positive impacts on visitor outcomes.

Theoretical implications

Past frameworks suggest several pathways to predict pro-environmental behaviour. This study's results support the norm activation model, but recommend moving beyond the study of guilt and pride, to explore the role of other emotions, especially positive emotions, to predict behavioural intentions (Coelho *et al.*, 2017). Increased appreciation of the role of emotions in decision-making and pro-environmental intentions and behaviours highlights the importance of incorporating emotions in pro-environmental messaging (Brosch and Steg, 2021).

Gender differences, while small in magnitude in this study, appear to align with a few interpretive studies demonstrating that females had a higher level of stewardship response than males (Stern *et al.*, 2012). An explanation for this trend may be found in previous studies that have documented higher levels of emotional empathy expressed towards wildlife and the environment (Arnocky and Stroink, 2010), environmental concern, attitudes and pro-environmental behaviours among women (Zeleny *et al.*, 2000). Such responses may be amplified when emotions are evoked (Barrett and Mowen, 2014) or woven into the content of the interpretive programme (Jacobs and Harms, 2014). However, we suggest cautious interpretation of these differences, since evidence of gender's effect is limited and debated in research on interpretation and pro-environmental behaviours (Peake *et al.*, 2009).

Practical implications

These results have some important programme marketing and design implications. Since most attendees are family-oriented, educated and affluent, marketing efforts can target these audiences to increase participation. However, attendees came from all demographic categories, so marketing and programme design should be inclusive for all groups. Since attendees gained immediate benefits of visitor satisfaction and learning, marketing can promote these outcomes to attract participants, while facilitating other outcomes that last longer and have benefits for the natural environment. Interpreters can incorporate the positive emotions of pride and inspiration into the planning and delivery of their programmes to help enhance visitor experiences and to improve resource management by targeting behavioural intentions. Integrating these emotions should be appropriate to the audience characteristics (e.g. age, group type) and park characteristics (e.g. resource types, threats and solutions). Overall, planners and interpreters should identify visitor motivations and benefits of visiting the site, clarify agency goals for visitor outcomes and decide on how best to use emotional responses to influence those preferred outcomes.

Limitations

This study used different scales for each visitor outcome and emotional response, which limited our ability to make comparisons. Moreover, all measures of outcomes and emotions depend on the ability of respondents to report their responses accurately (Lawton *et al.*, 2009). By focusing on behavioural intentions, and not measuring actual behaviour, our ability to estimate behaviour change from interpretation programmes is limited. Even though we measured some visitor outcomes after 1 year, we do not know if those outcomes are maintained for longer periods.

Future research

To improve estimates of visitor outcomes it would be helpful to determine actual behaviours. Future work should also determine which aspects of the interpretive programmes were influential on visitor outcomes (Cook *et al.*, 2021), in particular the role of emotion.

References

Ajzen, I. (2011) The theory of planned behaviour: reactions and reflections. *Psychology & Health* 26(9), 1113–1127. DOI: 10.1080/08870446.2011.613995.

Alberta Parks. (2016) *Alberta Parks Quick Facts*. Alberta Environment and Parks.

Arnocky, S. and Stroink, M. (2010) Gender differences in environmentalism: the mediating role of emotional empathy. *Current Research in Social Psychology* 16, Article 9. Available athttps://psycnet.apa.org/record/2014-37822-001accessed 22 October 2023

Bamberg, S. and Möser, G. (2007) Twenty years after Hines, Hungerford, and Tomera: a new meta-analysis of psycho-social determinants of pro-environmental behaviour. *Journal of Environmental Psychology* 27(1), 14–25. DOI: 10.1016/j.jenvp.2006.12.002.

Barrett, A. and Mowen, A.J. (2014) Assessing the effectiveness of artistic place-based climate change interpretation. *Journal of Interpretation Research* 19(2), 7–24. DOI: 10.1177/109258721401900202.

Baumeister, R.F. (1998) The self. In: Gilbert, D.T., Fiske, S.T. and Lindzey, G. (eds) *The Handbook of Social Psychology*. McGraw Hill, Maidenhead, UK, pp. 680–740.

Blye, C.J., Hvenegaard, G.T. and Halpenny, E.A. (2022) Personal interpretation starts with the interpreter: how do outcome priorities change over time among interpretive staff? *Journal of Park and Recreation Administration* 40(2), 20–41. DOI: 10.18666/JPRA-2021-10605.

Brosch, T. and Steg, L. (2021) Leveraging emotion for sustainable action. *One Earth* 4(12), 1693–1703. DOI: 10.1016/j.oneear.2021.11.006.

Brown, T.J., Ham, S.H. and Hughes, M. (2010) Picking up litter: an application of theory-based communication to influence tourist behaviour in protected areas. *Journal of Sustainable Tourism* 18(7), 879–900. DOI: 10.1080/09669581003721281.

Coelho, F., Pereira, M.C., Cruz, L., Simões, P. and Barata, E. (2017) Affect and the adoption of pro-environmental behaviour: a structural model. *Journal of Environmental Psychology* 54, 127–138. DOI: 10.1016/j.jenvp.2017.10.008.

Cook, K.J., Hvenegaard, G.T. and Halpenny, E.A. (2021) Visitor perceptions of the outcomes of personal interpretation in Alberta's provincial parks. *Applied Environmental Education & Communication* 20(1), 49–65. DOI: 10.1080/1533015X.2019.1693309.

Fishbein, M. and Manfredo, M.J. (1992) A theory of behavior change. In: Manfredo, M.J. (ed.) *Influencing Human Behavior: Theory and Applications in Recreation, Tourism, and Natural Resources Management*. Sagamore Publishing, Inc, Champaign, Illinois, pp. 29–50.

Government of Alberta (2009) *Plan for Parks: 2009–2019*. Available at: https://albertaparks.ca/media/123436/p4p.pdf (accessed 22 October 2023).

Harth, N.S., Leach, C.W. and Kessler, T. (2013) Guilt, anger, and pride about in-group environmental behaviour: Different emotions predict distinct intentions. *Journal of Environmental Psychology* 34, 18–26. DOI: 10.1016/j.jenvp.2012.12.005.

He, M., Blye, C.-J. and Halpenny, E. (2022) Impacts of environmental communication on pro-environmental intentions and behaviours: a systematic review on nature-based tourism context. *Journal of Sustainable Tourism* 31(8), 1–23. DOI: 10.1080/09669582.2022.2095392.

Hvenegaard, G.T. and Shultis, J. (2016) The role of interpretation. In: Dearden, P., Rollins, R. and Needham, M. (eds) *Parks and Protected Areas in Canada: Planning and Management*, 4th edn. Oxford University Press, Toronto, Canada, pp. 141–169.

Jacobs, M.H. and Harms, M. (2014) Influence of interpretation on conservation intentions of whale tourists. *Tourism Management* 42, 123–131. DOI: 10.1016/j.tourman.2013.11.009.

Lawton, R., Conner, M. and McEachan, R. (2009) Desire or reason: predicting health behaviors from affective and cognitive attitudes. *Health Psychology* 28(1), 56–65. DOI: 10.1037/a0013424.

Lemieux, C.J., Doherty, S.T., Eagles, P.F.J., Groulx, M.W., Hvenegaard, G.T. *et al.* (2016) Policy and management recommendations informed by the health benefits of visitor experiences in alberta's protected areas. *Journal of Park and Recreation Administration* 34(1), 24–52. DOI: 10.18666/JPRA-2016-V34-I1-6800.

Leung, Y.-F., Spenceley, A., Hvenegaard, G. and Buckley, R. Leung, Y.-F., Spenceley, A., Hvenegaard, G. and Buckley, R. (eds). (2018) *Tourism and Visitor Management in Protected Areas: Guidelines for Sustainability. Best Practice Protected Area Guidelines Series no. 27.* IUCN, Gland, Switzerland. DOI: 10.2305/IUCN.CH.2018.PAG.27.en.

National Association for Interpretation(2022) Mission, vision, and core values. Available at: https://www.interpnet.com/NAI/interp/About/About_NAI/What_We_Believe/nai/ (accessed 22 October 2023).

Peake, S., Innes, P. and Dyer, P. (2009) Ecotourism and conservation: factors influencing effective conservation messages. *Journal of Sustainable Tourism* 17(1), 107–127. DOI: 10.1080/09669580802276000.

Pelletier, L.G. and Sharp, E. (2008) Persuasive communication and proenvironmental behaviours: how message tailoring and message framing can improve the integration of behaviours through self-determined motivation. *Canadian Psychology/Psychologie Canadienne* 49(3), 210–217. DOI: 10.1037/a0012755.

Roberts, M., Mearns, K. and Edwards, V. (2014) Evaluating the effectiveness of guided versus non-guided interpretation in the Kruger National Park, South Africa. *Koedoe* 56(2), 1–8. DOI: 10.4102/koedoe.v56i2.1160.

Shipley, N.J. and van Riper, C.J. (2022) Pride and guilt predict pro-environmental behavior: a meta-analysis of correlational and experimental evidence. *Journal of Environmental Psychology* 79, 101753. DOI: 10.1016/j.jenvp.2021.101753.

Skibins, J.C., Powell, R.B. and Stern, M.J. (2012) Exploring empirical support for interpretation's best practices. *Journal of Interpretation Research* 17(1), 25–44. DOI: 10.1177/109258721201700103.

Stern, M.J. and Powell, R.B. (2021) Taking stock of interpretation research: where have we been and where are we heading? *Journal of Interpretation Research* 25(2), 65–87. DOI: 10.1177/1092587221 1023205.

Stern, M.J., Powell, R.B., Martin, E. and McLean, K. (2012) Identifying best practices for live interpretation programs in the United States National Park Service. Project Report for the U.S. National Park Service. Available at: https://media.clemson.edu/cbshs/prtm/research/interpret-environ-education/Stern-Powell-Interp-Study-Aug-6-2012.pdf (accessed 22 October 2023).

van Riper, C.J. and Kyle, G.T. (2014) Understanding the internal processes of behavioral engagement in a national park: a latent variable path analysis of the value-belief-norm theory. *Journal of Environmental Psychology* 38, 288–297. DOI: 10.1016/j.jenvp.2014.03.002.

Zeleny, L.C., Chua, P.-P. and Aldrich, C. (2000) Elaborating on gender differences in environmentalism. *Journal of Social Issues* 56, 443–457. DOI: 10.1111/0022-4537.00177.

Zeppel, H. (2008) Education and conservation benefits of marine wildlife tours: developing free-choice learning experiences. *The Journal of Environmental Education* 39(3), 3–18. DOI: 10.3200/JOEE.39.3.3-18.

19 The Contribution of Past Experiences on Revisit Intentions in National Parks: The Case of Tourists in Tanzania's Lake Manyara National Park

Erica J. Hando[1], Simon M. Thiong'o[2*], Rita Schulz[3] and Brendah N. Uluma[3]
[1]Nomad Tanzania Tours, Arusha, Tanzania; [2]School of Events, Hospitality and Tourism Management, Leeds Beckett University, UK; [3]School of Tourism, Hospitality and Events Management, Moi University, Eldoret, Kenya

Abstract

Studies of tourists' destination loyalty demonstrate that experiences during past visits to a destination influence future intentions to revisit the same destination. Similarly, literature on travel behaviour reveals that post-visitation variables significantly relate to revisit intentions. However, at the time of this study, there was a paucity of empirical studies exploring the correlation between 'during-visitation' variables and revisit intentions to parks. This study attempted to fill this gap by assessing the influence of on-site experiences on tourists' revisit intentions in Lake Manyara National Park (LMNP). The findings showed that experience (Mean = 1.876) influenced tourists' intentions to revisit LMNP. Further, Pearson's correlation coefficient depicted a relatively significant association between experience and tourist intentions to revisit the park ($r = 0.447^{**}$, $p = 0.000$). This study concludes that experience is a significant predictor of future tourists' behavioural intentions. Tourism destination managers should, therefore, strive to understand their target market(s) to create memorable experiences.

Introduction

Tourism products are usually intangible, and tourists rely on their individual feelings and interpretations, not just facts, to make travel decisions (Darnell and Johnson, 2001; Thiong'o *et al.*, 2022). In that regard, planning and designing products which fit the preferences of target markets along with understanding their experiences is an important task for tourism destination managers (Rop *et al.*, 2021). This means that a successful product or service experience requires an understanding of the likes, tastes and preferences of customers (Callahan and Lasry, 2004; Guimarães *et al.*, 2015).

Since 1998, when Pine and Gilmore came up with the phrase 'experience economy', the number of studies dedicated to the comprehension of diversified consumer experiences in tourism has been increasing. Further, and as postulated by Boonajsevee (2005), visitor experience is deemed the main construct in travel research. This is perhaps why it has attracted the curiosity of researchers over the last couple of decades (for instance, Crompton, 1979; Urry, 1990; Ryan, 2002; Pearce, 2013; and Tan *et al.*, 2021,

*Corresponding author: tm_simon@mu.ac.ke and s.thiongo2234@student.leedsbeckett.ac.uk

DOI: 10.1079/9781800621411.0019

among others). As indicated in many of these studies, tourists desire experiences that symbolize something meaningful. In most instances, this is defined through a network of relations between their individual responses and objective features of a product (Barlow and Maul, 2000).

Experience in the context of this study is viewed as a subjective psychological phenomenon which imparts an image representation on someone. In tourism research, for instance, tourists' experience during a safari (an expedition to natural wildlife areas, particularly in East Africa) focuses on visiting, discerning, discovering and appreciating different ways of life (Stamboulis and Skayannis, 2003). In the same vein, tourists' experiences could be perceived as that unique, emotionally charged and high personal value encounter in a tourism system's destination region (Mcintosh and Siggs, 2005). From a travel perspective, Chen and Tsai (2007) look at on-site experience as the perceived quality of a trip based on the comparison between pre-encounter expectations and performance after consumption.

While there is an abundance of literature on how visitation (push and pull) factors inform travel decisions, there is (at the time of this study) still a substantial paucity of empirical studies exploring how on-site or 'during-visitation' variables contribute to revisit intentions in case-specific (in the context of this study, wildlife) destinations in the region. In consideration of East Africa's world-class natural products endowment, this study sampled Lake Manyara National Park (LMNP), Tanzania, as a case reference to attempt to fill this gap.

Literature review

Tourists' experience and revisit intention

The relationship between tourists' experiences and their revisit intention in different touristic domains has been widely explored – wildlife viewing, cruise vacations, festival participation and consumption, parachuting over wildlife protected areas and experiences in sports events, among others (Chen and Funk, 2010). Most of these studies found a positive direct association between the two constructs. The study by Hosany and Witham (2010) on cruise tourism, for instance, showed that cruisers' experience confirmed intention to recommend the product to others. Other studies by Lee *et al.* (2005), Lee *et al.* (2012) and Weed (2005) on tourism experience and its relationship to revisitation point to the same conclusion.

Based on these references, it was generally hypothesized for this study that tourist on-site experiences would have an impact on their willingness and intent to revisit Lake Manyara National Park (or any other wildlife area in the region). In other words, tourists' intention to revisit the study area was hypothesized to be informed by their positive image and assessment of their experience during the visit at the time of this study (Petrick, 2004; Um *et al.*, 2006). It was also hypothesized that tourists' attitudes towards the touristic encounter were pertinent indicators of other post-encounter actions including the desire to recommend the product (Gnoth, 1997).

The 4Es tourism experience theory – a theoretical framework

The study was based on Pine *et al.*'s (1999) indicators of experience. Generally referred to as the 4Es, the illustration includes entertainment, esthetics, education and escapism. The 4Es are based on the level of tourist involvement along with the type of association (Fig. 19.1). Participation in this case is either *passive* – tourists are largely observers; or *active* – tourists delve into and become part of the occurrences and experiences at an event (Pine *et al.*, 1999). Regarding connection or environmental relationship, there is *absorption*, in which a tourist's mind is occupied by getting the experience into his/her attention; and *immersion*, where visitors become physically part of the experience itself. *Entertainment* and *esthetic* consumption of tourism represent passive participation such as immersing oneself in an event or sightseeing during a safari without necessarily influencing it (Oh *et al.*, 2007).

Active participation, on the other hand, is represented by *education* and *escapism*, where tourists are actively involved. Through the interaction, visitors are considered to acquire skills, knowledge and, in most instances, a learning experience. They also immerse themselves in events and environments such as mountain climbing, consuming theme parks and hunting.

This immersion is normally among the main motives of travel (Fig. 19.1).

The application and relevance of the 4Es tourism experience theory

In tourism literature, there are several researchers who have applied the 4Es model with significant results (for instance, Oh *et al.*, 2007; Jurowski, 2009; and Hosany and Witham, 2010). Most of them have suggested that the four realms can be a useful configuration to understand tourists' experiences in various destinations. For instance, Oh *et al.* (2007) used the structure to come up with indictors/measures of visitor experiences. Hosany and Witham (2010) went further and employed the indicators of Oh *et al.* (2007) to measure travel experience in cruise tourism. However, limited research has been done with respect to the application of the model to explore on-site tourist experiences, specifically in wildlife-protected areas, and the link between visitation experience and revisit intention to wildlife destinations in East Africa.

Regarding this topic and building on Pine and Gilmore's framework along with other related studies, measurement scales for experience variables in LMNP were derived (Table 19.1).

Revisit intention is one of either end-results of a 'touristic journey' that could be conceptualized into an *Expectations-Experience-Perception (EEP)* framework. *Expectations* are 'pre-visitation' attitudes regarding a destination or tourism product. These could be influenced by the brand image created by marketing efforts of destination marketing organizations (DMOs) and destination managers, or through referrals. The *experience*, which forms the basis of this study, is a 'during-visitation' encounter with the tourism product. This leads into a post-visitation *experience*, either negative or positive, after the consumption of the product. The outcome of the experience could be one of the following: 'met expectations', 'surpassed expectations',

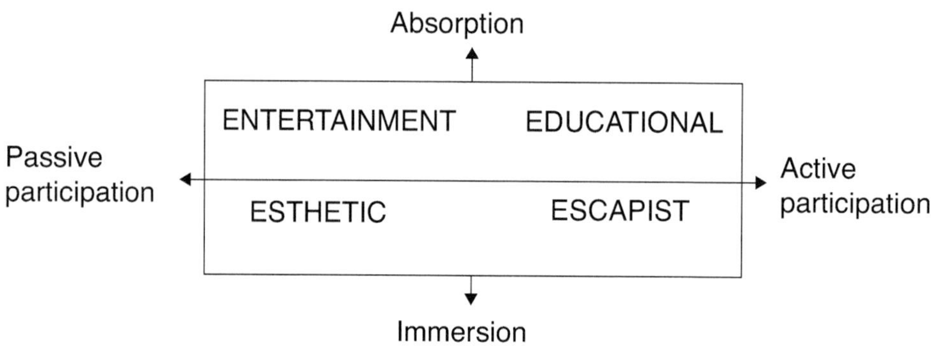

Fig. 19.1. The four realms of an experience.

Table 19.1. Elements used to measure experience in the study.

	Elements
EDUCATION	The experience has made me more knowledgeable The experience enhanced my skills
ESTHETIC	I experienced genuine peace and tranquility The settings really showed attention to protection and conservation
ENTERTAINMENT	I really enjoyed watching various tourism activities Watching others perform was captivating
ESCAPISM	I totally got away from the norm I thought I experienced distinct time and location

or 'under-met/did not meet expectations'. These outcomes later affect, either positively or negatively, intent to revisit the same destination, re-consume the product or even refer the product to others. This study hypothesized that the perception (either negative or positive) after the product consumption in the destination of study would have a direct relationship to intent to revisit the same destination.

Materials and Methods

Study area and data-collection procedure

The study was conducted in Lake Manyara National Park in Tanzania's northern circuit (Fig. 19.2). Famous for its spectacular landscape, nature and wildlife species, LMNP is accessible from Arusha and Dar es Salaam by road or air (Nkinda, 2016). The park covers an area of 330 km^2 (of which two-thirds is Lake Manyara) and receives well over 150,000 tourists per annum (World Bank, 2015).

In July and August 2016, we systematically sampled 500 international tourists (third-person- or group-selected) derived from a target population of 1001 international tourists visiting the park in 2015 (Tanzania National Parks Authority, 2020). We administered questionnaires to tourists at the exit gate of the park through an on-the-spot filling method. Research participants were also chosen based on their willingness to volunteer and participate in the

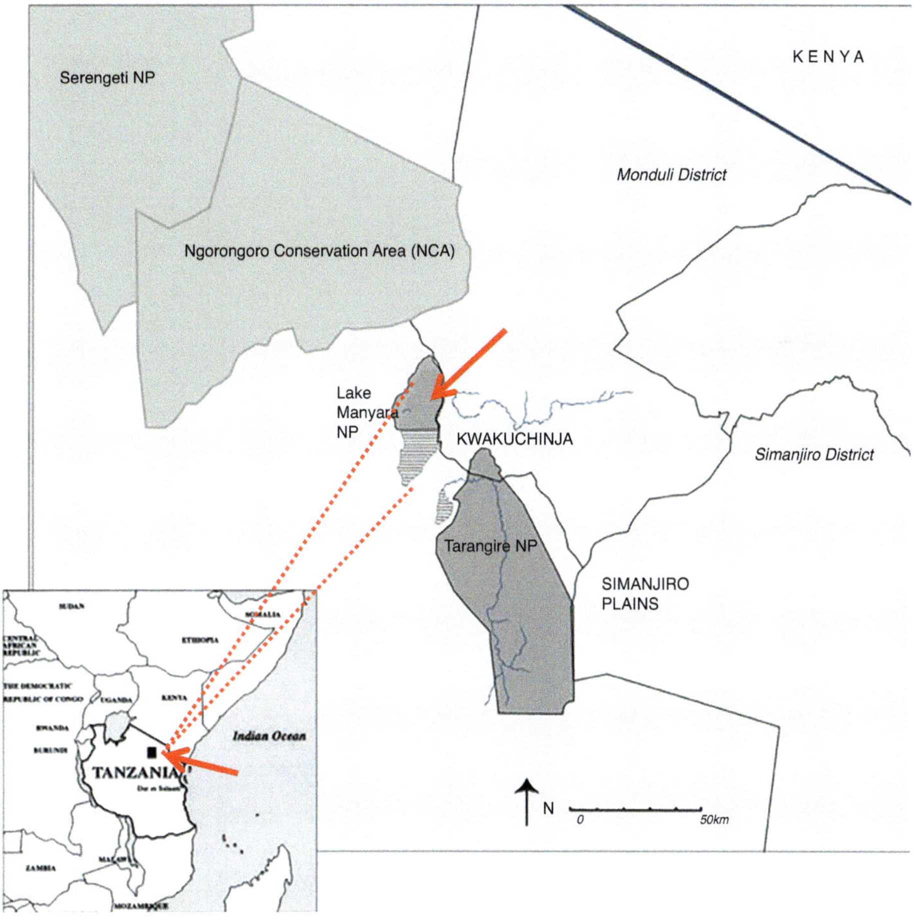

Fig. 19.2. Map showing location of Lake Manyara National Park.

study. Borrowing from the study by Oh *et al.* (2007), we adopted two items for each experience measurement dimension (Table 19.1). Respondents' opinions were sought on five-point Likert scales ranging from 'strongly agree' to 'strongly disagree'. The questionnaire was structured to gather details regarding tourists' experiences immediately after visiting Lake Manyara National Park. A total of 452 questionnaires were collected from the possible 500, giving a response rate of 90.4%. Fifteen questionnaires were discarded due to incompleteness and outliers, while the remaining 437 were used for the analysis.

Reliability of the research instrument

To establish the reliability of research instruments, the questionnaire used was first subjected to a pilot study and Cronbach's analysis (Tables 19.2 and 19.3). As a rule of thumb, a Cronbach's alpha coefficient value of $\alpha > 0.7$ was regarded as reliable for each of the datasets (Taber, 2018). The results from piloting were used in restructuring the questions that were not clear to the respondents.

From the results, the variables presented values greater than 0.7 and the questionnaire was, therefore, deemed reliable for the study.

Validity of the research instrument

A valid research instrument is one which accurately measures what is sought. To determine the content validity of the questionnaire, we consulted Tanzanian tourism officials (Tanzania Wildlife Research Institute and Tanzania National Parks Authority) who checked the instrument for its general content, analytical uniformity and clarity of individual elements. Based on their comments and suggestions, the questionnaire was modified, and the necessary review was made before administration.

Data analysis and ethical considerations

Descriptive statistics (mean and correlation coefficient) were used to analyse the data collected. To adhere to ethical procedures, consent to be involved in the study was sought

Table 19.2. Cronbach's analysis for piloting.

Scales	Items	Mean	Standard deviation	Cronbach's alpha
Experience				
Education	2	5.29	1.09	0.78
Esthetics	2	4.01	1.23	0.89
Entertainment	2	5.09	1.13	0.85
Escapism	2	3.69	1.27	0.89
Revisit intention	4	5.49	1.19	0.93

Table 19.3. Summary of Cronbach's analysis for piloting.

Scale	Items	Mean	Standard deviation	Cronbach's alpha
Experience	8	4.70	1.18	0.85
Revisit intention	4	5.49	1.19	0.93

Table 19.4. Role of tourists' experience in influencing their revisit intentions to LMNP. (Key: 1, Strongly agree; 2, Agree; 3, Undecided; 4, Disagree; 5, Strongly disagree)

Experience		1	2	3	4	5	Total	Mean
The experience has made me more knowledgeable	Freq.	221	165	23	19	9	437	1.62
	%	50.6	37.8	5.3	4.3	2.1	100	32.4
The experience has enhanced my skills	Freq.	191	175	48	13	10	437	1.80
	%	43.7	40	11	3	2.3	100	36
I felt genuine peace and tranquility	Freq.	123	173	74	41	26	437	2.25
	%	28.1	39.6	16.9	9.4	5.9	100	45
The settings really showed attention to protection and conservation	Freq.	143	195	45	27	27	437	2.08
	%	32.7	44.6	10.3	6.2	6.2	100	41.6
I enjoyed watching various tourism activities	Freq.	156	200	55	20	6	437	1.90
	%	35.7	45.8	12.6	4.6	1.4	100	38
Watching others perform was captivating	Freq.	205	169	50	8	7	437	1.69
	%	46.9	38.2	11.4	1.8	1.6	100	33.8
I completely escaped from reality	Freq.	179	197	37	23	1	437	1.79
	%	41	45.1	8.5	5.3	0.2	100	35.8
I thought I was in a different time and place	Freq.	161	205	42	22	7	437	1.88
	%	36.8	46.9	9.6	5	1.6	100	37.6

from the participants. Further, during data collection, the respondents were assured of anonymity and confidentiality, and were allowed to withdraw from the study at any time. No names or personal identification details were sought.

Results

Tourists' experience and revisit intentions

In assessing whether tourists' experiences at LMNP influenced their revisit intent, we measured each dimension of the 4Es using two items (see Table 19.1). The findings are presented in Table 19.4.

Tourists' revisit intentions

We measured tourists' revisit intentions using eight items that sought to identify their willingness to return to the park. The findings are illustrated in Table 19.5.

Correlation analysis

We used correlation analysis to establish whether there was a relationship between travellers' on-site experience and their intentions to revisit LMNP. Findings are depicted in Table 19.6.

Discussion

From the variables used to measure on-site visitation, experience had a mean ranging from 1.62 to 2.25 within the Likert scales. This suggests that tourists' experience was positively impacted. The overall mean for all the items under test was 1.87, which is very close to 2 (agreed). This has the implication that the respondents consented to all the items, affirming

Table 19.5. Tourists' revisit intentions. (Key: 1, Strongly agree; 2, Agree; 3, Undecided; 4, Disagree; 5, Strongly disagree)

Revisit intentions		1	2	3	4	5	Total	Mean
I would definitely recommend this attraction to my family members	Freq.	41	93	112	99	92	437	3.25
	%	9.4	21.3	25.6	22.7	21.1	100	65
I will recommend this attraction to my friends	Freq.	215	213	9	0	0	437	1.53
	%	49.2	48.7	2.1	0	0	100	30.6
I will definitely come back here in the next 12 months	Freq.	216	212	8	1	0	437	1.53
	%	49.4	48.5	1.8	0.2	0	100	30.6
I will spend my next holiday in these attractions	Freq.	27	78	113	92	127	437	3.49
	%	6.2	17.8	25.6	22.7	21.1	100	69.8
I will come next time with my family members	Freq.	60	130	149	44	54	437	2.78
	%	13.7	29.7	34.1	10.1	12.4	100	55.6
I enjoyed the game drive and will try it again	Freq.	105	205	83	25	19	437	2.19
	%	24	46.9	19	5.7	4.3	100	43.8
I enjoyed the services offered	Freq.	148	219	59	5	6	437	1.86
	%	33.9	50.1	13.5	1.1	1.4	100	37.2
I will definitely want to experience the services again	Freq.	102	183	108	24	20	437	2.26
	%	23.3	41.9	24.7	5.5	4.6	100	45.2

Table 19.6. Tourist experience and revisitation in wildlife destinations.

Correlations	Revisitation in wildlife destinations	
Tourists' experience	Pearson correlation	.447*
	Sig. (2-tailed)	.000
	N	.437

*Correlation is significant at the 0.01 level (2-tailed).

that experience is significantly related to revisit intention. Generally, Pine *et al.*'s (1999) variables were confirmed of their applicability in the study.

On the specific items measuring revisit intentions, most tourists agreed that they would recommend LMNP to their friends and 'definitely consume the park offerings within 12 months' (mean = 1.53). Additionally, most of them agreed that they enjoyed services offered (mean = 1.86), enjoyed game drives and would try the product again (mean = 2.19), and that they would want to experience the services package again (mean = 2.26). Most tourists (mean = 3.49) were close to agreeing that they may not spend their holiday in LMNP and had no opinion about coming to LMNP with their family members (mean = 2.79) or recommending the park to family members (mean = 3.25). These are strong areas for LMNP services marketers and destination managers to capitalize on with a view to enhance service delivery.

The study further demonstrated a strong positive correlation between tourists' experience and revisitation, $r(437) = 0.447^*$, $p = 0.000$. The positive result could have been attributed to the quality of service, availability of a variety of attractions in the park such as game drives (Figs 19.3, 19.4 and 19.5), hiking and birdwatching, among others (Tanzania National Parks Authority, 2020). This study had related results with those of Som and Badarneh (2011), who suggested that destination image strongly influenced tourist

Fig. 19.3. Tourists viewing a herd of African elephants in Lake Manyara National Park.

Fig. 19.4. Tourists viewing a giraffe, vultures and hyenas in Lake Manyara National Park.

Fig. 19.5. Tourists viewing the giraffes during a safari in Lake Manyara National Park's savannah grassland.

satisfaction on certain destinations – a positive image of LMNP could lead to successive tourist visits. This is further supported by the findings of Chang (2013), Beerli and Martin (2004) and Um *et al.* (2006), who assert that experiences at a destination are a major predictor of revisitation.

Additionally, the results are consistent with studies on diversified consumer experiences in various tourism research fields including gastronomy (e.g. Bruwer, 2014; Bruwer and Kelley, 2015; Chang *et al.*, 2018); tourist experience during cruise vacations (e.g. Parola *et al.*, 2014; Ozturk and Gogtas, 2016; Shahijan *et al.*, 2018); festivals (e.g. Shen, 2014; Lee and Kwon, 2021); and sporting events (e.g. Liu, 2013; Markus *et al.*, 2019; Vassiliadis *et al.*, 2021), among others. In the context of this study, a handful of specific case studies on the experiential revisit value of ecotourism/nature destinations corroborate the findings of this study (see Chien, 2016; Ramukumba, 2018; Ariya *et al.*, 2020).

This study, informed by the background literature cited and the data collected, aids in understanding wildlife destination attributes that influence revisitation in Tanzanian and, by extension, East African national parks. The findings from this study emphasize the place of experience in the tourism journey and offer a basis for further investigation on other variables (pre-, during and post-visitation) that influence visitation in destinations offering not only wildlife but also other tourism products. It also offers an outside-in perspective and insights for the Park Management Authority in Tanzania (and, by extension, similar authorities throughout East Africa) on tourists' travel motivation.

Conclusion

As with related studies on tourist behaviour, the study established that during-visitation experience was not only a decisive construct but also a predictor of visitor revisitation behaviour in wildlife destinations. This means that a destination's potential to create positive experiences through its products will influence favourable outcomes. It is, thus, destination managers' responsibility to cultivate unique customer experiences by developing value-added and distinct wildlife products. In this regard, it is also imperative that they understand their market in an attempt to develop service blueprints to meet, or perhaps surpass, the expectations of clientele.

References

Ariya, G., Sempele, C. and Wishitemi, B. (2020) An evaluation of wildlife tourism destinations' attributes, overall satisfaction and revisit intentions: a tourist guides' perspective. *African Journal of Hospitality, Tourism and Leisure* 9(6), 1220–1237. DOI: 10.46222/ajhtl.19770720-79.

Barlow, J. and Maul, D. (2000) *Emotional Value: Creating Strong Bonds with Your Customers*. Berrett-Koehler Publishers, Oakland, California.

Beerli, A. and Martin, J.D. (2004) Factors influencing destination image. *Annals of Tourism Research* 31(3), 657–681. DOI: 10.1016/j.annals.2004.01.010.

Boonajsevee, B. (2005) *Relationship Marketing: Loyalty Intentions in New Era of Thai Bank Marketing*. Nova Southeastern University, Fort Lauderdale, Florida.

Bruwer, J. (2014) Service quality perception and satisfaction: buying behaviour prediction in an Australian festivalscape. *International Journal of Tourism Research* 16(1), 76–86. DOI: 10.1002/jtr.1901.

Bruwer, J. and Kelley, K. (2015) Service performance quality evaluation and satisfaction in a USA wine festivalscape. *International Journal of Event and Festival Management* 6(1), 18–38. DOI: 10.1108/IJEFM-04-2014-0009.

Callahan, J. and Lasry, E. (2004) The importance of customer input in the development of very new products. *R&D Management* 34(2), 107–120. DOI: 10.1111/j.1467-9310.2004.00327.x.

Chang, L.L. (2013) Influencing factors on creative tourists' revisiting intentions: the roles of motivation, experience and perceived value. Doctoral thesis, Clemson University, Clemson, South Carolina. Available at: https://tigerprints.clemson.edu/all_dissertations/1084/ (accessed 23 October 2023).

Chang, M., Kim, J.H. and Kim, D. (2018) The effect of food tourism behavior on food festival visitor's revisit intention. *Sustainability* 10(10), 3534. DOI: 10.3390/su10103534.

Chen, C.F. and Tsai, D. (2007) How destination image and evaluative factors affect behavioral intentions? *Tourism Management* 28(4), 1115–1122. DOI: 10.1016/j.tourman.2006.07.007.

Chen, N. and Funk, D.C. (2010) Exploring destination image, experience and revisit intention: a comparison of sport and non-sport tourist perceptions. *Journal of Sport & Tourism* 15(3), 239–259. DOI: 10.1080/14775085.2010.513148.

Chien, M.C. (2016) An empirical study on the effect of attractiveness of ecotourism destination on experiential value and revisit intention. *Applied Ecology and Environmental Research* 15(2), 43–53. DOI: 10.15666/aeer/1502_043053.

Crompton, J.L. (1979) Motivations for pleasure vacation. *Annals of Tourism Research* 6(4), 408–424. DOI: 10.1016/0160-7383(79)90004-5.

Darnell, A.C. and Johnson, P.S. (2001) Repeat visits to attractions: a preliminary economic analysis. *Tourism Management*, 22(2), 119–126.

Gnoth, J. (1997) Tourism motivation and expectation formation. *Annals of Tourism Research* 24(2), 283–304. DOI: 10.1016/S0160-7383(97)80002-3.

Guimarães, M.H., Nunes, L.C., Madureira, L., Santos, J.L., Boski, T. *et al.* (2015) Measuring birdwatchers' preferences: a case for using online networks and mixed-mode surveys. *Tourism Management* 46, 102–113. DOI: 10.1016/j.tourman.2014.06.016.

Hosany, S. and Witham, M. (2010) Dimensions of cruisers' experiences, satisfaction, and intention to recommend. *Journal of Travel Research* 49(3), 351–364. DOI: 10.1177/0047287509346859.

Jurowski, C. (2009) An examination of the four realms of tourism experience theory. In: *International CHRIE Conference-Refereed Track.* University of Massachusetts Amherst 2009 Conference.
Lee, C.K., Lee, Y.K. and Lee, B. (2005) Korea's destination image formed by the 2002 World Cup. *Annals of Tourism Research* 32(4), 839–858. DOI: 10.1016/j.annals.2004.11.006.
Lee, M., Han, H. and Lockyer, T. (2012) Medical tourism: attracting japanese tourists for medical tourism experience. *Journal of Travel & Tourism Marketing* 29(1), 69–86. DOI: 10.1080/10548408.2012.638564.
Lee, W. and Kwon, H. (2021) The influence of personal involvement on festival attendees' revisit intention: food and wine attendees' perspective. *Sustainability* 13(14), 7727. DOI: 10.3390/su13147727.
Liu, D. (2013) Major sports events, destination image and intention to revisit from the foreign tourist's perspective. *International Journal of Sports Marketing and Sponsorship* 14(3), 178–189. DOI: 10.1108/IJSMS-14-03-2013-B003.
Markus, Z., Perovic, D., Pekovic, S. and Popovic, S. (2019) Assessing tourist revisit intention through the sports and recreational services offered. *Business Systems Research: International Journal of the Society for Advancing Innovation and Research in Economy* 10(2), 141–150. DOI: 10.2478/bsrj-2019-023.
Mcintosh, A.J. and Siggs, A. (2005) An exploration of the experiential nature of boutique accommodation. *Journal of Travel Research* 44(1), 74–81. DOI: 10.1177/0047287505276593.
Nkinda, P.D. (2016) Factors influencing tourism competitiveness of northern tourist circuit nature-based destinations in Tanzania from the tourists' perspective. *UONGOZI Journal of Management and Development Dynamics* 27(2).
Oh, H., Fiore, A.M. and Jeoung, M. (2007) Measuring experience economy concepts: tourism applications. *Journal of Travel Research* 46(2), 119–132. DOI: 10.1177/0047287507304039.
Ozturk, U.A. and Gogtas, H. (2016) Destination attributes, satisfaction, and the cruise visitor's intent to revisit and recommend. *Tourism Geographies* 18(2), 194–212. DOI: 10.1080/14616688.2015.1124446.
Parola, F., Satta, G., Penco, L. and Persico, L. (2014) Destination satisfaction and cruiser behaviour: the moderating effect of excursion package. *Research in Transportation Business & Management* 13, 53–64. DOI: 10.1016/j.rtbm.2014.11.001.
Pearce, P.L. (2013) *The Social Psychology of Tourist Behaviour*, International series in experimental social psychology, Vol. 3. Elsevier.
Petrick, J.F. (2004) The roles of quality, value, and satisfaction in predicting cruise passengers' behavioral intentions. *Journal of Travel Research* 42(4), 397–407. DOI: 10.1177/0047287504263037.
Pine, B.J., Pine, J. and Gilmore, J.H. (1999) *The Experience Economy: Work Is Theatre and Every Business a Stage*. Harvard Business Press, Boston, Massachusetts.
Ramukumba, T. (2018) Tourists' revisit intentions based on purpose of visit and preference of the destination: a case study of Tsitsikamma National Park. *African Journal of Hospitality, Tourism and Leisure* 7(1), 1–10.
Rop, W., Akunja, L. and Thiong'o, S. (2021) Leveraging on place brand architecture for competitive advantage: a mediated regression analysis. *Journal of Tourism Quarterly* 3(3), 130–138.
Ryan, C. (2002) From motivation to assessment. *The Tourist Experience* 2, 58–77.
Shahijan, M.K., Rezaei, S. and Amin, M. (2018) Qualities of effective cruise marketing strategy. *International Journal of Quality & Reliability Management* 35(10), 2304–2327. DOI: 10.1108/IJQRM-07-2017-0135.
Shen, S. (2014) Intention to revisit traditional folk events: a case study of Qinhuai Lantern Festival, China. *International Journal of Tourism Research* 16(5), 513–520. DOI: 10.1002/jtr.1949.
Som, A.P.M. and Badarneh, M.B. (2011) Tourist satisfaction and repeat visitation: toward a new comprehensive model. *International Journal of Human and Social Sciences* 6(1), 1106–1111.
Stamboulis, Y. and Skayannis, P. (2003) Innovation strategies and technology for experience-based tourism. *Tourism Management* 24(1), 35–43. DOI: 10.1016/S0261-5177(02)00047-X.
Taber, K.S. (2018) The use of Cronbach's Alpha when developing and reporting research instruments in science education. *Research in Science Education* 48(6), 1273–1296. DOI: 10.1007/s11165-016-9602-2.
Tan, K.K.H., Fu, Y.H. and Lee, S.E. (2021) A new statue to proclaim an old culture: reactions to and intentions of the statue of 'Christ the redeemer', Melaka. *Journal of Tourism and Cultural Change* 19(3), 295–314. DOI: 10.1080/14766825.2020.1718159.
Tanzania National Parks Authority (2020) Tourism activities. Available at: https://www.tanzaniaparks.go.tz/ (accessed 12 May 2022).

Thiong'o, S., Rop, W. and Akunja, L. (2022) Insecurity impacts, effects and implications for the tourism industry: a survey of selected incidences of terrorism, and safety issues during election years in Kenya. *Journal of Tourism Quarterly* 4(1–2), 21–37.

Um, S., Chon, K. and Ro, Y. (2006) Antecedents of revisit intention. *Annals of Tourism Research* 33(4), 1141–1158. DOI: 10.1016/j.annals.2006.06.003.

Urry, J. (1990) *The Tourist Gaze: Leisure and Travel in Contemporary Societies*. Sage, London.

Vassiliadis, C.A., Mombeuil, C. and Fotiadis, A.K. (2021) Identifying service product features associated with visitor satisfaction and revisit intention: a focus on sports events. *Journal of Destination Marketing & Management* 19, 100558. DOI: 10.1016/j.jdmm.2021.100558.

Weed, M. (2005) Sports tourism theory and method: concepts, issues and epistemologies. *European Sport Management Quarterly* 5(3), 229–242. DOI: 10.1080/16184740500190587.

World Bank (2015) Tanzania's tourism futures: harnessing natural assets: a World Bank environment and natural resources global practice policy note no. 96150. Available at: https://documents1.worldbank.org/curated/en/204341467992501917/pdf/96150-REVISED-PN-P150523-PUBLIC-Box393206B.pdf (accessed 26 June 2023).

20 The Potential of Coasteering as a Coastal Touristic Product: The Case of the Azores

Paulo Torres[1]*, Andrea Zita Botelho[1,2], Ana Cristina Costa[1,2], Maria Anunciação Ventura[1,2] and Francisco Silva[3]

[1]CIBIO, Centro de Investigação em Biodiversidade e Recursos Genéticos, InBIO Laboratório Associado, BIOPOLIS Program in Genomics, Biodiversity and Land Planning; UNESCO Chair – Land Within Sea: Biodiversity & Sustainability in Atlantic Islands, Ponta Delgada, Portugal; [2]Faculdade de Ciências e Tecnologia, Universidade dos Açores, Ponta Delgada, Portugal; [3]Escola Superior de Hotelaria e Turismo do Estoril, Centro de Estudos Geográficos, IGOT, Universidade de Lisboa CiTUR, Estoril, Portugal

Abstract

Nature-based tourism plays a vital role in promoting socio-economic and environmental sustainability on many islands worldwide. Its continued success requires the promotion of sustainable tourism practices. Over recent decades, there has been a sharp increase in the adventure tourism segment, associated with a growth in visitor experiences associated with some level of participant risk, as in the case of coasteering. Potential growth for this recently emerged activity can be significant in predominantly rocky and irregular coastal regions like those in the Azores archipelago. Considering the huge gap in knowledge and information available, this chapter characterizes the activity, analyses its potential as a tourist product and identifies its main features. The chapter is supported by a thorough literature review on the subject. Based on interviews with coasteering service providers in the Azores, where the coasteering product is still in its infancy, this case study demonstrates tremendous growth potential for this emerging nature-based tourism activity.

Introduction

Tourism is a sector with increasing economic relevance on a global scale, particularly important where rural, natural and insular territories have unique conditions and resources that allow nature tourism to be considered as an emerging and strategic product (e.g. IPDT, 2016; Turismo de Portugal, 2017). Within this industry, coastal nautical tourism constitutes a major source of income for European economies, especially those with extensive coastlines such as Portugal.

The attraction of activities linked to nature tourism is based on the exploration of remote and pristine locations, which have contributed to highlighting the importance of protected areas (Ceballos-Lascuráin, 1996; Hardiman and Burgin, 2010). Over recent decades, there has been a sharp increase within the adventure tourism segment, associated with a growth in experiences and sports activities among a large

*Corresponding author: biol.paulo@gmail.com

DOI: 10.1079/9781800621411.0020

segment of the population (Cunha, 2003; Taylor *et al.*, 2013).

Today, the search for adventure, thrills, risk and 'adrenaline rushes' is increasingly a key factor that motivates and attracts tourists and is also associated with recreation, sports and education (Taylor *et al.*, 2013). Millington *et al.* (2001) and Bentley *et al.* (2007) argue that adventure tourists' expectations coincide with experiences of excitement and tranquillity, tests of a personal nature and different levels of perceived risk, defining the adventurers themselves as explorers of exotic and virgin territories, looking for personal challenges.

At the same time, there has been growing research interest focusing on 'blue space' or aquatic encounters, human health and wellbeing, with prescriptions of nature-based activities to stimulate physical activity, mental health and social interactions (Juster-Horsfield and Bell, 2022). These are particularly relevant concerning what is referred to as 'blue prescriptions' (Maund *et al.*, 2019), which are water-based activities such as kayaking, canoeing, sailing, stand-up paddle-boarding, surfing, wild swimming and coasteering.

Coasteering is an emerging nautical, nature and adventure tourism activity with socio-economic relevance for both service providers and regions where it occurs. Potential growth can be significant in predominantly rocky and irregular coastal regions like the ones that can be found in the Azores archipelago.

The Azores, with its notable natural beauty, high cliffs and steep, rocky volcanic coast, provides constant challenges and moments of adrenaline surge, ensuring memorable emotional experiences. As such, this is an activity with a high potential for future development in the archipelago, given the growing number of tourists, the potential of resources associated with the sea and the trend towards valuing experiences of high emotional value.

Considering the huge gap in knowledge and information available, this chapter analyses the potential of coasteering in the Azores as a tourist product and identifies its main features, challenges and potential sustainability at all levels, emphasizing risk, environmental management, conservation and impacts.

The chapter is supported by a thorough literature review and is based on fieldwork and interviews with Azores tourism operators. Interviews were conducted on four islands: Flores, Pico, Santa Maria and São Miguel, selected within the scope of the SCAPETOUR project (Promotion of Coastal and Marine Landscapes for the Diversification of Tourist Products). Sustainability concepts (economic, environmental and social) are specifically addressed, considering the need to promote local natural resources, according to each island's specificities.

Coasteering

The word 'coasteering' is a combination of 'coast' and 'mountaineering'. This activity allows the participant to explore a rocky coast, particularly in the intertidal zone, through scrambling, climbing, walking, swimming and cliff jumping, with the possibility of snorkelling, abseiling, zip-lining and cave-exploring (Fig. 20.1; Silva and Almeida, 2011; NE and MMO, 2017). It attracts a variety of people, from those whose main motivation is to visit places practically inaccessible and of great natural beauty, to those looking for strong emotions and adrenaline rushes. Usually, it does not employ any kind of boat, although these can, potentially, support the activity.

Key features include the exploration of pristine places, with rich biological and geodiversity, cliff jumping, swimming between the waves, and exploring caves, gullies and small coves. Main equipment (Fig. 20.2) used includes a wetsuit (3–5 mm), life jacket, helmet, footwear and a whistle, with the possibility of including gloves, a diving mask, a headlamp (if caves are visited) or a waterproof camera (Silva and Almeida, 2011). Some emergency gear should also be considered, such as a first-aid kit, mobile phone or VHF radio, and a rescue tow line.

Practitioners recognize three levels of difficulty depending on sea conditions, terrain features, temperature, precipitation, wind and route length: blue line (calm sea); white line (1–2 m waves); and red line (rough sea, waves > 2 m) (Silva and Almeida, 2011).

Considering the adventurous outdoor nature of this activity, risk factors are mostly related to the lack of technical skills, physical capabilities,

Fig. 20.1. Coasteering; skills and abilities involved. (Photo: Francisco Silva)

Fig. 20.2. Coasteering equipment. (Photo: SCAPETOUR)

psychological profile, lack of communication, level of experience, inadequate or uncertified equipment, meteorology, terrain and sea conditions (Almeida and Silva, 2009, 2014).

The environmental impacts of this activity are, so far, poorly studied. Nevertheless, the main impact is, essentially, trampling, causing the crushing of invertebrates and macroalgae communities, leading to habitat degradation (Boaventura *et al.*, 2002; Tyler-Walters, 2005; Davenport and Davenport, 2006; Thomas, 2007).

A brief history

People have always explored the coast, walking, jumping and swimming, to reach more remote and isolated places along the coastal cliffs, particularly fishermen and later climbers and mountaineers (Pyatt, 1978). *Sea Cliff Climbing* (Cleare and Collomb, 1973) was probably the first work referring to the term 'coasteering'. Although this term was used 'to cover the activities of climbing on the coastal cliffs' (Pyatt, 1978), eventually the activity and definition evolved to refer to moving along and exploring the coast, using a mix of swimming, walking, jumping and climbing.

It was in the 1990s, in the UK, that this activity began to flourish as an autonomous and recreational sporting activity, specifically on the rocky coast of Pembrokeshire (Rogers, 2011). In 2000, the activity grew, with several companies adopting it, initially to other regions of the UK (Cornwall, Dorset, Anglesey and the Highlands and Islands of Scotland), and it also began to be featured in magazine ads and travel journals. Over recent decades, coasteering potential has been recognized worldwide, expanding to many international destinations such as Turkey, the Canary Islands, Ireland, Greece, Mallorca, Corsica, Portugal, the UK, Cape Town in South Africa and the Ngari Capes Marine Park in Australia, with upward trends over the years (Gonçalves, 2017; Silva and Almeida, 2018; Arkenford, 2019; Red Bull, 2019).

In 2004, this activity emerged in Portugal through the Associação Desportos de Aventura Desnível, with a first formal experience in the Sintra region, but it was only in 2009 that it began to be commercialized as a tourist product. In 2011, only nine companies offered it (Silva and Almeida, 2011; Turismo de Portugal, 2011), increasing to 15 companies currently. In Portugal, coasteering is expected to grow significantly, expanding to more tour operators and regions, justified by the factors previously mentioned and by specific natural conditions (climate, sea and coast) and the existence of an enthusiastic young group of promoters of the activity, who seek to diversify and expand this tourism sector (Silva *et al.*, 2020).

The Azores Archipelago

The Azores archipelago (36° 45' to 39° 43' N and 24° 45' to 31° 17' W; Fig. 20.3) is composed of nine volcanic islands, clustered in three groups, emerging from the Mid-Atlantic Ridge. The archipelago includes a maritime territory with an Exclusive Economic Zone (EEZ) of around 1,000,000 km^2 (Menezes, 2003).

The islands were populated by the Portuguese in the 15th century (Santos *et al.*, 1995), who established the first settlements on the coast (Leite, 2017). Tourism, despite a timid start in the 19th century (Machado, 2019), had a great expansion from the 1990s onwards and, today, coastal recreational activities such as boat trips, sailing, hiking, whale and dolphin watching, sport fishing and diving attract an increasing number of tourists (Calado *et al.*, 2011). These activities have great growth potential, particularly since the creation and promotion of the 'Azores' brand, especially focused on ecotourism and tourist products linked to the coastal zone (IPDT, 2016).

The Azores are internationally recognized as a tourist destination with a strong focus on nature. This status comes from the region's development model, since 1998, promoting a strategic focus on tourism, not neglecting the need to articulate the natural characteristics with tourist activities (Ponte, 2010). The biodiversity and geodiversity of the Azores, enhanced by the geographical setting of the islands and their volcanic origin, require careful and strategic management to sustain them (Alves and Silva, 2018). In 2014 and 2015, the

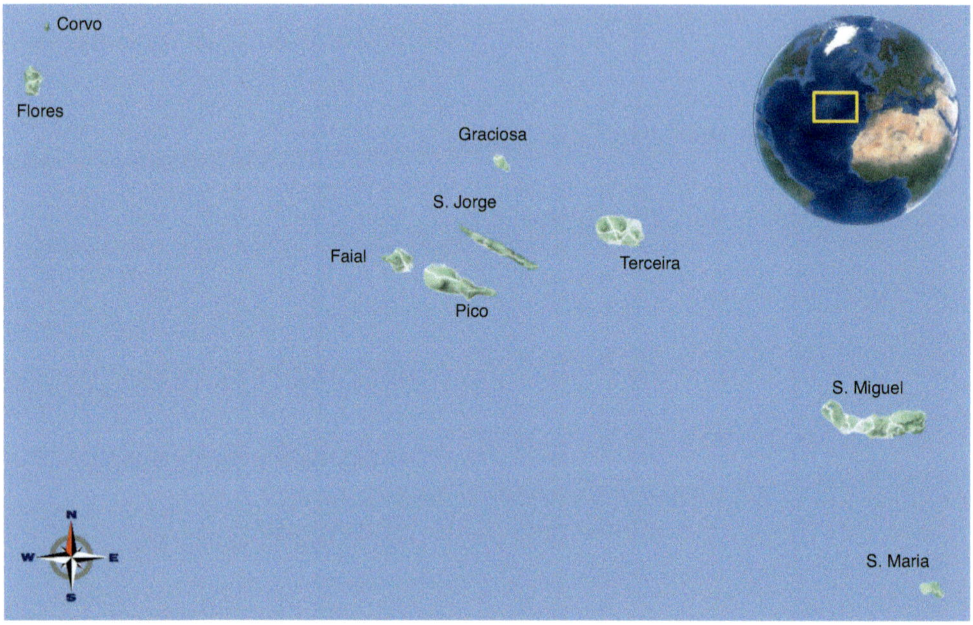

Fig. 20.3. Location of the Azores archipelago and its islands in the North Atlantic Ocean.

Azores won the Quality Coast Award Gold and, in 2021, the Quality Coast Award Platinum, an international award that distinguishes destinations for their coastal environmental quality for sustainable tourism. Also in 2021, the Azores were given an award, supported by the European Commission, for Europe's leading adventure tourism destination.

Recreational activities practised in natural areas grew quickly in the Azores from 2003 onwards. However, as with many recreational adventure activities, there are still important challenges to be addressed. There is an absence of baseline information, including essential statistics and development trends. It is still necessary to ensure visitor and environmental security through proper access and equipment management. Much can still be done to promote the tourist product and entrepreneurship, responsible access to the sport and greater community involvement.

Since the COVID-19 pandemic, people are searching for safer destinations (psychological, physical and food safety levels), pristine environments, privacy (less massified destinations), comfort, unique and memorable experiences (local culture), longer vacations, and accessible and affordable destinations. All these features indicate a continued growth in tourism in the Azores, driving the development and diversification of ecotourism products and services based on local natural resources.

Coasteering in the Azores

Biodiversity

Coasteering is mainly performed in the intertidal zone, which is inhabited by a wide variety of organisms, which should be valued as participants cross the selected trails and many times observe and interact with these species. The biological richness of the Azores is well known (Morton *et al.*, 1998) and varies according to each island's context and ecological zoning patterns among algae, invertebrates (crabs, barnacles, gastropods, sea urchins, ascidians), fish, coastal birds, and coastal plants. Figures 20.4–20.6 show some of the most common coastal species observed during coasteering. The coastal

Fig. 20.4. Common coastal algae, fish and invertebrate species of the Azores. (Photos: *Ophioblennius atlanticus* – Jorge Fontes; *Parablennius parvicornis* – Miguel Pais; *Gobius paganellus* – Gui Costa; *Pachygrapsus marmoratus* – Fun Activities; all other photos – SCAPETOUR.)

habitats include many endemic plant species, though they are threatened due to the increase of invasive species and trampling of vegetation.

Geodiversity

All islands are volcanic, having emerged from the seafloor due to the build-up of submarine volcanic eruptions 36 million years ago (Nunes, 2014). This process produced 27 volcanic systems with 16 central volcanoes, 11 volcanic ridges and about 1750 monogenetic volcanoes (Nunes and Lima, 2008). Thus, and despite its small land surface (2324 km^2), the region has a wide variety of landforms and different types of rocks and structures, such as historical eruptive centres and products, pillow lava, prismatic outcrops, hydrothermal fields, volcanic caves, pyroclastic deposits, tectonic structures (fault scarps, sag ponds) and sedimentary deposits (Nunes *et al.*, 2011; Nunes and Lima, 2013; Nunes, 2014). These diverse and abundant features that comprise the Azorean geo-landscapes have exceptional educational, scientific, sociocultural, economic and scenic values, reflecting a huge potential for tourism, especially in activities such as coasteering (Fig. 20.7).

Activity development and trail selection and evaluation

The initial development of coasteering as a tourism activity in the Azores was at São Jorge, in 2011. However, since 2006, companies offering canyoning also started to explore more coastal areas, particularly in Santa Maria, São Miguel, São Jorge and Terceira. According to Silva (unpublished data), the number of coasteering customers was very low until 2019, with an estimated value of approximately 500

Fig. 20.5. Common coastal bird species of the Azores. (Photo: Roberto Resendes)

participants in 2021, which are still low figures, but expected to grow. In 2022, 15 companies directly sold coasteering trips. Nevertheless, the offer continues to be very concentrated in São Miguel and Terceira (Table 20.1).

The island of Santa Maria is one of the choicest venues, while Flores and Pico have some potential considering the coastal geodiversity. However, it is in São Miguel and Terceira that the offer is most relevant, as these islands already concentrate most tourists (Tiago *et al.*, 2016, 2021), particularly during summer. Nevertheless, the activity is still somewhat unknown and practically all demand is from adventurous participants who want to explore the coastline in a more radical way, searching for unique and different experiences. The infancy of the activity is also reflected in the very few trails available so far and the inexistence of trail diversity in quantity or quality/difficulty.

The offer is practically absent in Pico, Flores and even Graciosa (where some coasteering expeditions are organized), though operators are trying to expand the activity and discover new and exciting trails. In Santa Maria, the offer is slightly differentiated, with a mix of canyoning and coasteering in São Lourenço and a more challenging trail in the south, for coastal and cave exploration (Fig. 20.8).

As part of the SCAPETOUR project, companies from the four target islands (Flores, Pico, Santa Maria and São Miguel) were asked to classify each trail according to several features, including access, difficulty, geodiversity and biodiversity (Fig. 20.9). As expected, difficulty does not vary greatly across the several trails, and geodiversity, along with cliffs and seascapes, is the most valuable asset. In terms of biodiversity, several groups are severely under-rated, such as algae or plants. The trail with the highest score was the one from Santa Maria (south), even accounting for tough access and a higher difficulty level. This route starts very close to the island's wind park and ends in the marina and includes two caves. The experience starts at Pedreira do Campo, an old quarry about 260m long where underwater

Fig. 20.6. Common coastal flora of the Azores. (Photo: Roberto Resendes)

volcanic rocks and fossils can be seen, followed by a thrilling combination of swimming, cliff jumping and rock climbing along the coastline. São Miguel trails have a high negative score regarding the presence of litter, which could impair activity sustainability in the long term.

It is interesting to note that some of the most highly rated trails in terms of overall ranking, as scored by the tour operators, were not the destinations with the highest scores in terms of the 'natural' or 'technical' components. In general, there is a clear over-estimation of these routes, not revealed in the set of their characteristics, either of technical or natural elements, substantiated in the opinions collected.

It should be noted that this initial assessment was carried out only with the input and viewpoints of the tour operators, so it is a limited assessment. To overcome subjectivity, an analysis of these components from the point of view of users/participants is already underway, which will allow a more 'impartial' assessment.

Information provided to participants and awareness

In nature and adventure tourism activities, it is common to have a brief conversation (briefing) between the company guides and the customers, providing information about (i) the activity (technical); (ii) standards of behaviour and safety; (iii) codes of conduct; and (iv) characteristics of the site (Oliveira, 2013). To get a sense of the importance of the information provided, each company was asked to assign a score from one to ten regarding the importance of the information they provide to their customers. It should be noted that all companies mentioned that, in addition to the information given in the briefing, they also usually provide information during

Fig. 20.7. Examples of the coastal marine geodiversity in the Azores. (Photo: SCAPETOUR)

Table 20.1. Companies and coasteering trails on each island.

Island	First year	Company	Main trails	Time	Extension
Santa Maria	2014	Bootlá	Lourenço (east) Figueiral (south)	1 h 4 h	Canyon + 600 m 2.2 km
São Miguel	2014	Azorean Blueberry Azores Adventure Islands Fun Activities	Caloura, 2 trails (south) Ribeirinha (north)	3 h 2 h	1.8 km 1 km
Terceira	2014	Rope Adventures	Contendas (south) Baía 4 Ribeiras (north)	2 h 2.5 h	420 m 600 m
Pico	2019	Natur Factor	Cachorro	2 h	500 m
São Jorge	2011	AvenTour Discover Experience	Calheta (south) Ouvidor (north)	4 h 3 h	1.8 km 1 km
Flores	2022	Ocidental Adventures	South of the airport	2 h	500 m

the activity, as well as at the end of it. The most important component of the briefing concerns technical issues and safety rules, emphasizing the interest and concern for safeguarding and wellbeing of customers (Fig. 20.10).

The information provided by tour operators regarding the itinerary to be carried out follows. It should be noted that the natural component is indicated as one of the most addressed issues, especially geodiversity.

Fig. 20.8. Santa Maria trail in the south (*left*) and cave (*right*). (Photo: SCAPETOUR)

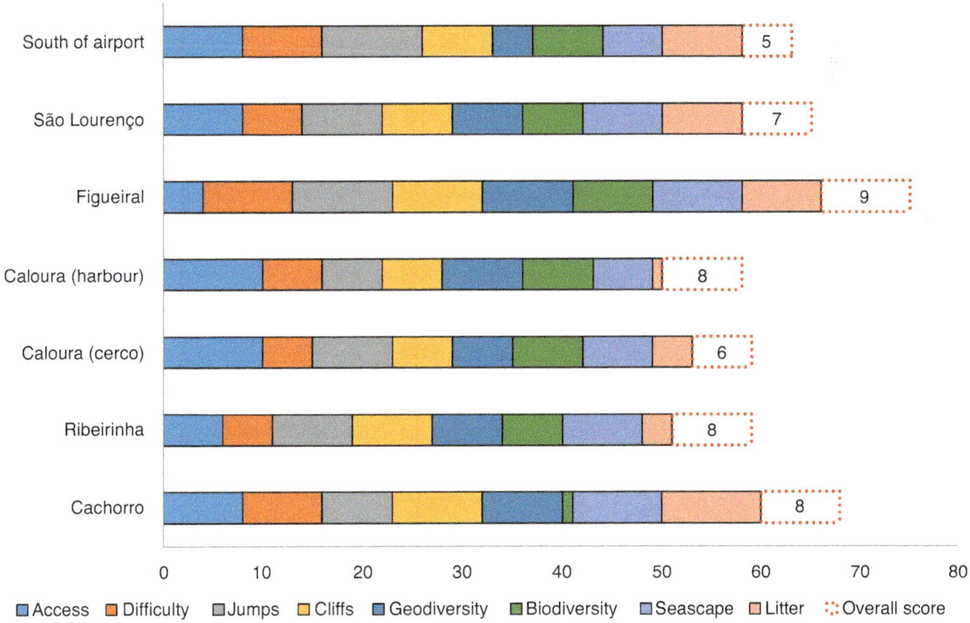

Fig. 20.9. Score (1–10) of main elements in each trail. Geodiversity includes boulders, presence of pillow lavas, caves, arches, dykes, prismatic disjunctions and giant's kettles. Biodiversity includes birds, fish, invertebrates, terrestrial flora and algae. Overall score is given by operators considering all elements.

Invasive species, protected areas and biodiversity are not aspects considered by companies to be high priority. However, operators are sensitive to the importance of protected areas and their conservation, as well as preserving biodiversity. Other topics that are not mentioned very often, such as history and culture or protected species, despite being considered important, are not explored due to the lack of basic knowledge.

Sustainability and challenges

The impacts on the routes are limited by technical and operational issues. As for changes in the

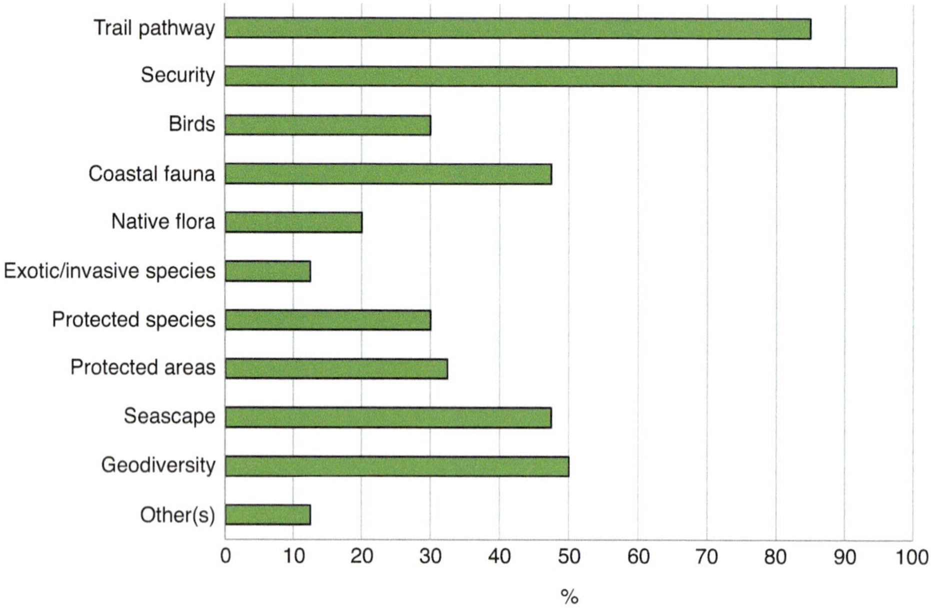

Fig. 20.10. Degree of importance attributed by coasteering operators in the Azores to information provided before, during or after coasteering.

environment, although small, as any wave or tide 'erases' any traces of the passage of visitors, to date, this issue lacks in-depth study and impact assessment. There are no elements introduced and the quality of the water is not significantly affected. Still, there are always some impacts, most of which can be avoided or mitigated with responsibility and good tour operator and environmental education practices. Practitioners and companies that commercialize this product are aware that the environment is an indispensable asset for the quality of the product, so they are also focused on its conservation. According to our field prospection and interviews, the main impact of coasteering is trampling of vegetation, and in some cases nest disturbance of coastal birds. This issue should be addressed and carefully monitored, and the extent and magnitude of these impacts evaluated, especially considering the expected increase in the number of participants and companies offering this activity.

One of the concerns pointed out by the operators is the presence of litter, especially in São Miguel. It is necessary to raise awareness among residents about this issue as the maintenance of environmental quality of coasteering destinations is essential for the activity to take place safely, to provide a good experience and to continue to be a source of income to the region. Of the islands with the greatest potential for coasteering, two have lower levels of economic development and attract fewer tourists (São Jorge and Flores) so further development of coasteering on these islands could have good economic potential.

High-quality tour guides are also essential for the development of coasteering tourism, connecting with their clients while ensuring high safety standards (NCC, 2015; TIA, 2016; Gonçalves, 2017). In Portugal, despite having already acceptable standards for guides and tour companies, it is still necessary to continue to conduct more training and higher-level and more thorough certifications, establishing demanding standards for the whole the industry, as in the UK and New Zealand (Tiago *et al.*, 2016).

Conclusions

Coasteering in the Azores is on the rise, thanks to the promotion of the region as an excellent destination for its practice, and because there is a growing number of tour companies offering coasteering tours. The increase in demand and supply has been sustainable and gradual, and companies have been focusing on ensuring quality service and a product capable of providing unforgettable experiences.

Although operators seek to provide a unique, adrenalin-filled experience, there is also concern regarding the presentation of detailed information about the environment. Pre-activity briefings by guides focus on technical issues so that the activity can be carried out safely, but along the way, some of the characteristics of the places visited are mentioned. This additional information is considered of special importance for promoting environmental awareness, as well as for enriching knowledge about the natural component. It was noted, during interviews of tour operators, that the respondents began to have the perception that, in the briefings and information provided along the routes, some interesting and dissemination information about the visited places was not adequately addressed (e.g. protected areas and biodiversity).

The need to obtain financial and human resources for trail maintenance was mentioned several times, since most of this is carried out by the companies. Another issue mentioned relates to the need to improve the training and emergency response capacity of rescue teams in places of difficult access by fire brigades, civil protection authorities and the guides themselves.

The focus on training of guides and certification of the activity has promoted the quality of the service provided. Companies are starting to increasingly feel the need to build their teams due to the increase in demand for this type of activity.

Nature tourism activities are usually related to sustainability standards that attest to an awareness of environmental preservation and conservation issues. Management authorities have a leading role in the application of necessary management measures to guarantee favourable environmental outcomes.

Thus, given the still embryonic offer of coasteering and its challenges, it is essential to develop an interactive and inclusive approach to bringing stakeholders, scientists and managers together to develop management solutions for coasteering tourism that promote economic, social and environmental sustainability.

Acknowledgements

This work was supported by FEDER (85%) and regional funds (15%) through Programa Operacional Açores 2020, under SCAPETOUR project (SeaSCAPEs promotion to diversify TOURistic products – Ref. ACORES 01 0145 FEDER 000083). This work is funded by National Funds through FCT – Foundation for Science and Technology under the project UIDB/50027/2020. The authors also would like to thank all the companies that participated in the project for their availability and support.

References

Almeida, M.C. and Silva, F. (2009) Gestão do risco em animação turística. In: Instituto Superior Técnico, (ed.) *Riscos Industriais e Emergentes. Edições Salamandra, Lda Lisboa*. pp. 309–322.

Almeida, M.C. and Silva, F. (2014) Segurança e gestão do risco em desportos na natureza e turismo ativo. In: Carvalhinho, L. (ed.) *Desporto de Natureza e Turismo Ativo. Contextos e Desenvolvimento.* Rio Maior. Escola Superior de Desporto de Rio Maior/Instituto Politécnico de Santarém.

Alves, P.H. and Silva, F. (2018) Geoturismo e desporto aventura nos açores: aplicação ao caso do canyoning. *Vulcânica- Revista Portuguesa de Vulcanologia* 2, 483–484. Available at: http://hdl.handle.net/10400.9/3100 (accessed 26 June 2023).

Arkenford (2019) *Watersports Participation Survey 2018*. Summary Report. Arkenford, Guildford, UK. Available at: chrome-extension://oemmndcbldboiebfnladdacbdfmadadm/https://www.nationalwa

tersafety.org.uk/media/1150/watersports_participation_survey_2017_summary.pdf (accessed 25 October 2023).

Bentley, T.A., Page, S.J. and Macky, K.A. (2007) Adventure tourism and adventure sports injury: the New Zealand experience. *Applied Ergonomics* 38(6), 791–796. DOI: 10.1016/j.apergo.2006.10.007.

Boaventura, D., Ré, P., Cancela da Fonseca, L. and Hawkins, S.J. (2002) Intertidal rocky shore communities of the continental portuguese coast: analysis of distribution patterns. *Marine Ecology* 23(1), 69–90. DOI: 10.1046/j.1439-0485.2002.02758.x.

Calado, H., Borges, P., Phillips, M., Ng, K. and Alves, F. (2011) The Azores archipelago, Portugal: improved understanding of small island coastal hazards and mitigation measures. *Natural Hazards* 58(1), 427–444. DOI: 10.1007/s11069-010-9676-5.

Ceballos-Lascuráin, H. (1996) *Tourism, Ecotourism, and Protected Areas: The State of Nature-based Tourism around the World and Guidelines for its Development*. International Union for the Conservation of Nature. DOI: 10.2305/IUCN.CH.1996.7.en.

Cleare, J. and Collomb, R. (1973) *Sea Cliff Climbing in Britain*. Constable, London.

Cunha, L. (2003) *Perspectivas e tendências do turismo*. Edições Universitárias Lusófonas, Lisbon.

Davenport, J. and Davenport, J.L. (2006) The impact of tourism and personal leisure transport on coastal environments: a review. *Estuarine, Coastal and Shelf Science* 67(1–2), 280–292. DOI: 10.1016/j.ecss.2005.11.026.

Gonçalves, B.P. (2017) Inovação no Turismo de Experiências e de Aventura: coasteering na empresa Vertical Descents. MSc Thesis, Escola Superior de Hotelaria e Turismo do Estoril, Estoril, Portugal.

Hardiman, N. and Burgin, S. (2010) Canyoners' perceptions, their evaluation of visit impacts and acceptable policies for canyon management in the blue mountains (Australia). *Managing Leisure* 15, 264–278. DOI: 10.1080/13606719.2010.508667.

IPDT (Instituto de Planeamento e Desenvolvimento do Turismo) (2016) *Plano Estratégico e de Marketing para o Turismo dos Açores (PEMTA)*. Instituto do Turismo.

Juster-Horsfield, H.H. and Bell, S.L. (2022) Supporting 'blue care' through outdoor water-based activities: practitioner perspectives. *Qualitative Research in Sport, Exercise and Health* 14(1), 137–150. DOI: 10.1080/2159676X.2021.1879921.

Leite, A.R. (2017) Os caminhos da construção do território dos Açores nos séculos XV e XVI. *Genius Loci: Lugares e Significados* 2, 599–610.

Machado, M.V.D.R. (2019) A literatura de viagens e os olhares femininos sobre os Açores setecentistas e oitocentistas. In: *Memória e Identidade Insular: Religiosidade, Festividades e Turismo Nos Arquipélagos Da Madeira e Açores*. pp. 277–290.

Maund, P.R., Irvine, K.N., Reeves, J., Strong, E., Cromie, R. *et al.* (2019) Wetlands for wellbeing: piloting a nature-based health intervention for the management of anxiety and depression. *International Journal of Environmental Research and Public Health* 16(22), 4413–4430. DOI: 10.3390/ijerph16224413.

Menezes, G.M. (2003) Demersal fish assemblages in the Atlantic Archipelagos of the Azores, Madeira, and Cape Verde. PhD thesis, University of the Azores, Portugal.

Millington, K., Locke, T. and Locke, A. (2001) Occasional studies: adventure travel. *Travel and Tourism Analyst* 4, 65–97.

Morton, B., Britton, J.C. and Martins, A.M.F. (1998) *Coastal Ecology of the Azores*. Azores, Portugal: Sociedade Afonso Chaves, Ponta Delgada, São Miguel.

NCC (2015) *Safety Advice for Coasteering Providers*. National Coasteering Charter. Available at: chrome-extension://oemmndcbldboiebfnladdacbdfmadadm/https://www.celticquestcoasteering.com/Safety-Advice-for-Coasteering-Providers-V3-Nov-2015.pdf (accessed 25 October 2023).

NE and MMO (2017) *Marine Recreation Evidence Briefing: Coasteering*. Natural England and Marine Management Organisation. UK Government.

Nunes, J.C. (2014) The Azores Archipelago: islands of geodiversity. In: Erfurt-Cooper, P. (ed.) *Volcanic Tourist Destinations*. Springer, Berlin, Heidelberg, pp. 57–67. DOI: 10.1007/978-3-642-16191-9.

Nunes, J.C. and Lima, E.A. (2008) Paisagens Vulcânicas dos Açores: Valor Intrínseco Enquanto Recurso Natural e Património Geológico. Livro de Resumos – IV Congresso Nacional de Geomorfologia. Braga, Portugal.

Nunes, J.C., Lima, E.A., Ponte, D., Costa, M.P. and Castro, R. (2011) *Azores Geopark application*. Available at: www.azoresgeopark.com. Available at: (accessed 25 October 2023).

Nunes, J.C. and Lima, E.A. (2013) Ranking and classifying the Azores Islands geosites. Rendiconti Online della Società Geologica Italiana. In: *Special Issue for the 12th European Geoparks Conference*, Ascea, Italy.

Oliveira, C.P. (2013) Caracterização do mercado de actividades de Turismo de Natureza em Portugal. MSc thesis, Universidade Nova de Lisboa, Faculdade de Ciência e Tecnologia.

Ponte, J.B. (2010) Caracterização do grenn-business no sector do turismo na Região Autónoma dos Açores. MSc thesis, University of the Azores, Portugal.

Pyatt, E. (1978) Some pioneers of coasteering. *Alpine Journal* 83(327), 191–198.

Red Bull (2019) The 12 most breathtaking coasteering spots in the world. Available at: https://www.redbull.com/mea-en/worlds-best-coasteering-spots-locations (accessed 25 October 2023).

Rogers, A. (2011) An investigation into perceptions of coasteering in the Pembrokeshire Coast National Park amongst participants in other forms of recreation. MSc thesis, Sheffield Hallam University.

Santos, R.S., Hawkins, S.J., Monteiro, L.R., Alves, M. and Isidro, E.J. (1995) Case studies and reviews: marine research, resources and conservation in the Azores. *Aquatic Conservation: Marine and Freshwater Ecosystems* 5(4), 311–354. DOI: 10.1002/aqc.3270050406.

Silva, F. and Almeida, M.C.T. (2011) Caracterização e potencialidades do coasteering em Portugal. *Journal of Tourism and Sustainability* 1, 27–35.

Silva, F. and Almeida, M.C. (2018) Empreendedorismo e inovação na animação turística: o caso do canyoning e do coasteering nos Açores. I Congresso internacional de 'Turismo activo, desporto, natureza e sustentabilidade – CITADNS. Setúbal, Portugal.

Silva, F., Almeida, M. and Gonçalves, B. (2020) Products in maritime and costal tourism: the case of coasteering. In: Brito, M., Dias, A. and Patuleia, M. (eds) *Managing, Marketing, and Maintaining Maritime and Coastal Tourism*. IGI Global, Hershey, Pennsylvania, pp. 187–203. DOI: 10.4018/978-1-7998-1522-8.

Taylor, S., Varley, P. and Johnston, T. (2013) *Adventure Tourism: Meaning, Experience and Learning*. Fort William, UK. DOI: 10.4324/9780203120095.

Thomas, D. (2007) The impact of coasteering on Fucus vesiculosis at St Nons Bay. BSc thesis, Pembrokeshire College, Haverfordwest, UK.

TIA (2016) *Activity Safety Guideline: Coasteering*. Tourism Industry Aotearoa, New Zealand.

Tiago, F., Gil, A., Stemberger, S. and Borges-Tiago, T. (2021) Digital sustainability communication in tourism. *Journal of Innovation & Knowledge* 6(1), 27–34. DOI: 10.1016/j.jik.2019.12.002.

Tiago, T., Dias Faria, S., Luís Cogumbreiro, J., Pedro Couto, J. and Tiago, F. (2016) Different shades of green on small islands. *Island Studies Journal* 11(2), 601–618. DOI: 10.24043/isj.369.

Turismo de Portugal (2011) *Turismo de Portugal. Registo Nacional de Agentes de Animação Turística (RNAAT).*

Turismo de Portugal (2017) *Estratégia Turismo 2027*. Liderar o turismo do futuro, Turismo de Portugal I.P.

Tyler-Walters, H. (2005) *Assessment of the Potential Impacts of Coasteering on Rocky Intertidal Habitats in Wales*. Countryside Council for Wales, Plymouth, UK.

21 Wellbeing Tourism in Costarainera, Italy: The Giacomo Filippo Novaro Park

Adriana Ghersi*
Department of Architecture and Design, University of Genoa, Italy

Abstract
This chapter illustrates a case study in the Ligurian Western Riviera, which has a historical vocation to be a therapeutic landscape-based tourism set, starting from the beginning of the 20th century. It describes the Costarainera Sanatorium (built to cure respiratory sickness) and its park, which has been disused in the last decades. Then it illustrates the recently realized landscape project for the enhancement of a pilot area, giving birth to the Novaro Wellbeing Park, with its 'therapeutic rooms' in nature (based on different theories about natural therapies). It focuses on the role of this opening as a strategic phase of a further project for the valorization of the whole complex, inserted in the tourist circuit of the San Lorenzo Valley, and along the coastal bicycle pathway Sanremo-San Lorenzo al Mare. Finally, it shows the value of the case study, the future outlook of the intervention and its managerial challenges.

Landscape and Wellbeing in the Ligurian Western Riviera

The invention of the Riviera (Blume, 1922; Astengo *et al.*, 1982), as a therapeutic landscape of beauty and wellbeing for elite winter tourism, dates from the 19th century. A great influence was the book by Giovanni Ruffini, *Doctor Antonio*, written in English in 1855, a bestseller that promotes the image of the Ligurian Western Riviera and of the landscape of Bordighera, in particular. Starting from the climatic character of the hotter Riviera, this landscape is proposed as an ideal place for the treatment of various respiratory diseases, for the European elites, during the mild winter (Zanini, 2012).

The success of the Riviera dei Fiori as a luxury tourism destination determined the development and transformation of the coastal landscape between the end of the 19th century and the 1930s. A complete transformation of the image of the Ligurian coast was accomplished. The seaside walks are now identified by the presence of palm trees (like the ones for the promenade in Sanremo, gifted by the Russian Empress, wife of Tsar Alexander II, in 1874/1875) or other particular species (such as the row of *Araucaria* along the seaside in Bordighera). To suggest a relationship with exotic countries, some species have been introduced from distant warm countries, also through forms of acclimatization (like in the famous Hanbury Gardens, in La Mortola, near Bordighera) (Mazzino, 1997). Many visitors came to stay in the Riviera, to treat respiratory diseases and also to enjoy the beauty of the places and the sweetness of the climate during the winter: first many foreign nobles (in particular English, Russian and German) and Italians, then international wealthy *bourgeoisie*.

*adriana.ghersi@unige.it

DOI: 10.1079/9781800621411.0021

With the building of large hotels, for some international communities who chose the Riviera for long winter rests, after the construction of the Genoa–Nice railway, which reached Sanremo in 1872, the tourist appeal extended to the international middle-upper classes.

At the beginning of the 20th century, different sanatoriums and heliotherapy colonies for children were built in the Riviera, linked to a wider Italian network of coastal healing structures, to treat chronic health problems affecting a large portion of the European population (Labò and Podestà, 1941; Frabboni, 1971; Pitzurra, 1989; Sabbatani, 2005).

After the World War II, the railway became the most important means to 'take to the sea' for thousands of Italians: the beginnings of mass tourism in a rapidly growing country.

This last change – accompanied by a conflicting development of residences, hotels and second homes in the villages, with the opening to mass tourism – was signalled by the transition from winter tourism to summer seaside tourism (Cigolini and Croce, 1997; Maniglio Calcagno, 2009).

Costarainera Sanatorium and Its Therapeutic Park

Plants and a direct contact with nature have positive psychic and physical effects that can stimulate our bodies and calm some symptoms (physical and mental weakness), helping to find a natural equilibrium (Kaplan, 1995; Hartig *et al.*, 1991; Berger and McLeod, 2006).

The Novaro-Barellai ensemble (Costarainera, Imperia), built in 1932–1934 by Muratorio for the treatment of tuberculosis (a page in the history of the fight against tuberculosis during the Fascist era, and a struggle between ideological and social questions), and mainly abandoned in recent decades, is today a green oasis (45,000 m^2) filled with vegetation within the surrounding rural terraces (Fig. 21.1; Pirero, 2019).

The complex was created to host a large community of tuberculosis patients, with outdoor areas for healing sessions in the sun, and with a farm in the surrounding rural areas to ensure food sufficiency. The continuity between

Fig. 21.1. Novaro-Barellai heliotherapy complex in Costarainera: the park and the agricultural surrounding areas, with glasshouses and terraced cultivation.

the sanatorium, the formal park and the rural area is fundamental.

Two different buildings hosted different patients: the Barellai building, facing the sea, located under the Aurelia Road, was to cure bone tuberculosis by sunbathing (prolonged exposure to sunlight and salubrious sea air); the Novaro is located on a higher terrace, to cure consumption and pulmonary tuberculosis, by air-bathing (extended exposure to the salty breeze coming from the sea, enriched by the balsamic essences of the park). The park is structured suiting composition axes in connection with the landscape: an axis connects the Novaro sanatorium with the sea, and a second axis links the formal garden with the cultivated landscape (solar thermal axis), towards a scenic point of view with pines, along the road climbing to the hamlet of Costarainera. The park features a series of spaces and terraces for heliotherapy and thalassotherapy, and extends through a system of stairs and terraces. Three monumental stairs characterize the lower part of the park, called Boschetto (one of them directly connects the Barellai building). This part is separated from the main formal garden by the internal driveway (with rows of pines and *Quercus ilex*) leading to the Novaro building.

The park contains both Mediterranean and exotic plants in groups and rows, with olive groves also, to underline the link between the sanatorium and the park (Fig. 21.2).

The Project for the Enhancement of the Park: The Novaro Wellbeing Park and the 'Therapeutic Rooms' in Nature

A park enhancement has been started, thanks to an Alcotra-Jardival European cross-border project (involving local administration,

Fig. 21.2. The park of the Novaro-Barallai complex and its connection with the surrounding landscape, with olive groves and territorial axes.

stakeholders and the Department of Architecture and Design at the University of Genoa, under the author's scientific responsibility), in 2017, laying out different strategies to highlight its therapeutic role and potentialities. The main aim was to restore the therapeutic role of the park, helping to read the story of the heliotherapic complex, proposing new activities for psychophysical rehabilitation and wellbeing, especially by the use of plants, respecting the original structure and linking it to its landscape and social context (Farina and Ghersi, 2007).

Thanks to a wider process of revaluation, a first experimental 'starting tassel' has been opened to public fruition, near the entrance, offering a variety of new compatible uses in a green area (Fig. 21.3). The renovation of the Boschetto ($5500\,m^2$) has given rise to a new treatment area, filled with sensory inspirations, showing different open-air natural therapies, to underline the healing history of the whole complex (Nieri and Mencagli, 2017; Ghersi, 2020). Flowerbeds with 'healing' plants have been created, with colours, perfumes and flavours that can have positive influences on the physical and psychological health of visitors (Ulrich *et al.*, 1991). The actions respected the structure of the garden, both in the architectural artefacts (monumental stairs, stone-paved paths), and in the plant structure, adding richness and colours to the park.

There is evidence of the benefits of the contact with living plants for all kinds of people, whether they are suffering from ailments or diseases, or whether they simply need to recover from the wear and tear of life in urban metropolises and the distance from a natural environment (Heath, 2004; Borghi, 2011; Murroni *et al.*, 2021).

The path of medical plants proposes a series of plants with healing principles and leads to the the centre of the 'Breathing Garden' where it is possible to see the main axis of the formal garden at the other side of the road. Some intervention on the vegetation has re-opened the view to the sea, offering 'sun-/air-bathing'. The breath releases our vital energy, trapped in defences, in resistances, in the need for control, in modes of behaviour, in our rigid schemes, in the armour of tensions transformed in our second skin. Some plants deal with good breathing (rosemary, mint, thyme, together with existing pines and eucalyptus). From this area an almost flat path reaches a second, higher entrance to this public area, which makes it possible for everyone to use the outdoor spaces, including disabled and sight-impaired people (with guide). Along this path there is a sequence of spaces dedicated to specific treatments (Fig. 21.4). These different nature-based therapies are narrated in some panels, in three languages and Braille texts (Fig. 21.5), with QR codes and links to an audio guide, located near the specific curative plants. To understand these approaches, we need a holistic view, like in Einstein's theory, where the human body is not the assemblage of chemical parts, but a total, complete system, in harmony with the electromagnetic/energy system of the universe, in contrast to a Newtonian mechanistic approach. If our energy levels are blocked or depleted then our body cannot function properly, and this can lead to a variety of problems. Nature has a deep effect on us; the practice of green and biophilic design brings people closer to the healing power of nature's sights (Wilson, 1984).

Alongside a sensory path and a children's playground, there is the Garden for Aromatherapy, with collections of plants from which we can get essential oils, to defend, communicate and heal us, such as lavender, rosemary, honeysuckle and other fragrant plants. These and other fragrant species compose the 'Garden to Smell', and the 'Bach Flower Garden', showing some plants of the famous curative theory through flower extracts (Fig. 21.6). The famous homeopathic doctor Edward Bach (1886–1936) believed that the power of flowers could restore the inner harmony of our spirit, since disease is, in essence, 'the result of a conflict between Soul and Mind, and will never be eradicated except by spiritual and mental effort' (Bach and Wheeler, 1997).

The Chromotherapy Garden refers to the possibility to stimulate our inner reactions by different colours, recreating our psychophysical balance (Ghersi, 2018). There is an influence of each colour on different parts of the human body, different glands, different chakra (e.g. the orange colour stimulates the metabolism and lungs, cures spasms and cramps, restores the balance of nervous system to overcome anxieties and fears). Natural therapists use

Fig. 21.3. The steep part of the park that the Municipality of Costarainera has opened to the public, as 'Wellbeing Park' (1, five-colours-diet; 2, chromotherapy garden; 3, edible flowers; 4, garden to smell; 5, Bach's flowers; 6, aromatherapy; 7, good breath; 8, medical plants path; 9, art/vegetable garden therapy).

chromotherapy to recover stroke victims and patients affected by chronic depression.

In gardens and parks, the living colours of flowers and foliage can impress our visual memory and interact with different parts of our body, helping to assess our inner balance and to mend physical decline (Clark, 2004).

In the design of some gardens, landscape architects use colours (articulating nuances and tones of different colours) to interact with the theory of chromotherapy and other alternative medicine and esoteric studies, from alchemy to numerology (Winterbottom and Wagenfeld, 2015). These concepts are utilized in the composition of healing gardens but also of public parks, as in some important examples. The connection among our 5 + 1 senses, six colours, six states of water, six planets and six metals is shown in the series of six gardens of the well-known Citroen Park, by Gilles Clément and Alain Provost/Jean Paul Viguier team, in Paris, opened in 1992. The theory of colours was at the base of the famous temporary realization of the Optical Garden by Bernard Lassus, in Niort, in 1993, by a sequence of mirrors and colours, of the plants and of the wooden painted slats (Lassus, 2004). In the Leichtag Family Healing Garden, by Topher Delaney, at the San Diego Children's Hospital, vivid colours of plants, walls and objects are protagonists (Cooper Marcus and Barnes, 1999).

In the Novaro Park, again with the help of colours, some principles of food education come from the Garden of the Five-colour Diet, an easy way to make children get the habit to eat fruits and vegetables daily. The Edible Flowers Garden, according to the lesson by Libereso Gugliemi (Calvino's famous gardener), suggests

Fig. 21.4. A view of the original stone path, along the playground with the different therapeutic 'green rooms'.

the tasting of flowers, that can enrich our feeding, both in the fresh (spiced lavender and calendula, acid begonia, cucumber flavoured borage, minty violet) and cooked versions, like in traditional Ligurial cuisine: filled zucchini flowers (Marchisio and Porchia, 2009). A little educational orchard with local varieties has also been added.

Around the cement basement of a demolished glasshouse, the Art/Horticultural Laboratory takes place, becoming an open-air classroom for children (Chermaz, 2003). On that pavement, some little benches have been built, recently covered by an attractive mosaic by the artist Judith Török (realized with the help of a volunteer group of apprentices), narrating all the therapeutic themes expressed in the park. The playful aspect is emphasized by a series of slides, adhering to the slope, which allow children to descend quickly, returning to the start of the monumental staircase.

The park, equipped with a series of fitness devices, located in shadowed areas, is also a free outdoor gym, for the visitors' wellness.

An Opportunity for Residents' and Tourists' Wellbeing

The park opened to the public in 2018, after just 18 months of work, and represented a significant moment for the stakeholders who contributed to its making and management. The park is open to everyone, families and sportspeople, a bench offers free recharging for electric bikes and opening is guaranteed every day thanks to the local Association Gli Amici dei Piani (which is also entrusted with the management of the small building of the former pharmacy, located at the entrance to the park). The ancient vocation of this health

Fig. 21.5. Panel about the Chromotherapy Garden.

resort is restored, allowing everyone to rest and breath in scenic areas, in a green balsamic setting.

Many people have chosen this place for diverse outdoor activities, from tai chi to yoga, and local stakeholders have been animating events and providing guided tours to schools, building up an affective relationship with this special park, its surrounding landscape and all the people working as a team for its valorization. Meeting the physical, psychological, emotional and social needs of a wide range of user groups, the park has become a pleasant point for socialization for the local inhabitants, but also an attractive place to visit.

Linked to the cycle-pedestrian path along the disused railway Area 24, the Novaro Wellbeing Park is linked with the tourism circuit in the San Lorenzo Valley. Thanks to its location, it becomes a real connecting node from the seaside promenade towards the internal valley, promoting cultural and natural interesting elements.

The cycle-pedestrian path of the Riviera dei Fiori Coastal Park is located on the 24 km old railway line that served western Liguria from 1872 to 2001. Since 2004, the Area 24 company has revived this enchanting path, which fascinated artists and intellectuals from all over Europe, through the management and co-ordination of the services necessary for the promotion, enhancement and development of the environmental and tourist potential of the former railway site.

The track is connected with other coastal cycle routes and with more challenging routes also, with a wide choice of excursions in the splendid Ligurian hinterland, along the many off-road itineraries for mountain bikes. The cycle-pedestrian path, starting from Ospedaletti, crosses Sanremo, Arma di Taggia, Riva Ligure and Santo Stefano and arrives at Costarainera and San Lorenzo al Mare. It maximizes the potential of this coastal path and returns to the community a space occupied by the railway,

Fig. 21.6. *Salvia* with smell of pineapple and *Lavandula*, in the Garden to Smell.

proposing an experience on a human scale in the heart of the Riviera, designed and built to provide comfort, wellbeing and fun. Leaving the tunnel that passes under the sanatorium, you will find the signs to go up to the Novaro Park, crossing the Aurelia Road.

In the Novaro Wellbeing Park many elements are tied to typical production and agriculture, such as the aromatic plants (of which a specialized nursery cultivates a large collection in Pietrabruna, Fig. 21.6). Other references are the lavender (once a largely cultivated plant, which is now reproposed in some areas like the upper part of the versants near Pietrabruna) and the edible flowers (commercialized by some producers in the Albenga cultivated plain, and more and more proposed by local chefs to enrich and characterize the dishes of traditional Ligurian cuisine). The ancient species of fruit trees (the brindle apricot from Valleggia, the jujube and other forgotten varieties) in the little orchard, have also a didactic aim.

The artistic mosaic (Fig. 21.7) is a direct reference to the two private visitable parks about sculpture that characterize the San Lorenzo Valley: one is Villa Biener in Costarainera, and the other is the Sculpture of the Worlds Park, in Lingueglietta. The park of monumental sculptures and contemporary art, Among the Worlds, below the historic town of Lingueglietta, presents a series of bronze sculptures by the German artist Carin Grudda, arranged starting from 2001, among the olive groves and the ancient terraces of the valley. Under the village of Cipressa there is instead the Villa Biener, a park of contemporary art and the Museum of Migrant Cultures. Here the Hungarian-born artist Judith Török, together with her partner and artist Carlo Maglitto, were able to shape, expand and enrich a series of *en plein air* installations, works created by young emerging artists, splendid ceramics and gaudy polychrome and multi-material mosaics. With a spirit of welcome and sharing, Judith Török wanted to participate in the enhancement of the

Fig. 21.7. A detail of the mosaic by the artist Judith Török, realized with the help of a volunteer group of apprentices, narrating all the therapeutic themes expressed in the park.

Novaro Park, building the mosaic dedicated to the themes of therapy in nature (Fig. 21.7).

The hope is that the positive energy (Nieri, 2009; Strimmer, 2019) and the emotional relationship with the park established during the works and the visits can drive the restoration of the entire compound in the near future.

Conclusion

The intervention promoted by the Mayor of Costarainera and the qualifying project, created for the reactivation of the pilot area, have given way to the reopening of the complex to the public, starting from the public Novaro Wellbeing Park and its connections with the coastal cycle path. This intervention has enriched the tourist system of the San Lorenzo Valley with a new excellence, attracting visitors and offering meeting places and activities for the inhabitants. The Wellbeing Park remains an important and rare example of 'staging' and explanation of the different therapies based on nature, interesting for many visitors. In the park, the inhabitants can find a place to reconnect with nature, with the history of this area and the sanatorium complex, which has held great importance for the community in the past. They can take advantage of the outdoor fitness equipment, use the spaces to practise yoga and group sports, as well as meet and enjoy the play areas with children in the cool shade of the park.

The limits of the intervention concern the management of the public park and the possible interactions between the public administrations and the possible operators interested in the purchase of the Novaro building and in the reconstruction of its curative therapeutic role.

The recent sale of the Barellai building, located downstream of the Via Aurelia, could also be an opportunity to recover the meaning of the entire therapeutic complex, for example with the arrangement of the garden overlooking the sea, allowing the possibility to appreciate its view from the external public path.

Like any public park, the maintenance and management of activities that can enliven the space are very important and must always be confirmed among the objectives of the municipal administration to prevent a further degradation phase of the recovered part.

References

Astengo, D., Duretto, E. and Quaini, M. (1982) *La scoperta della Riviera*. Sagep, Genova, Italy.

Bach, E. and Wheeler, F.J. (1997) *The Bach Floral Remedies*. McGraw-Hill, New York.

Berger, R. and McLeod, J. (2006) Incorporating nature into therapy: a framework for practice. *Journal of Systemic Therapies* 5(2), 80–94. DOI: 10.1521/jsyt.2006.25.2.80.

Blume, M. (1922) *Côte d'Azur: Inventing the French Riviera*. Thames and Hudson, London.

Borghi, C. (2011) *Un giardino per stare bene*. Apogeo, Milan, Italy.

Chermaz, A. (2003) *Terapia orticolturale*. Civico Orto Botanico di Trieste, Trieste, Italy.

Cigolini, M.C. and Croce, M.R. (1997) *Il turismo sulla costa ligure. Urbanistica e architettura dalla metà '800 a oggi*. Erga, Genova, Italy.

Clark, L. (2004) *Cromoterapia. Gli effetti curativi dei colori e della luce*. Red, Como, Italy.

Cooper Marcus, C. and Barnes, M. (1999) *Healing Gardens: Therapeutic Benefits and Design Recommendations*. John Wiley & Sons, New York.

Farina, A. and Ghersi, A. (eds) (2007) Therapeutic landscapes (special issue). *Journal of Mediterranean Ecology* 8.

Frabboni, F. (1971) *Tempo libero infantile e colonie di vacanza*. La Nuova Italia, Firenze, Italy.

Ghersi, A. (2018) Chromotherapy in the wellness park G.F. Novaro in Costarainera (IM). In: Pellegri, G. (ed.) *De_Sign: Environment Landscape City 2018*. GUP, Genova, Italy, pp. 370–378.

Ghersi, A. (2020) Di sole e d'Azzurro, Parco del Benessere (Costarainera, IM). *Acer* 2(2020), 57–62.

Hartig, T., Mang, M. and Evans, G. (1991) Restorative effects of natural environment experiences. *Environment and Behavior* 23(1), 3–26. DOI: 10.1177/0013916591231001.

Heath, Y. (2004) Evaluating the effect of therapeutic gardens. *American Journal of Alzheimer's Disease and Other Dementias* 19(4), 239–242. DOI: 10.1177/153331750401900410.

Kaplan, S. (1995) The restorative benefits of nature: toward an integrative framework. *Journal of Environmental Psychology* 15(3), 169–182. DOI: 10.1016/0272-4944(95)90001-2.

Labò, M. and Podestà, A. (1941) Colonie marine, montante, elioterapiche. *Casabella-Costruzioni* 167.

Lassus, B. (2004) *Couleur, Lumière, Paysage: Instants d'une pédagogie*. Monum, Ed. du Patrimoine, Paris.

Maniglio Calcagno, A. (2009) *Paesaggio costiero, sviluppo turistico sostenibile*. Gangemi, Rome.

Marchisio, O. and Porchia, C. (2009) *Mangiare il giardino.La lezione di Libereso Guglielmi*. Socialmente, Bologna, Italy.

Mazzino, F. (1997) *An Earthly Paradise: The Hanbury Gardens at La Mortola*. Sagep, Genova, Italy.

Murroni, V., Cavalli, R., Basso, A., Borella, E., Meneghetti, C. *et al.* (2021) Effectiveness of therapeutic gardens for people with dementia: a systematic review. *International Journal of Environmental Research and Public Health* 18. DOI: 10.3390/ijerph18189595.

Nieri, M. (2009) *Bioenergetic Landscape*. SE, Milan, Italy.

Nieri, M. and Mencagli, M. (2017) *La terapia segreta degli alberi*. Sperling & Kupfer, Milan, Italy.

Pirero, S.G. (2019) *Il 'Sanatorio' di Costarainera e il suo Parco fra storia e attualità*. Typ. San Giuseppe, Taggia (IM), Italy.

Pitzurra, M. (1989) *Igiene del turismo*. Franco Angeli, Milan, Italy.

Ruffini, G. (1855) *Doctor Antonio*. Constable & Co, Edinburgh, UK.

Sabbatani, S. (2005) La nascita dei sanatori e lo sviluppo socio-sanitario in Italia e in Europa. *Le Infezioni in Medicina* 2, 123–132.

Strimmer, U. (2019) Bathing between trees. *Topos* 106, 94–99.

Ulrich, R.S., Simons, R.F., Losito, B.D., Fiorito, E., Miles, M.A. *et al.* (1991) Stress recovery during exposure to natural and urban environments. *Journal of Environmental Psychology* 11(3), 201–230. DOI: 10.1016/S0272-4944(05)80184-7.

Wilson, E.O. (1984) *Biophilia: The Human Bond with Other Species*. Harvard University Press, Cambridge, Massachusetts. DOI: 10.4159/9780674045231.

Winterbottom, D. and Wagenfeld, A. (2015) *Therapeutic Gardens: Design for Healing Spaces*. Timber Press, Portland, Oregon.

Zanini, A. (2012) *Un secolo di turismo in Liguria. Dinamiche, percorsi, attori*. F. Angeli, Milan, Italy.

22 Decomposition of the Dimension of Wine Destination for the Experiences of Tourists: The Concept of 'Winescape'

Patrik Kubát*, Andrea Králiková and Kateřina Ryglová
Mendel University, Faculty of Business and Economics, Department of Marketing and Trade, Brno, Czech Republic

Abstract
A wine destination offers not only vineyards and landscapes but also wine attractions, bike trails and a comprehensive wine experience for every visitor. The concept of 'winescape' is another layer of the many attributes that attract visitors to a destination and appeal to the human senses, as, of course, wine can do. In the Czech Republic, there is still a lack of utilization and acceptance of the winescape concept. Therefore, it is essential to focus on this goal and consider adoption of this concept as an important step for further development of Czech wine destinations. This chapter focuses on the attributes that are present in a destination and thus constitute a wine destination. The dimensions of the winescape are considered from the perspective of the visitors. The aim is to initially define the concept of winescape and its attributes in the Czech environment, thus laying the foundations for its conceptualization and broader utilization.

Introduction

Relationships between hosts, guests and the tourism industry are often strained as they compete for access to, control over and use of resources. In this context, some forms of tourism can contribute to tourist experiences while respecting the natural environment and improving the socio-economic wellbeing of local people, accompanied by sustainable development. Wine tourism is one of these forms. Recently, there have been different perspectives on how to understand wine tourism, its range of services and products and its place of origin. The concept of 'winescape' is one of them. This concept is one among others in wine tourism research that fully explains the supply side of wine tourism (Terziyska, 2020; Quintal *et al.*, 2021). In the Czech Republic (Fig. 22.1), there is a lack of utilization of the concept of winescape, especially in the initial stage and subsequently in its introduction and development. Therefore, it is appropriate to address this issue and consider this concept as an important step for wine destinations and their further development, focusing on the values associated with wine and related aspects in the destination.

As mentioned by do Valle *et al.* (2012), visitor satisfaction with products and services influences consumer loyalty. Since satisfaction can be used as a measure to evaluate the products and services offered at the destination (do Valle *et al.*, 2012), the winescape concept evaluates a wine destination based on its attributes. The destination

*Corresponding author: patrik.kubat@mendelu.cz

DOI: 10.1079/9781800621411.0022

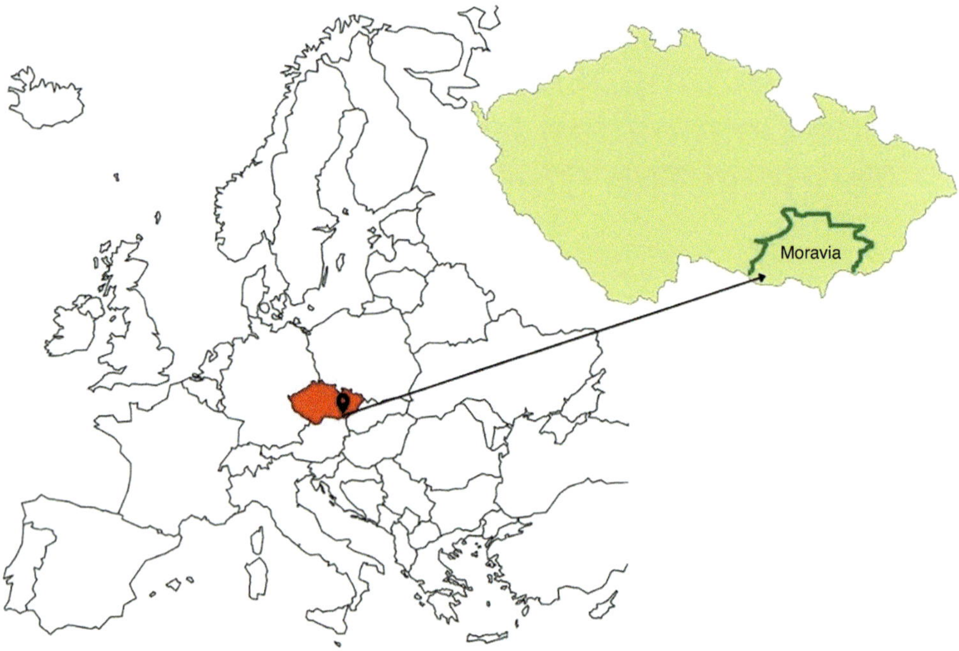

Fig. 22.1. Location of the Czech Republic and its main wine region, Moravia, in the south-eastern part.

attributes studied were used to determine visitors' perception and satisfaction with the wine destination as part of the winescape concept. Therefore, for this chapter, the attributes of the wine destination were selected as variables for the perception of the wine destination (destination image) as a dependent variable to conduct a principal component analysis (PCA). Satisfaction from, and perception of, the wine destination with its attributes as explanatory variables may increase tourists' return intentions and consequently their loyalty to the destination. The study is complemented by (i) an enumeration of selected attributes of the Moravian wine destination; (ii) a list of the most important attributes according to the mean value based on a five-point Likert scale; and (iii) the attributes that visitors consider most representative of the wine destination image.

Background

Wine tourism is categorized as part of rural tourism because of its place of development and origin, and especially due to the agricultural production of the main product, wine. However, today there are various forms of wine tourism that go beyond the boundaries of rural areas (Králiková and Kubát, 2022). The promotion of this type of tourism is about preserving traditions, promoting local products and supporting small-scale artisanal production based on sustainability principles. It can be considered as one of the current territorial engines of regional development that have a tendency to promote socio-economic and environmentally sustainable solutions (Mihaila, 2011). As Szolnoki states, wine tourism is a special form of tourism. Since there is an agricultural product with a solid connection to the environment behind this type of tourism, there are huge differences compared to general tourism. Wine tourism can adopt some sustainable practices from the general tourism industry; however, the fact that we are dealing with wineries producing an alcoholic beverage makes sustainable wine tourism unique (Szolnoki *et al.*, 2022).

Several research studies on wine tourism examine which of a number of attributes have the greatest influence on factors such as the following: decisions to visit a wine destination (Byrd *et al.*, 2016); loyalty intentions and satisfaction with wine destination attributes (do Valle *et al.*, 2012; Králiková *et al.*, 2021); typology and segmentation of wine visitors (Terziyska, 2020; Quintal *et al.*, 2021); evaluation of destination quality (Vajčnerová *et al.*, 2012); use of the winescape scale (Quintal *et al.*, 2015; Thomas *et al.*, 2016). The concept of the winescape is another layer of diverse attributes that attract visitors to a destination and appeal to the human senses, as wine can do naturally. This concept dates back to the 1970s and is connected to concepts such as landscape, cultural landscape and servicescape. In wine tourism literature, a winescape is also closely related to a destination's image. Therefore, we can assume that the image of a wine destination is, in the minds of wine visitors, viewed as a projection (Terziyska, 2020). The image of a destination and its perception as such represents a summary by the individual visitor's awareness of the destination. Each destination is characterized by changing attributes, but a certain portion of them, as also evidenced by the studies, is permanent and retained, e.g. the natural character, the traditions and the culture. As do Valle *et al.* (2012) state, the tourist image and attributes (quality variables) of the destination influence the intention to recommend and revisit the place. As noted by Mihaila (2011), the charm of the place makes people feel comfortable, especially in the case of wine destinations. Wine, wineries, vineyards and natural landscapes are the background elements of a wine destination's tourism offer.

Considering the previous research in this region (Králiková and Kubát, 2022), understanding visitors' perceptions and the image of the destination based on certain factors is essential for individual wineries and wine destinations to be able to define policies and strategies that favour development. There are two main approaches to the evaluation of winescape attributes – through expert assessment and by taking the customer perspective (Terziyska, 2020). Through research, approaching a section of the winescape concept with selected attributes of the wine destination, the retrieved principal components (factors) found lead to a grasp of the image of the wine destination through the most representative attributes.

Methods and Data

In order to define the critical attributes of the wine destination, primary data were collected by interviewing tourists in 2021. The survey was distributed with the participation of university students on courses who collected data through an online questionnaire (Google Forms) which included a filter question about the last visit to one of the Czech wine subregions. Thus, only respondents who had visited one of the four subregions of the Moravian wine region – the superior Czech wine region (Fig. 22.1) – were included (N = 271). A controlled quota selection based on gender was performed (Table 22.1). To determine the

Table 22.1. Sample of correspondents (N = 271).

Category	Subcategory	N	N (%)
Gender	Male	116	42.8
	Female	155	57.2
Age	18–23	122	45.0
	24–30	41	15.1
	31–40	22	8.1
	41–50	42	15.5
	51–60	25	9.2
	61+	19	7.1
Education	Elementary	20	7.4
	High school	169	62.4
	University	82	30.3
Economic activity (primary)	Unemployed	5	1.8
	Employed	116	42.8
	Student	120	44.3
	Retired, maternity, etc.	30	11.0
Income in CZK	Less than 20,000	143	52.8
	20,001–30,000	66	24.4
	30,001–40,000	39	14.4
	Above 40,000	23	8.5

groups of most-similar factors, a tool of multidimensional statistical analysis – principal component analysis – was used. PCA is most often used to reveal the internal structure of the data in a way that best explains the variance in the data (Vajčnerová *et al.*, 2012; Hair *et al.*, 2019). PCA was conducted to assess the dimensionality of the 18 selected factors (Table 22.2). These factors were adopted from previous research on this topic (Quintal *et al.*, 2015; Byrd *et al.*, 2016; Thomas *et al.*, 2016; Králiková *et al.*, 2021). The analysis was performed by creating a correlation matrix of the variables and examining the correlation coefficients between them. The Pearson correlation coefficient was used to measure the strength of correlation between two variables. All correlation coefficients were statistically significant. Because of the use of PCA, the percentage of variance criterion was used to determine the proportion of variance explained by the derived factors. Thus, at least 60% of the total variance had to be obtained (Hair *et al.*, 2019). Together with the scree test performed (Fig. 22.2) and the latent root criterion, factors with eigenvalues greater than 1.0 were selected (Table 22.2).

Bartlett's test of sphericity was significant (Chi-square = 3603.092; p<000) and the Kaiser-Meyer-Olkin measure of sampling adequacy (KMO) was 0.959, indicating that the data considered were suitable for PCA (Brata *et al.*, 2022). To assess the degree of consistency between variables, Cronbach's alpha was selected as the most commonly used measure. An acceptable reliability coefficient for Cronbach's alpha was >0.7 (Hair *et al.*, 2019). The internal consistency of each principal component (PC) was determined with an overall value of 0.957. The mean value of the inter-item correlation was measured as 0.556, indicating solid internal consistency. The item-to-total correlation was >0.5 for each variable, which is also adequate (Thomas *et al.*, 2016; Hair *et al.*, 2019). Consequently, a varimax rotation was applied to rotate the PCs for interpretive purposes (Hair *et al.*, 2019). Eigenvalues > 1.0 were considered significant and then varimax rotated factors (VFs) representing the new groups of variables were generated. The exception was F14, which was close to the lower bound with a value of 0.505. PCA results were obtained using SPSS Statistics 25 software.

Results

A total of 271 respondents participated in the study, all of whom were considered to be wine visitors to a specific wine subregion in Moravia. Thus, for the purposes of the study, both residents and non-residents were considered, as the perception of the image of a wine destination and the evaluation of its attributes are perceived by each visitor to the destination. Most wine visitors (40%) gave their answers based on a visit to the Mikulovská wine subregion, followed by the Velkopavlovická wine subregion (34%), the Znojemská wine subregion (16.6%) and the Slovácká wine subregion (9.4%).

The eigenvalues represent the importance of the characteristics. Two components were recorded and the others can be considered insignificant. It is obvious that the most important first principal component (PC1) explains more than half of the overall variability of the variables (58.6%). The total information (variance) contained in the data interpreted into two principal components is 64.446%. The higher the PC1 value, the greater the perception that a factor indicates for wine destination (the mean perception is symbolized by 0). On the contrary, the lower the PC1 value, the lower is the perception of wine destination by a factor.

Discussion and Conclusion

Based on PCA with varimax rotation, the original 18 factors were reduced to two new representative components explaining 64.446% of the total variance. The highest factor loading had, in order, F8 as nature (scenery), rural environment, F5 as relaxation and recreation, and then, almost equally, F3 and F18 as wine quality and as sustainability and ecology. Factors F8 and F5 also received the highest scores of 4.55 and 4.50, respectively. These two factors play a crucial role in the minds of wine visitors in perceiving the image of a wine destination. The photos in Fig. 22.3 illustrate the highly evaluated factors in the Moravian wine region.

Table 22.2. Principal component analysis of wine destination perception.

Component (PC)	Eigenvalue	Variance %	Factors/ variables	Rotated factor loadings	Mean (1–5)	SD
			Purchase of wine (F1)	0.681	4.28	1.056
			Wine tasting (F2)	0.687	4.28	1.077
			Wine quality (F3)	0.786	4.40	1.002
			Relaxation and recreation (F5)	0.812	4.50	1.003
			Nature (scenery), rural environment (F8)	0.866	4.55	1.006
1	10.552	58.624	Friends and acquaintances (F9)	0.755	4.41	1.074
			Wine culture and tradition, local atmosphere (F10)	0.771	4.37	1.037
			Travel (F12)	0.657	4.14	1.023
			Cultural monuments and other attractions in the destination (F13)	0.652	4.26	1.032
			Participation in wine events (open wine cellars, vintage, etc.) (F4)	0.557	3.91	1.044
			Visit a winery and meet the winemaker (F6)	0.717	4.01	1.046
			Interest in learning more about wine (education, training) (F7)	0.693	3.78	1.013
			Social interaction (F11)	0.555	4.07	0.990
2	1.048	5.823	Transport accessibility (quality of infrastructure, frequency of transport links, etc.) (F14)	0.553	3.69	1.064
			Friendly welcome by the local population (F15)	0.690	4.11	1.013
			Vineyard and winery tour (F16)	0.680	4.10	1.041

Continued

Table 22.2. Continued

Component (PC)	Eigenvalue	Variance %	Factors/ variables	Rotated factor loadings	Mean (1–5)	SD
			Gastronomic specialties in the restaurant at/close to the winery (F17)	0.623	3.97	1.011
			Sustainability and ecology (F18)	0.784	3.78	0.942
	Total variance %	64.446				

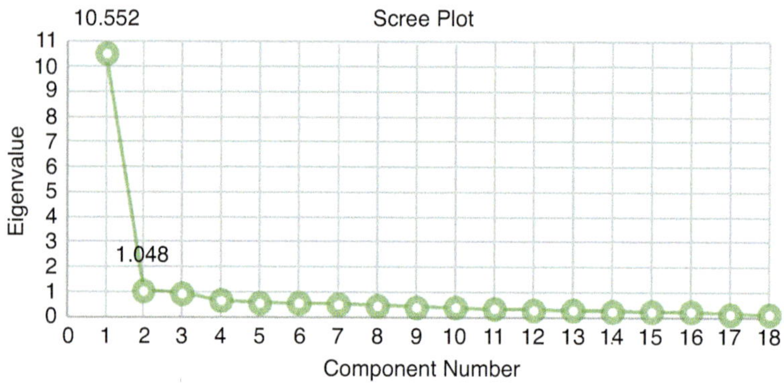

Fig. 22.2. Scree test of the squared loadings of the components.

Considering the nature of wine tourism and its rootedness in rural tourism, the rural setting is the logical conclusion of the study, as well as the admiration for local nature and its components. As Szolnoki *et al.* (2022) pointed out in the three-pillar model of sustainable wine tourism, the wineries surveyed from around the world consider the environmental pillar to be the most important. Thus, the environmental dimension of sustainability seems to outrank the other two dimensions and their aspects. Thus, the natural environment seems to be a fundamental building block for wine tourism and, in a way, an important prerequisite for its further development. This factor could be categorized as a localization factor for tourism development. As Mihaila (2011) adds, making a wine is the activity around which everything revolves and which provides a comprehensive dialogue with the land. To emphasize the importance, not only from visitors, the wine region of Moravia includes several protected areas (Kubát, 2021): national parks, protected landscape areas (Fig. 22.4), etc. Nature, landscape and wine production are an interconnected stimulus and are therefore considered as a section of the winescape concept researched attributes (Quintal *et al.*, 2021).

The connection of landscape (as natural and rural environment), servicescape (as tourism industry) and winescape (as wine as the main motivation for any activity related to it) includes different types of tourism and, increasingly, sought-after experiences and wine destinations. As Hansen (2018) adds, it is often a combination of several factors at the same time, but always with the natural environment as an active part or setting for conducting wine tourism activities. Similarly, in previous research

Fig. 22.3. (a) The wine quality factor takes its place in a winery after the grapes are harvested in the vineyards. A picture shows the typical entrance to a small family cellar. (b) Wine culture and tradition play an important role in the development of winemaking and wine tourism in the Moravian region. Folkloric patterns can be seen wherever wine is involved. Own accommodation above the wine cellar is another additional service offered. (c) Local atmosphere with wine tasting and purchasing is still offered in the wine cellar in many Moravian wineries. A picture shows a characteristic view of the family wine cellar and its equipment.

in the Moravian wine region (Králiková and Kubát, 2022), the factor of natural attractions was found to be very influential in recommending the wine destination and the intention to revisit it. Furthermore, Hansen (2018) describes that the natural environment plays a dual role in creating feelings of health and wellbeing. This connection between health, wellbeing and the natural environment is an important component of nature-based recreation and tourism, that is, recreation and tourism that focus on nature and experiences in nature. The combination of nature, winemaking and wine tourism thus represents elements that are interdependent. In addition to the highly evaluated factor of relaxation and recreation, the wine destination of Moravia is a year-round destination for short trips or longer stays as a vacation. This factor was already crucial in an earlier study by Byrd *et al.* (2016), as they placed it in the category of ancillary services. Similarly, the factors of wine quality and wine tasting were contrived to be a core attribute for wine visitors at a wine destination, as this study also proves for the wine region of Moravia.

In order for the wine region of Moravia to be perceived by every visitor as a wine destination as such, international as well as national awareness is essential. In order to strengthen the position of this destination, it participates in international wine competitions, where the competition is held on the level of wine and its quality in comparison with other wines from other wine regions. At the tourist level, for example, at the national scale, the largest Czech portal about viti/viniculture and wine events is

Fig. 22.4. Landscape view of the Pálava protected landscape area from the Sonberk winery in Popice (Mikulovká wine subregion).

VinazMoravyVinazCech.cz (*Wines from Moravia, Wines from Bohemia*), and at the international scale, mainly the tourism platform TripAdvisor or the online booking platform WineTourism.com, which brings together wineries and winemakers from many wine countries on a voluntary basis and offers wine tastings and tours accessible to visitors. The wine region of Moravia began integrating with this booking platform in the autumn of 2021. This will make a visit to this destination even more accessible to foreign visitors and make it easier to book any type of wine experience. The development of wine tourism in rural areas follows a greater understanding among wineries and other stakeholders in the sector, as well as a growing awareness and demand from visitors. As a result of this progress, domestic wine tourism is clearly gaining a profile compared to other activities in rural areas. Regardless of what COVID-19 caused to this sector, it remains the most visited region in the country, with the exception of the capital (Kupčíková, 2022). As a result of the pandemic, new wine services and products were introduced, strengthening its position. The resulting components prove the importance of the wine destination's attributes. For business-like purposes, the two extracted components could be named: PC1 as *wine and relaxed atmosphere* and PC2 as *more than/about wine in the environment*.

In conclusion, the results enable a review and evaluation of the attributes of the Moravian wine destination and make it possible to break them down into components. Moreover, they confirm the key attributes, and their values thus indicate the most important aspects of the destination for its image and design in the space as a wine destination. The identified components can be used by destination management organizations and winemakers to define (or improve) key products and services, provide a basis for conceptualizing the winescape concept in the Moravian wine region, and contribute to further expand the list of attributes that make up a winescape-based destination. Subsequently, more comprehensive and profound research should reflect a larger number of (wine) visitors to the destination in order to substantiate a concept of the winescape in the Moravian wine region with the expansion and addition of existing, but also new, dimensions. At the same time, or after these processes, a winescape could be better implemented in the regional development strategies and their management and marketing efforts.

A certain shortcoming is the generalizability of the results, which in Table 22.1 indicate a preponderance of respondents of lower age and income, which explains the high proportion of students in the sample as they were co-operating with the data collection. A quota selection of more demographically and economically diverse respondents by economic activity would be useful for further proceeding with the conceptualization of the winescape in the Czech environment.

Acknowledgements

This chapter collaborated with projects of the Internal Grant Agency of the Faculty of Business and Economics of Mendel University in Brno in 2022 (IGA-PEF-DP-22–014 and IGA-PEF-DP-22–018).

References

Brata, A.M., Chiciudean, D.I., Brata, V.D., Popa, D., Chiciudean, G.O. *et al.* (2022) Determinants of choice and wine consumption behaviour: a comparative analysis between two counties of Romania. *Foods (Basel, Switzerland)* 11(8), 1110. DOI: 10.3390/foods11081110.

Byrd, E.T., Canziani, B., (Jerrie) Hsieh, Y.-C., Debbage, K. and Sonmez, S. (2016) Wine tourism: motivating visitors through core and supplementary services. *Tourism Management* 52, 19–29. DOI: 10.1016/j.tourman.2015.06.009.

do Valle, P.O., Silva, J.A., Mendes, J.D. and Guerreiro, M. (2012) Tourist satisfaction and destination loyalty intention. In: Seba, J.A. (ed.) *Tourism and Hospitality: Issues and Developments*. Apple Academic Press, Palm Bay, Florida, pp. 54–75. DOI: 10.1201/b12231-7.

Hair, J.F., Black, W.C., Babin, B.J. and Anderson, R.E. (2019) *Multivariate Data Analysis*, 8th edn. Cengage Learning EMEA, Boston, Massachusetts.

Hansen, A.S. (2018) The visitor: connecting health, wellbeing and the natural environment. In: Azara, I., Michopoulou, E., Niccolini, F., Taff, D.B. and Clarke, A. (eds) *Tourism, Health, Wellbeing and Protected Areas*. CAB International, Wallingford, UK, pp. 125–137. DOI: 10.1079/9781786391315.0000.

Králiková, A. and Kubát, P. (2022) Wine tourism as an alternative of sustainable development of the rural region: a case study of the Czech Republic. In: Vujicic, M.D., Kasim, A., Kostopoulou, S., Olmo, C.J. and Aslam, M. (eds) *Cultural Sustainable Tourism (Advances in Science, Technology & Innovation)*, 2nd edn. Springer, pp. 25–32. DOI: 10.1007/978-3-031-07819-4.

Králiková, A., Kubát, P. and Ryglová, K. (2021) Visitors' happiness and loyalty in the Moravian wine region. *European Countryside* 13(4), 750–767. DOI: 10.2478/euco-2021-0040.

Kubát, P. (2021) Wine tourism and its application in the context of localization and implementation factors. In: *VI International Scientific Conference Challenges of Globalization in Economics and Business: Proceedings*, Ivane Javakhishvili Tbilisi State University Press, pp. 181–189. Available at: https://dspace.tsu.ge/handle/123456789/930 (accessed 26 October 2023).

Kupčíková, T. (2022) Návštěvnost HUZ – podrobná data 2012 – 2022 Q. *Tourdata*. Available at: https://tourdata.cz/data/navstevnost-huz-2012-2022/ (accessed 26 October 2023).

Mihaila, M. (2011) Wine tourism – architecture and landscape. *Urban Space – Architectural Space – Interior Space* 41–48. DOI: 10.13140/RG.2.1.4505.8087.

Quintal, V.A., Thomas, B. and Phau, I. (2015) Incorporating the winescape into the theory of planned behaviour: examining 'new world' wineries. *Tourism Management* 46, 596–609. DOI: 10.1016/j.tourman.2014.08.013.

Quintal, V.A., Thomas, B., Phau, I. and Soldat, Z. (2021) Segmenting hedonic wine tourists using push–pull winescape attributes. *Australasian Marketing Journal* 30(3), 237–245. DOI: 10.1177/1839334921999478.

Szolnoki, G., Ridoff, N., Nilsson, C., Tafel, M. and Stelter, A.-C. (2022) *Sustainable Wine Tourism – A Global Survey*. Geisenheim University and WineTourism.com. Available at: https://www.winetourism.com/

files/2022/01/Presentation_sustainable_wine_tourism_Geisenheim_winetourismcom.pdf (accessed 25 October 2023).

Terziyska, I. (2020) *Wine Tourism: Critical Success Factors for an Emerging Destination*. Gea Libris, Sofia.

Thomas, B., Quintal, V.A. and Phau, I. (2016) Wine tourist engagement with the winescape: scale development and validation. *Journal of Hospitality & Tourism Research* 42(5), 793–828. DOI: 10.1177/1096348016640583.

Vajčnerová, I., Šácha, J. and Ryglová, K. (2012) Using the principal component analysis for evaluating the quality of a tourist destination. *Acta Universitatis Agriculturae et Silviculturae Mendelianae Brunensis* 60(2), 449–458. DOI: 10.11118/actaun201260020449.

23 Commodifying Clear Waters and Green Mountains through Tourism: A Case Study of Fazhan Village in Giant Panda National Park

Yulei Guo*
Chengdu Research Base of Giant Panda Breeding, Chengdu, China

Abstract
As the nation faces the burden of an ageing population, an increasing number of senior Chinese tourists are establishing a new way of life through their extended stays in rural areas. This case study explores how nature-based tourism helps retired visitors (re)integrate into social life in Fazhan, Sichuan, a rural village with abundant natural resources. The study demonstrates that retirees develop a bond with nature and share this experience with their friends, family and the local community. The study concludes, through interviews with 34 Chinese retirees, that nature experiences help seniors build a unique social network with the host community and their existing life community.

Introduction

This case study examines how nature-based tourism helps elderly Chinese tourists (re) integrate into social life in the rural village of Fazhan, Sichuan, China. In 2018, 17.9% of China's population was comprised of individuals aged 60 and older. Sichuan province, located in western China, has the highest rate of population ageing. Several recent studies have documented the rapid growth of senior tourism in China, labelling this new way of life for the Chinese elderly 'long-stay in rural China' (Lyu *et al.*, 2021), 'rural stay' (Cheng *et al.*, 2020), 'retirement-tourism' (Deng *et al.*, 2022), and 'retirement migrants' (Liang *et al.*, 2021). This case study contributes to the investigation of the participation of Chinese senior citizens in rural long-term tourism stays. As demonstrated by Lyu *et al.* (2021), Chinese senior citizens' long trips to rural China are both a response to the nation's rural revitalization programme and a desire for a new way of life. Sedgley *et al.* (2011) suggest that 'more individualized, subjective research that explores the intricacies of older people's lives' is required because the majority of these studies have focused on the Chinese elderly's participation in rural tourism as a social phenomenon on a large scale.

This study examines, using semi-structured interviews conducted in Fazhan Village and based on the notion of social integration, how nature-based tourism in rural China has (re) integrated the elderly into social life after retirement. The researcher interviewed retirees face-to-face and spent time listening, speaking, laughing and engaging with each individual. The semi-structured interviews enabled the researcher to investigate the research question posed by this case study. Specifically, the study investigates how time and experiences spent with nature in rural China assist the formation

*yulei.guo@panda.org.cn

DOI: 10.1079/9781800621411.0023

of social ties that promote the wellbeing and health of elderly Chinese.

Literature Review

According to Otoo and Kim (2020), senior visitors will be the largest segment of the tourism industry during the next three decades due to the increasing number of seniors in modern society. Tourism has helped elders reconstitute their social roles and identities, according to researchers (Nimrod, 2007; Buse, 2009; Scherger *et al.*, 2011; Dorfman, 2013; Lee *et al.*, 2019). Studies (Carter and Cook, 1995; Gustafson, 2002; Heaven *et al.*, 2013; Damman *et al.*, 2015; Eismann *et al.*, 2019) indicate that engagement in these activities can be essential to the health of the elderly, whereas a few researchers (Kim and Merriam, 2004; Istance, 2015; Jamieson, 2016; Tam, 2018) have demonstrated that retirement can be a challenging transition for a substantial portion of the working population. It is considered that tourism and leisure activities minimize transitional burdens and challenges. Therefore, Nimrod (2008, p. 874) highlights the need to appreciate how tourism also provides the elderly with a 'serious' path for reconstructing 'structure, affiliation, duty, challenge, and a sense of essentiality'.

This study applies the notion of social integration to investigate how elderly Chinese tourists establish a segmented social position through their interactions with nature and rural host communities. Specifically, the notion of social integration refers to 'the web of roles and relationships' (Pillemer *et al.*, 2000, p. 1) that Chinese seniors attempt to bind together through leisure time spent in nature in rural China, according to a path that is distinctively theirs. Tourism can improve social inclusion, according to Minnaert *et al.* (2009, p. 413). Kastenholz *et al.* (2018) indicate that rural tourists have the opportunity to engage with other tourists and hosts, which has a positive effect on their life experiences. These studies suggest that tourism and travel experiences can contribute to the social integration of the elderly by encouraging the formation of various social networks.

Nature supports the social reintegration of the elderly in its own unique way. Nature can be therapeutic (Finlay *et al.*, 2015; Wen *et al.*, 2018; Marques *et al.*, 2019), nurturing (Rissanen, 2013; Dreyer *et al.*, 2018; Taylor *et al.*, 2018) and strengthen social bonds (Taylor *et al.*, 2018). Orr *et al.* (2016) demonstrate that nature can provide older persons with a great deal of pleasure and delight, while memories and recollections of places and people can enrich their everyday life and sense of self. In this light, time spent in nature helps to reintegrate the elderly into society.

Many of the anticipated benefits of social integration are realized when senior Chinese tourists engage in rural tourism. This bond is strengthened via nature-based activities. Tourism enables the elderly to rebuild social networks through connection with nature. In exchange, these social networks give senior participants respect and acknowledgement. It also illustrates that, rather than being a passive experience, nature may actively change the memory, experience and social ties of senior travellers.

Method

This study employed semi-structured interviews, which allow researchers to address research questions with 'good listening skills and emotional control' (Adams, 2010, p. 18) and gain insight into the experiences of elderly individuals. Before doing fieldwork, the researchers developed a questionnaire on the Chinese website Wenjuanxing, which specializes in questionnaire design.

Considering the elderly's difficulty in accessing mobile devices, the questionnaire was collected in two methods. The first method targeted individuals with adequate online literacy. Participants were instructed to scan a QR code in order to access the online survey, which they could complete on their own mobile devices. The researcher analysed the questions and provided additional inquiries as individuals reviewed the questionnaire. The second method enabled the researcher to establish rapport with individuals. On the basis of the questionnaire, the researcher conducted conversations with visitors and was able to collect answers. In this light, the researcher was able to collect information from

Table 23.1. Summary of interviewees.

Measure	*N*	%
Age	34	Mean = 64.9; SD = 6.91
Gender		
Female	16	47.06
Male	18	52.94
Place of residence		
Chengdu	21	61.76
Other places in Sichuan	13	38.24
Occupation before retirement		
Business	12	35.29
Government	13	38.24
Public service	2	5.88
Self-employed	5	14.71
Housewife/husband	2	5.88
Group type		
Individual	1	2.94
Friends	14	41.18
Family	19	55.88
Number of group members		
< 3	6	17.65
3–5	7	20.59
> 5	21	61.76
Length of stay (days)		
< 4	18	52.94
4–7	9	26.47
> 7	7	20.59
First/returned visitors		
1	13	38.24
2	6	17.65
> 2	15	44.12

participants who said they lacked internet literacy, had bad eyesight or low energy. In August 2022, the study team spent one week in Fazhan Village. Thirty-four senior visitors took part in the interview (Table 23.1). All interviews were recorded with participants' consent. The research group transcribed the audio recordings and conducted additional theme analysis.

The researcher conducted the investigation in Fazhan Village, Yinjing, Sichuan. Deep in the Longcan Valley, Fazhan Village is near Giant Panda National Park, which is home to giant pandas, dove trees, and a variety of protected species and plants. The relationship between Fazhan Village and nature has moved from exploitation to conservation and protection during the past two decades. In 2021, 65 rural stays were constructed in Fazhan Village to satisfy the increasing demand from tourists. Summer temperature differences of 5°C between Fazhan Village and metropolitan Chengdu make Fazhan Village attractive to urban inhabitants, as witnessed by its high occupancy rates during peak seasons. Fazhan Village's abundant natural resources are the primary draw for visitors. Since 2010, the average annual revenue generated by tourism has increased more than sixfold, from \$7899 to \$48,005. Senior tourists and retirees, who frequently spend weeks in Fazhan Village throughout the summer, have considerably contributed to this increase.

Discussion

Tourist-nature

Summertime migration of senior residents from the city to Fazhan Village was driven mostly by the village's refreshingly cold and clean air. Over 90% of survey respondents identified Chengdu and its neighbouring cities as their place of origin. Fazhan Village was a place to 'escape the heat of Chengdu' (Interviewees 4, 11, 18). Steiger *et al.* (2016) discovered that mountain visitors who were older were more vulnerable to heat. The older inhabitants of Chengdu, Meishan and Ya'an viewed Fazhan Village as a viable alternative to their urban existence. Fifty per cent of visitors indicated that they frequently visit the area annually or seasonally.

The majority of tourists decided to spend their time surrounded by hilly landscapes. When asked how his friends and he spent the day in Fazhan Village, Interviewee 5 gave the following description:

> We rise in the morning and walk down the route. We return to the rural guesthouse where meals are supplied after 30 minutes. After

> breakfast, my buddies and I drive to nearby scenic spots and spend the day there. We return at noon for lunch. After a nap, we will continue along the route. Then it is time for dinner.

At Fazhan Village, senior visitors have often appreciated the opportunity to interact with the natural environment. Fazhan Village serves as a connection between nearby visitor attractions. Although senior tourists have more time to interact with the natural environment, their activities have been limited to strolling and sightseeing.

The 'naturalness' and 'wildness' of Fazhan Village, however, were not appealing to younger Chinese tourists, according to Interviewee 28:

> The government has made a modest investment in Fazhan Village. Clearly, the public sector is insufficient. The walking track is insufficiently long, and there are no alternative entertainment options. Teenagers will become bored in this environment. They have no activities to occupy their time.

Fazhan Village is geographically, culturally and socially isolated due to its location in the mountains. The unique natural features and cultural 'rawness' of Fazhan Village attracted senior travellers. The elderly have accepted the dearth of urban leisure venues such as shopping malls and sports centres that are appealing to younger generations. According to Interviewee 28, the retro aspect of Fazhan Village in comparison to the industrialized megacity Chengdu is 'how these seniors lived when they were young'.

Tourist-host-nature

A woman in her late thirties who owned a homestay recounted her 'family-like' relationship with her older visitors in a lengthy conversation:

> They adored the idea that I can provide them with medical treatment. My clients are aware that I am the local clinic's physician and that I am more knowledgeable about their health. They have confidence in my ability to give superior service. In addition, as a result of my medical training, I have learned to communicate with and listen to the elderly.

For senior tourists, establishing intimate ties with local hosts has been a vital experience. Hosts and guests concur that the tourists' long and frequent visits in Fazhan Village have been motivated by their close interaction with the locals. While hosts have learned to profit from intimacy through their hospitality company, senior citizens believe they have received a different return. Interviewee 9 stated that she aided the host in the kitchen since she believed the host lacked cooking expertise. Interviewee 16 stated that he provided 'great guidance' to the local host at the outset of the rural stay business. The relationship between hosts and senior visitors has been a mutually beneficial investment.

By establishing relationships with local hosts, elder visitors have better access to natural resources. The fourth interviewee states that local hosts 'know the finest places to shoot photos and discover the tastiest meals'. All visitors to Fazhan Village, where the tap water was pure and the food was organic, remarked on the high quality of the food. The rural community provided them with great cuisine that was rarely available in urban supermarkets, according to the fourth interviewee.

Tourist-friends/family-nature

Over 50% of those questioned stayed in Fazhan Village with more than five other tourists. Family and friends make up the membership of the group. In the study, a large friendship network consisting of co-workers, classmates and interest groups was observed. A few interviewees stated that they stayed in Fazhan Village because family or friends wanted them to, but the vast majority underlined the importance of spending quality time with family or friends in nature. As Interviewee 15 notes:

> We are on vacation this summer. Our dance class is over, and I eagerly anticipate the September commencement of the second semester. We're here to snap photographs. We are dancing in front of the flowers in this photo. And the mountain is covered in luxuriant vegetation, and the river is crystal clear.

The bulk of elderly visitors to Fazhan Village brought along their own social networks. Due to the natural setting, the elderly in Fazhan Village engaged in new conversations and exchanges.

The appreciation of nature was a common topic of conversation among older vacationers. Meanwhile, environmental experiences fuelled the recreational activities of senior travellers. They practised dance, playing instruments, photography and calligraphy. A bird lover (Interviewee 11) showed the researcher his photos of birds and claimed that some of these were his 'classic collection'.

Discussion and Conclusion

The value of social integration for health, happiness and ageing has been well-investigated (Pillemer *et al.*, 2000). This case study examines how elderly Chinese visitors' experiences with nature-based tourism in Fazhan Village have helped them retain and build social networks. The study indicates that, via their encounters with nature, retirees have formed a unique network of contact with the host community, their own social circle and themselves. Seasonal rural stays in Fazhan Village develop a distinct tourism culture that benefits both the locals and senior citizens. This culture is the outcome of a successful dialogue between the host community and the Chinese retirees, which addressed the various needs of each group. Tourism experiences at Fazhan Village (re)integrate Chinese retirees into both Fazhan Village's new social network and their current social circles.

Nature has played a major role in the (re) integration of the elderly into Chinese society. First, nature, and specifically the cool climate of Fazhan Village, has sparked the interest of retirees who can afford an extended stay with the local community to escape the metropolis. For seniors who grew up in a less developed China in the 1950s, the natural setting of Fazhan Village also evokes memories. Through assistance and direct contact with the locals, senior travellers forge a deep connection with the local environment. Thirdly, nature revitalizes the retirees' own social network through the interchange of photographs with family and friends, renewed interest in a new environment, and the co-creation of new stories.

This study's focus on Chinese seniors is a limitation because it does not explain how the retirees' relationship with nature differs from that of the younger generation, despite the assertion of Clayton *et al.* (2019) that older people have a closer relationship with and affinity for nature. More research is required to understand whether older adults' appreciation for and interest in nature have affected their travel experiences.

The implication of this study is that nature-based tourism facilitates the development of a social network that is good for the wellbeing and health of Chinese retirees. Nature has played a significant role in attracting senior tourists and bridging the gap between locals and tourists. Meanwhile, nature preserves the existing social networks of retirees. Tourism offers Chinese retirees the means to accomplish social integration by connecting tourists to nature, the local community and the tourists' own social networks.

References

Adams, E. (2010) The joys and challenges of semi-structured interviewing. *Community Practitioner* 83(7), 18–21.

Buse, C.E. (2009) When you retire, does everything become leisure? Information and communication technology use and the work/leisure boundary in retirement. *New Media & Society* 11(7), 1143–1161. DOI: 10.1177/1461444809342052.

Carter, M.A.T. and Cook, K. (1995) Adaptation to retirement: role changes and psychological resources. *The Career Development Quarterly* 44(1), 67–82. DOI: 10.1002/j.2161-0045.1995.tb00530.x.

Cheng, H., Yang, Z. and Liu, S.-J. (2020) Rural stay: a new type of rural tourism in China. *Journal of Travel & Tourism Marketing* 37(6), 711–726. DOI: 10.1080/10548408.2020.1812467.

Clayton, S., Bexell, S.M., Xu, P., Tang, Y.F., Li, W.J. *et al.* (2019) Environmental literacy and nature experience in Chengdu, China. *Environmental Education Research* 25(7), 1105–1118. DOI: 10.1080/13504622.2019.1569207.

Damman, M., Henkens, K. and Kalmijn, M. (2015) Missing work after retirement: the role of life histories in the retirement adjustment process. *The Gerontologist* 55(5), 802–813. DOI: 10.1093/geront/gnt169.

Deng, T., Zhao, W. and Hu, Y. (2022) Retirement and household tourism consumption—a case study in China. *Tourism Economics* 29(4), 1055–1073. DOI: 10.1177/13548166221090170.

Dorfman, L.T. (2013) Leisure activities in retirement. In: Wang, M. (ed.) *The Oxford Handbook of Retirement*. Oxford University Press, Oxford, UK, pp. 339–353.

Dreyer, B.C., Coulombe, S., Whitney, S., Riemer, M. and Labbé, D. (2018) Beyond exposure to outdoor nature: exploration of the benefits of a green building's indoor environment on wellbeing. *Frontiers in Psychology* 9, 1583. DOI: 10.3389/fpsyg.2018.01583.

Eismann, M., Verbeij, T. and Henkens, K. (2019) Older workers' plans for activities in retirement: the role of opportunities, spousal support, and time perception. *Psychology and Aging* 34(5), 738. DOI: 10.1037/pag0000377.

Finlay, J., Franke, T., McKay, H. and Sims-Gould, J. (2015) Therapeutic landscapes and wellbeing in later life: impacts of blue and green spaces for older adults. *Health & Place* 34, 97–106. DOI: 10.1016/j.healthplace.2015.05.001.

Gustafson, P. (2002) Tourism and seasonal retirement migration. *Annals of Tourism Research* 29(4), 899–918. DOI: 10.1016/S0160-7383(01)00084-6.

Heaven, B., Brown, L.J.E., White, M., Errington, L., Mathers, J.C. *et al.* (2013) Supporting well-being in retirement through meaningful social roles: systematic review of intervention studies. *The Milbank Quarterly* 91(2), 222–287. DOI: 10.1111/milq.12013.

Istance, D. (2015) Learning in retirement and old age: an agenda for the 21st century. *European Journal of Education* 50(2), 225–238. DOI: 10.1111/ejed.12120.

Jamieson, A. (2016) Retirement, learning and the role of higher education. *International Journal of Lifelong Education* 35(5), 477–489. DOI: 10.1080/02601370.2016.1224036.

Kastenholz, E., Carneiro, M.J. and Eusébio, C. (2018) Diverse socializing patterns in rural tourist experiences - a segmentation analysis. *Current Issues in Tourism* 21(4), 401–421. DOI: 10.1080/13683500.2015.1087477.

Kim, A. and Merriam, S.B. (2004) Motivations for learning among older adults in a learning in retirement institute. *Educational Gerontology* 30(6), 441–455. DOI: 10.1080/03601270490445069.

Lee, Y., Chi, I. and Palinkas, L.A. (2019) Retirement, leisure activity engagement, and cognition among older adults in the United States. *Journal of Aging and Health* 31(7), 1212–1234. DOI: 10.1177/0898264318767030.

Liang, Z., Luo, H. and Hui, T. (2021) Moving for a good life: tourism mobility and subjective well-being of Chinese retirement migrants. *Tourism Geographies* 1–9. DOI: 10.1080/14616688.2021.1990385.

Lyu, J., Huang, H. and Mao, Z. (2021) Middle-aged and older adults' preferences for long-stay tourism in rural China. *Journal of Destination Marketing & Management* 19, 100552. DOI: 10.1016/j.jdmm.2020.100552.

Marques, B., McIntosh, J. and Kershaw, C. (2019) Healing spaces: improving health and wellbeing for the elderly through therapeutic landscape design. *International Journal of Arts and Humanities* 3(2), 20–34. DOI: 10.26686/wgtn.12413336.v1.

Minnaert, L., Maitland, R. and Miller, G. (2009) Tourism and social policy: the value of social tourism. *Annals of Tourism Research* 36(2), 316–334. DOI: 10.1016/j.annals.2009.01.002.

Nimrod, G. (2007) Expanding, reducing, concentrating and diffusing: post retirement leisure behavior and life satisfaction. *Leisure Sciences* 29(1), 91–111. DOI: 10.1080/01490400600983446.

Nimrod, G. (2008) Retirement and tourism themes in retirees' narratives. *Annals of Tourism Research* 35(4), 859–878. DOI: 10.1016/j.annals.2008.06.001.

Orr, N., Wagstaffe, A., Briscoe, S. and Garside, R. (2016) How do older people describe their sensory experiences of the natural world? A systematic review of the qualitative evidence. *BMC Geriatrics* 16(1), 1–16. DOI: 10.1186/s12877-016-0288-0.

Otoo, F.E. and Kim, S. (2020) Analysis of studies on the travel motivations of senior tourists from 1980 to 2017: progress and future directions. *Current Issues in Tourism* 23(4), 393–417. DOI: 10.1080/13683500.2018.1540560.

Pillemer, K., Moen, P., Glasgow, N. and Wethington, E. (2000) *Social Integration in the Second Half of Life*. Johns Hopkins University Press, Baltimore, Maryland.

Rissanen, S. (2013) Wellbeing and environment: concepts in the elderly care home context. In: Hujala, A., Rissanen, S. and Vihma, S. (eds) *Designing Wellbeing in Elderly Care Homes*. Aalto University publication series, Helsinki, pp. 20–35.

Scherger, S., Nazroo, J. and Higgs, P. (2011) Leisure activities and retirement: do structures of inequality change in old age? *Ageing and Society* 31(1), 146–172. DOI: 10.1017/S0144686X10000577.

Sedgley, D., Pritchard, A. and Morgan, N. (2011) Tourism and ageing: a transformative research agenda. *Annals of Tourism Research* 38(2), 422–436.

Steiger, R., Abegg, B. and Jänicke, L. (2016) Rain, rain, go away, come again another day. Weather preferences of summer tourists in mountain environments. *Atmosphere* 7(5), 63. DOI: 10.3390/atmos7050063.

Tam, M. (2018) Retirement and learning: a longitudinal qualitative approach. *Educational Gerontology* 44(1), 54–63. DOI: 10.1080/03601277.2017.1389486.

Taylor, L., Hahs, A.K. and Hochuli, D.F. (2018) Wellbeing and urban living: nurtured by nature. *Urban Ecosystems* 21(1), 197–208. DOI: 10.1007/s11252-017-0702-1.

Wen, C., Albert, C. and Von Haaren, C. (2018) The elderly in green spaces: exploring requirements and preferences concerning nature-based recreation. *Sustainable Cities and Society* 38, 582–593. DOI: 10.1016/j.scs.2018.01.023.

24 Sustainable Tourism Recovery in Nature-based Tourism Destinations of Nepal

Sitaram Dahal* **and Kabindra Bhatta**
Nepal Tourism Board, Kathmandu, Nepal

Abstract
Tourism has proved a boon for the highlands given the limited livelihood options. Sagarmatha National Park, a home to the highest mountain in the world, is a perfect example of a transformative journey that tourism brings to the highlands. This chapter highlights how tourism has brought sustainable development to the region.

Introduction

Nepal is one of the most famous tourism destinations in the world, especially among those seeking to experience mountains and associated features. However, the highlands, which are otherwise very rich in natural beauty and diversity, have minimal options for peoples' livelihoods. The highlands are a barren land with very few possibilities for agriculture. The barren land and tundra vegetation hinder local communities from practising commercial farming. However, the region is gifted with beautiful landscapes, with mountains as a backdrop. After the formal start of tourism in Nepal in the 1950s (Shrestha and Shrestha, 2012), the country witnessed new opportunities in the form of tourism. This chapter discusses the opportunities tourism brought to the region and for the livelihood of the local communities in the highlands.

Background context

Nepal is a geographically diverse country. Though having only an area of 147,181 km^2, it is geographically classified into terai/lowlands (<60 m above sea level [masl]), mid-hills (60–3000 masl) and highlands with an altitude of >3000 masl. The highlands of Nepal make up 16% of the total land mass and have a population of 6.73% of the total population, with a very low contribution to the economy (Government of Nepal, 2022). However, tourism is seen as a 'ray of hope' to accelerate the pace of development in the highlands. 'The Grandeur of the natural scenery from the high mountains to the tropical lowland, is coupled with a population of diverse, friendly ethnic and caste groups' (Thapa, 1998, p. 227). The first tourists came to climb Nepal's mountains, the single biggest draw being Mount Everest, called Sagarmatha in the local language (Thapa, 1998). Nepal has somehow

*Corresponding author: dahalsram@gmail.com

DOI: 10.1079/9781800621411.0024

developed itself as an adventurous destination and mountains have played a pivotal role in that process. Tourism has become the main business in the highlands. Sagarmatha National Park, an official protected area since 1973, is listed as a UNESCO World Heritage Site for its rich ecosystem, providing habitats to different endangered species of flora and fauna. Sherpa culture, and the brand associated with them as guides to the Himalayas, ensures a steady flow of visitors. The protected area covers an area of 1148 km^2 (Sagarmatha National Park, 2022).

In addition to Everest, the biggest pulls in the region are Gokyo Lake, Dudh Kosi Valley, Mount Amadablam and the Namche region. Tourism is the major occupation of the people. In 2018/19, the park collected revenue of USD 1,611,340. In 2019, Sagarmatha National Park welcomed 59,179 international tourists (Sagarmatha National Park, 2022).

Literature Review

Tourism in Nepal

Nepal was restricted to very few tourism-related activities during the Rana regime before the 1900s. 'In 1949, the Ranas decided to allow mountaineers to explore as long as scientists accompanied them' (Mahato *et al.*, 2021, p. 15). Similarly, Rana rulers of Nepal hosted Prince Edward in 1890 for tiger hunting in Nepal. Later, Nepal was popular among high-end adventure tourists looking for big safaris and hunting. In the late 1920s, a British expedition team attempted the summit of Mount Everest. Nepal gradually opened to foreigners in the 1950s and was popular as a destination for mountaineers. As trekking tourism gained popularity among adventurers in the world, Nepal became their favourite. Similarly, in the 1960s, Nepal, especially Kathmandu, was famous among hippies. It is said that the hippies came chasing enlightenment and preaching peace as the trade of marijuana was legal at that time. However, as the government of Nepal banned such trade, slowly the number of tourists seeking such pleasure plunged. It is worth mentioning here that pilgrimage tourism (temples and monasteries) has been attracting Hindu and Buddhist visitors in large numbers from ancient times.

Tourism as a driver of development: a late realization of the government

Nepal is primarily an agro-based country where the sector engages 66% of the total population and contributes approximately one-third of the total GDP (FAO and WTO, 2023). Other industries accompanying agriculture are manufacturing, service and tourism. The WTTC (2022) shows the direct contribution of travel and tourism in Nepal to GDP as USD 292.2 billion, which is 7.5% of the total economy (2019). Here, direct contribution means economic activities generated by industries like hotels, travel agencies, aviation industries and others. Though the data show a high contribution of tourism to the economy, the country had not seriously considered tourism as a priority area in its national plans and policies until the 2010s.

The country's actions towards making Nepal a perfect destination were not compatible with the vision it made through Tourism Vision 2020. The infrastructures and the environment are still not tourism-friendly. Deepak Raj Joshi, former CEO of the Nepal Tourism Board, during interview, stated:

> Though having immense possibilities of tourism in Nepal and having policies in writing, the practices of the government and the concerned authorities are still insufficient. There is a scope that tourism can itself become a means of development of a place for which some basic infrastructures are necessary and concerned authorities have to work on it. However, it is beyond possibility. The co-ordination and co-operation which we lack among different stakeholders is a major challenge we have not been able to overcome.

If we analyse this in line with the stakeholder theory considering the tourism industry as a company, there will be a range of actors, such as local leaders, government, NGOs, and when their roles and co-ordination do not synchronize, then development is not possible (Aguera, 2013).

With time, business communities and leaders realized the potential tourism has had for the country. However, no significant emphasis by the policy-makers was visible until the 1970s, when the Fourth National Plan of Nepal (1970–1975) emphasized national policy and international recognition of Nepal through tourism (Mahato *et al.*, 2021). Consequently, the first tourism master plan was launched in 1972. Different agencies were established including the Nepal Tourism Board. The government of Nepal seems to prioritize tourism in the form of policies and strategies; however, the practices are still not enough.

In conclusion, it can be said that the government has realized quite late that tourism is a driver of development.

Theoretical review

There are very few theories that are directly about tourism and there are plenty of theories that can be related to tourism as it is an interdisciplinary subject. So, different theories were reviewed.

The theory of the tragedy of the commons of Hardin (1968) is applicable when we talk about exploitation of nature and natural resources by humans. Even in mountains, the cases of grassland, as with Hardin (1968), can be referred to the land, forests and natural resources. As stated earlier, had there not been any tourism, the major source of livelihood for people in the mountains would be agriculture, which is often subsistence farming. People tend to exploit nature and natural resources as rational choices. This theory can be referred to in the case of the mountain community in Nepal.

Whenever one talks about sustainable development and sustainability, one needs to talk about resilience. Theory of resilience can be referred to in tourism as a capacity of the industry and communities to bear shocks and bounce back (Pappas *et al.*, 2023). In other words, the theory can be relevant to the study as it describes resilience by three characteristics: the amount of disturbance a system can absorb and still remain the domain of attraction; the degree to which the system is capable of self-organization; and the ability to build and increase the capacity for learning and adaptation (Carpenter, 2001), as cited in ESDN (2012). Another relevant theory for the study is the theory of community-based tourism. The concept of community-based tourism is said to have emerged in the mid-1990s (Henema *et al.*, 2019) and is a form of tourism pursued and supported by local communities.

Tourism and Sustainable Development in Sagarmatha National Park

There is no dilemma about the contribution tourism can make in development. The more interesting fact is its multiplier effects in a place, which is depicted in Fig. 24.1. The tourism sector is directly linked with other sectors (infrastructure, agriculture, food production etc.) and if backward linkages are well utilized it can help develop other sectors as well.

Besides talking about an ideal situation, as discussed in different reports, we have to look at the ground reality and context of the rural highlands of Nepal. To be specific to the Sagarmatha region, most of the park (69%) comprises barren land above 5000 m with 28% being grazing land and almost 3% forest. The local inhabitants are Sherpas whose primary occupation is agriculture; however, due to the limited amount of land on which they can practise agriculture, they are dependent on the limited trade-based activities directly or indirectly related to tourism (UNESCO, 2022).

Sustainable development is defined as 'development which meets the needs of current generations without compromising the ability of future generations to meet their own needs' (WCED, 1987, p. 41). The concept, which was first used by the Brundtland Commission in 1987, is now a global agenda. Now, 193 countries have agreed common goals, called sustainable development goals. Sustainable development does not limit economic progress; it blends other dimensions, including socio-cultural and environmental. This concept makes tourism, especially in rural areas, an opportunity to

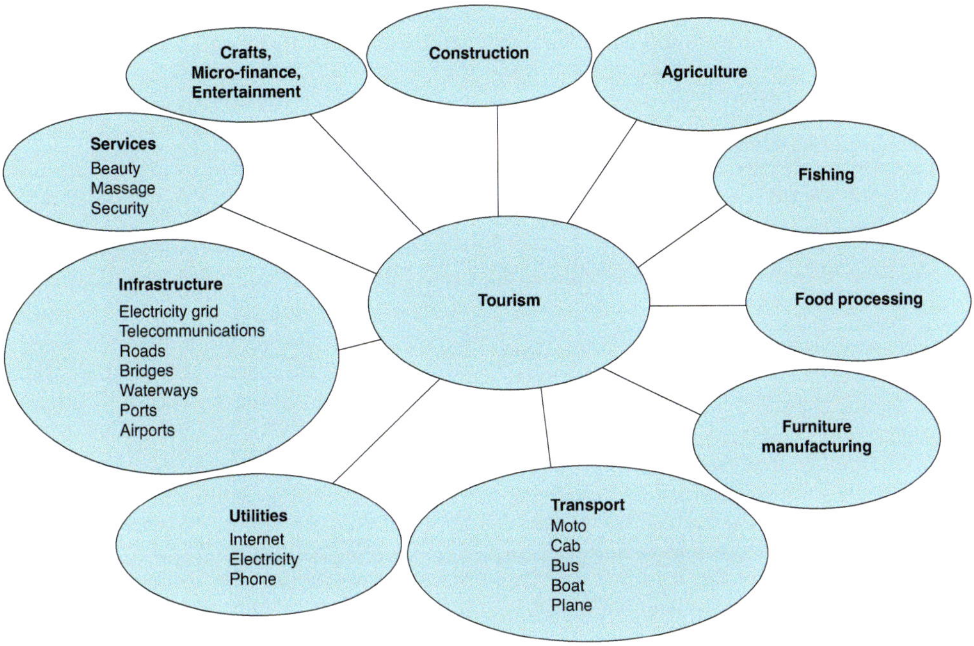

Fig. 24.1. Links between the tourism sector and other sectors. (OECD and UNWTO, 2013).

work towards sustainable development, which is discussed below.

Tourism: people first

Tourism is a labour-intensive sector and so it provides employment opportunities. To be specific, when a trekker goes trekking in the highlands, they directly contribute to the daily income of guides, porters, small lodges, retail shops, etc. There is a direct connection to the banking institutions, helicopter service providers, medical teams, etc. Moreover, the multiplier effects also result in the development of other amenities like health and education (Forum on Environment and Development, 1998; DEVCO and UNWTO, 2013; UNWTO, 2022).

Tourism has also helped reduce inequalities in the region (UNESCO, 2022). Tourism is a sector with high female participation rates and it caters for the daily needs of even unskilled human resources including women and young people. Braber *et al.* (2018) shared that tourism activities in protected areas of Nepal including Sagarmatha have lowered poverty. On a similar note, Pasang Sherpa, during interviews, said that there is no significant difference in the daily wages of female and male porters. Porters have to carry the stuff of the trekkers and it does not matter whether the porter is male or female. Actually, female trekkers are more comfortable with female porters. Therefore, it can be said that, looking at the Commission on Sustainable Development (CSD) theme indicator framework of sustainable development, tourism has contributed to promoting equality in the Sagarmatha region.

The analysis of the multi-dimensional poverty index of Province 1 (Province of Sagarmatha) shows the trend of progressive decrease of poverty and incremental progress of school enrolment and living standards, including health, sanitation, drinking water, cooking fuel, assets and electricity (Government of Nepal, 2022). Despite the absence of location-specific data, it can be assumed that there has been progress in the field of socio-cultural dimensions, which can be linked with tourism. During the key informant interview (KII), the ward chairperson stated:

> Tourism seems to be the only option for our daily livelihood, all industries here are directly/ indirectly related to tourism, and all the people, irrespective of their gender and age, are affected by the tourism. For instance, not only owners and workers of teahouses but small shops are also dependent on tourism.

Analysing the statement, it can be said that tourism is a business which is done by the community and community-based tourism theory's principle is that the major benefits remain in society, though in small amounts; in the region, most of the revenues of trekking are received by local communities in the form of accommodation, food, porter charge and others (Baniya *et al.*, 2018).

To conclude on the impacts tourism has had in different socio-cultural dimensions of the region, it can be summarized in the words of Mr Pasang as: 'Tourism is the sector which puts people in the first priority as the one who serves is also human and the one who gets served is also human.' Analysing all these, the tourism sector can be said to be socio-culturally sustainable.

Tourism: the viable source of economy

The tourism industry is widely regarded as the most dependable source of foreign currencies as the industry is based on visits by foreigners. Durbarry (2004) suggests tourism as an instrument of employment generation as it helps in earning through commodities trade. 'Tourism is the crucial growth sector for Least Developed Countries (LDCs)' (UNWTO,International Trade Centre & Enhanced Integration Framework, 2017, p. 1). To add to this statement, different scholars classify the impacts of tourism in the economy as direct, indirect and induced (Dwyer *et al.*, 2000). The same can be justified

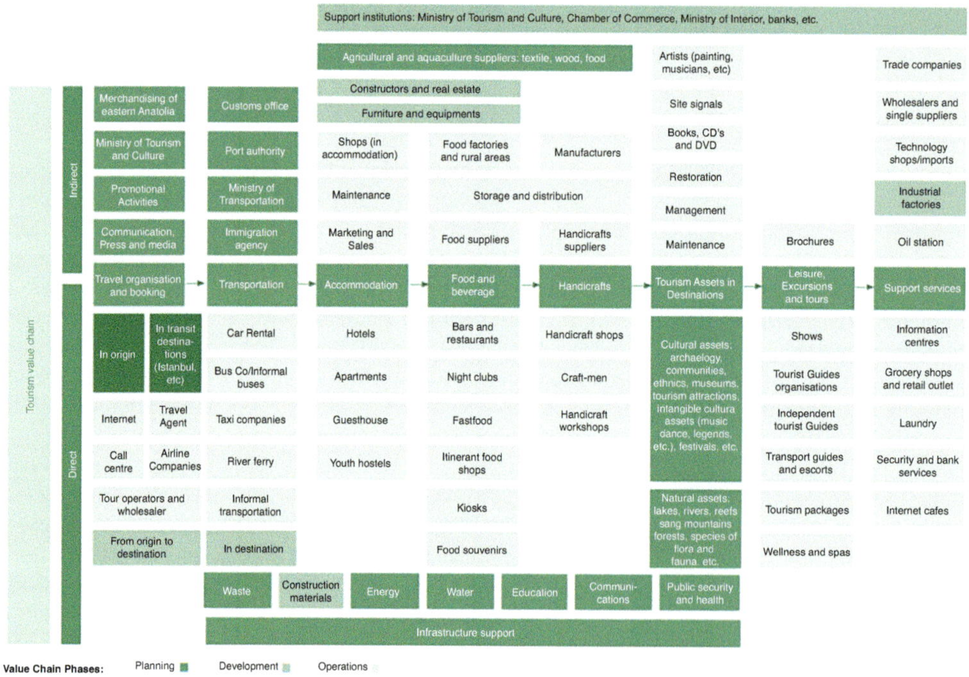

Fig. 24.2. Tourism value chain. (From DEVCO and UNWTO, 2013).

by the tourism value chain shared by DEVCO and UNWTO (2013) in Fig. 24.2.

The tourism value chain shown in Fig. 24.2 illustrates a wide range of activities dependent on tourism or that are results of the tourism industry. It blends different sectors like transportation, accommodation, food and beverage, and infrastructure Mahato *et al.* (2021), while describing the tourism value chain, states: 'As a result of interconnectivity of the outbound and inbound activities both directly as depicted in the value chain, the tourism sector can become a driving force for economic development by contributing to three high priority goals of developing countries' (p. 71). Therefore, the economic impact tourism industry brings to a place is widely agreed.

In the Sagarmatha region, there are many tangible results tourism has brought. The local Sherpas have got employment opportunities in the form of guides, tourists, support service providers and the like. Further, there has been exponential growth in tourism-related facility development. Sun and Watanabe (2021) state that lodges in the major villages in the Sagarmatha region have grown exponentially. For instance, there has been growth of 81.8% in lodges in Namche Bazar from 1997 to 2019; and there has been growth of 400% in Zamphute in the same period. Furthermore, the tourism industry also brings opportunities to small and micro-level enterprises as even low cost would add value (UNESCO, 2022).

Lakhpa Chiring, local leader, states:

> Since the only business local Sherpas can afford here is business of hotel, lodges and teahouses, and the number of visitors is also rising except during lockdown, the rise in number of hotels is very good. A single hotel can employ a minimum of five to ten people. So this region's economy is fully dependent on tourism.

During the KII with the rural municipality official, it was found that tourism has made lives possible in the region. The multiplier impacts of tourism are visible in the form of schools, banks and health posts being built in the region. In sum, the tourism industry has had positive impacts on the economy of the region and its local communities and it is the only viable source of economy in the region.

Tourism: conservation of nature and natural resources

To begin with the contribution tourism can make in conservation, an example of tragedy of commons would be relevant as tourism is all about common resources and its management. Hardin (1968) put a valid logic behind deterioration of the common resources due to lack of ownership by the locals and due to rational decisions a local makes while calculating possible gain and potential hazard while investing more for personal benefits. This could have been the fate of the region if it had not been declared a national park. Besides the national rules and regulations that restrict people from deteriorating the resources in the region, the values tourism has brought have made the local people more responsible. Mr Pasang Sherpa, during the interview, states: 'If there is no Everest and no greeneries, glaciers; why would tourists visit here? And what would we do to survive.' This shows the responsibility tourism has brought among local people. If we analyse this in line with the theory of 'governing the commons' of Ostrom (1990), it can be said that the whole region can be considered a common resource and the locals can be considered the people who eventually contribute in conservation as it also contributes to the economy of the people.

UNESCO on its website writes:

> The carefully preserved natural heritage and the dramatic beauty of the high, geologically young mountains and glaciers were recognized by UNESCO with the inscription of the park as a world heritage site in 1979. The property hosts over 20 villages with over 6000 Sherpas who have inhabited the region for the last four centuries. Continuing their traditional practice of culture and religion including the restriction of animal hunting and slaughtering, and reverence of all living beings. These practices combined with indigenous natural resource management practices, have been major contributing factors to the successful conservation of the SNP.
>
> (UNESCO, 2022)

This also justifies how tourism has contributed to conservation.

Being a protected area, Sagarmatha region helps in safeguarding different rare and endangered species of flora and fauna. Further, the security of the Nepal Army has nullified the threat of wildlife trade in the region as well. Thaman Gurung, a captain in the Nepal Army, during the interview stated that there is very little chance and incidents of wildlife trade and illegal hunting in the region. Furthermore, the region itself is a slaughter-free zone and local people do not believe in slaughter, which also helps in conservation of animals. There are regular national campaigns of mountain clean-up done by the Army, during which tonnes of waste, including dead bodies, are brought and managed properly. These all are outcomes of the tourism industry.

If we analyse the changes brought by tourism, we can relate them to the case of governing the commons of Ostrom (1990) as tourism is also about economic activities around common natural resources and tourism has made local people realize that those resources have to be conserved to yield the highest possible results. Therefore, from the theoretical perspective of community-based tourism, we can conclude that tourism in the highlands is actually a viable example of a sustainable development initiative.

Analysing all three dimensions – socio-cultural, economic and environmental – against the indicators of the CSD framework and theories of tragedy of commons and governing the commons, tourism in Sagarmatha has proved a driver of sustainable development.

Implications of the Study

Some of the major implications of the study are:

- The study provides a rationale for the policy-makers and leaders of the places like Sagarmatha region with huge potential to put tourism as a priority.
- The study shows that tourism can be a driver of development in the highlands and other regions where there is minimal infrastructure. Tourism strengthens facilities and supports other basic needs including education and safe drinking water.
- The tourism industry is an industry of local communities and people; therefore the local government has many opportunities to intervene to develop tourism and to strengthen the industry.

Conclusion

This chapter attempted to highlight how tourism can be one of the best options for the least developed countries like Nepal. Countries like Nepal have priorities to eradicate poverty and hunger and ensure basic education and health facilities. Tourism is usually an area often overlooked by the leaders. However, with proper orientation and planning, tourism can not only help in eradicating poverty but also contribute to achieving goals for the generations to come.

Sagarmatha region is an example of how tourism can transform a place and people's lives in a region. In other words, tourism can help a place develop socio-culturally, economically and environmentally. Tourists do not come alone to a destination; they bring foreign currency, employment opportunities and trade to the local area. Tourism has multiplier effects in any place. Tourism can contribute in equity and inclusive development, including educational development and health facilities, making it a valid contributor in socio-cultural development of a region.

The case study in this chapter shows how tourism became a sustainable replacement for subsistence farming in the Sagarmatha region. Tourism has given reasons not only to the authorities but also to the local communities to conserve nature and natural resources. Hence, the value of nature-based tourism for the wellbeing of the environment and the local communities is immense (Azara *et al.*, 2018; Clarke *et al.*, 2018). Sagarmatha region has successfully established itself as a demonstration of the idea of tourism for sustainable development.

References

Aguera, F. (2013) Stakeholder theory as a model for sustainable development in ecotourism. *TURyDES* 6(15). Available at: https://www.eumed.net/rev/turydes/15/stakeholders.html#:~:text=The%20concept%20of%20the%20stakeholder%20in%20a%20tourist%20company.&text=In%20this%20respect%2C%20the%20responsibility,stakeholder%20theory%20may%20be%20applied (accessed 28 October 2023).

Azara, I., Michopoulou, E., Niccolini, F., Taff, B.D. and Clarke, A. (2018) *Tourism, Health and Wellbeing in Protected Areas*. CAB International, Boston, Massachusetts. DOI: 10.1079/9781786391315.0000.

Baniya, R., Shrestha, U. and Karn, M. (2018) Local and community well-being through community based tourism – a study of transformative effect. *Journal of Tourism and Hospitality Education* 8, 77–96. DOI: 10.3126/jthe.v8i0.20012.

Braber, B., Evans, K.L. and Oldekop, J.A. (2018) Impact of protected areas on poverty, extreme poverty, and inequality in Nepal. *Conservation Letters* 1(9). DOI: 10.1111/conl.12576.

Carpenter, S.R. (2001) Alternate states of ecosystems: evidence and its implications. In: Press, M., Huntly, N. and Levin, S. (eds) *Ecology: Achievement and Challenge*. Blackwell, London.

Clarke, A., Azara, I., Michopoulou, E., Niccolini, F. and Taff, D. (2018) Reconsidering the relationship between tourism, health, wellbeing and protected areas. In: Azara, I., Michopoulou, E., Niccolini, F., Taff, D. and Clarke, A. (eds) *Tourism, Health, and Wellbeing and Protected Areas*. CAB International, Wallingford, UK. DOI: 10.1079/9781786391315.0000.

DEVCO and UNWTO (2013) *Sustainable Tourism for Development*. UNWTO, Madrid.

Durbarry, R. (2004) Tourism and economic growth: the case of Mauritius. *Tourism Economics* 10(4), 389–401. DOI: 10.5367/0000000042430962.

Dwyer, L., Forsyth, P., Madden, J. and Spurr, R. (2000) Economic impacts of inbound tourism under different assumptions regarding the macroeconomy. *Current Issues in Tourism* 3(4), 325–363. DOI: 10.1080/13683500008667877.

ESDN (European Sustainable Development Network) (2012) Resilience and sustainable development: theory of resilience, systems thinking, and adaptive governance. Available at: chrome-extension://oemmndcbldboiebfnladdacbdfmadadm/https://www.esdn.eu/fileadmin/ESDN_Reports/2012-September-Resilience_and_Sustainable_Development.pdf (accessed 29 October 2023).

FAO and WTO (2023) Understanding and quantifying mountain tourism. Food and Agricultural Organization of the United Nations. Rome.

Forum on Environment and Development (1998) *Tourism and Sustainable Development*. Forum on Environment and Development, Bonn.

Government of Nepal (2022) *Nepal Tourism Statistics 2021*. Government of Nepal.

Hardin, G. (1968) The tragedy of the commons. *Science* 162(3859), 1243–1248. DOI: 10.1126/science.162.3859.1243.

Henema, U.S., Strydom, A.J. and Mangope, D.F. (2019) A critique of the interface between tourism, sustainable development and sustainable tourism in community-based tourism heory. Available at: https://www.researchgate.net/publication/337423554_A_critique_of_the_interface_between_tourism_sustainable_development_and_sustainable_tourism_in_community-_based_tourism_theory (accessed 29 October 2023).

Mahato, G., Nyaupane, P., Lamichhane, M., Bhattarai, U., Subedi, D. *et al.* (2021) *Tourism in Nepal*. Nepal Mountain Academy, Kathmandu.

OECD and UNWTO (2013) Aid for Trade and Value Chains in Tourism. OECD and UNWTO.

Ostrom, E. (1990) *Governing the Commons: The Evolution of Institutions for Collective Action*. Cambridge University Press, Cambridge, UK. DOI: 10.1017/CBO9780511807763.

Pappas, N., Michopoulou, E. and Farmaki, A. (2023) Tourism innovation and resilience during uncertainty. *Tourism Planning & Development* 20(2), 135–137. DOI: 10.1080/21568316.2023.2176647.

Sagarmatha National Park (2022). Available at: https://snp.gov.np/tourism

Shrestha, H.P. and Shrestha, P. (2012) Tourism in nepal: a historical perspective and present trend of development. *Himalayan Journal of Sociology and Anthropology* 5, 54–75. DOI: 10.3126/hjsa.v5i0.7039.

Sun, Y. and Watanabe, T. (2021) Tourism-related facility development in sagarmatha (Mount Everest) National Park and buffer zone, Nepal Himalaya. *Land* 10(9), 1–20. DOI: 10.3390/land10090925.

Thapa, D. (1998) *Toni Hagen's Nepal: The Kingdom in the Himalaya*. Himal Books, Kathmandu.

UNESCO (United Nations Educational, Scientific and Cultural Organization) (2022) UNESCO World Heritage Convention. Available at: https://whc.unesco.org/en/list/120/ (accessed 28 October 2023).

UNWTO (2022) *Sustainable Development for Tourism Guidebook*. Available at: https://www.e-unwto.org/doi/book/10.18111/9789284415496 (accessed 29 October 2023).

UNWTO (United Nations World Tourism Organization), the ITC (International Trade Centre), and EIF (Enhanced Integrated Framework) (2017) *Tourism for Sustainable Development in the Least Developed Countries –Leveraging Resources for Sustainable Tourism with the Enhanced Integrated Framework*. UNWTO, ITC, EIF, Geneva.

WCED (World Commission on Environment and Development) (1987) *Report of the World Commission on Environment and Development: Our Common Future*. Brundtland Commission.

WTTC (World Travel and Tourism Council) (2022) *Nepal 2022 Annual Research: Key Highlights*. WTTC, London.

25 Nature-based Tourism Potential in Ukraine: Strengthening the Industry's Postwar Recovery through Cluster Development

Nataliia Kochkina[1*], Iryna Trunina[2], Myroslava Bosovska[3] and Olena Sushchenko[4]

[1]Marketing and Business Administration Department, Taras Shevchenko National University of Kyiv, Ukraine; [2]Business Administration, Marketing and Tourism Department, Kremenchuk Mykhailo Ostrohradskyi National University, Kremenchuk, Ukraine; [3]Hotel and Restaurant Business Department, Kyiv National University of Trade and Economics, Kyiv, Ukraine; [4]Tourism Department, Simon Kuznets Kharkiv National University of Economics, Kharkiv, Ukraine

Abstract

This chapter aims to provide a more comprehensive understanding of green tourism clusters by defining them according to 12 semantic features. To facilitate the development of green tourism clusters, a matrix of strategies is suggested that considers the interconnections between members and their potential to benefit from mutual support and interaction. The research highlights Ukraine's significant potential for rural green tourism, thanks to its rich natural and cultural heritage. This study examines the challenges and prospects of the rural green sector in the Kharkiv region, located in the east of Ukraine, as a part of the national recovery plan following the full-scale Russian invasion in 2022.

Introduction

In the 21st century, clusters have emerged as a key factor influencing socio-economic phenomena and driving humanitarian and economic development. They represent an integral component of the knowledge economy, entrepreneurial environment and informational society, built on principles of consolidation, strategic partnership and co-competition. In Ukraine's tourism industry, clusters have enabled tourism players to acquire unique competitive advantages and organizational competencies while optimizing the resource potential and strategic opportunities to yield synergistic and multiplicative benefits (Boiko *et al.*, 2017). Furthermore, clusters have been crucial in promoting the national Ukrainian tourism brand (Melnychenko *et al.*, 2021). Developing clusters is a national priority for nature-based tourism, as it integrates over 30 spheres of the Ukrainian economy, including tourism, the agricultural sector, adjacent and auxiliary industries, territorial communities,

*Corresponding author: nataliakochkina@knu.ua

DOI: 10.1079/9781800621411.0025

and regions. In this context, cluster studies should consider the specificity of nature-based tourism in Ukraine, with rural green tourism emerging as the most popular form.

The Ukrainian tourism industry has been one of the hardest-hit sectors following the full-scale Russian invasion in 2022. The closure of Ukrainian airspace, cessation of tour operator activities and millions of people becoming refugees rather than tourists had a profound impact. Of course, discussing tourism during military conflicts is challenging. However, in the postwar reconstruction phase, the cluster approach could contribute to the redevelopment of the tourism industry by facilitating collaboration among competitors who benefit by achieving common goals.

Clusters have been recognized as valuable for managing highly competitive economies through co-competition based on co-operation and mutual benefit (Porter, 1980; Roelandt and den Hertog, 1999; Sushchenko, 2009; Brandenburger and Nalebuff, 2011). Clusters help increase competitiveness and drive innovative economic development, especially during times of crisis. Consequently, world-leading economies frequently adopt a cluster approach to enhance national and regional competitiveness (Sokolenko, 2004; Boiko *et al.*, 2017; Gorina *et al.*, 2020). This trend has become even more prominent during the COVID-19 pandemic and is expected to continue after the war between Russia and Ukraine. The cluster approach is particularly effective in tourism due to its synergistic effect, leading to higher profitability of competing players within the same cluster.

Marshall (1964) states that implementing a dynamic regional policy requires cluster development. Wolfgang (1999) emphasized that co-operation among enterprises is indispensable in enhancing industry competitiveness. Cluster members mutually support each other in exchanging information, deploying innovative activities, and adopting and acquiring new technologies. These relationships foster the creation of new competitive advantages and opportunities for success in the global market. Unlike business networks, clusters involve a larger range of members, including manufacturers, suppliers, distributors, engineering and consulting firms, research institutes, educational institutions, banks, business associations, regional administrations and public organizations.

A cluster typically refers to a group of manufacturers and public institutions collaborating to enhance their competitiveness (Drevinh, 2008). However, when it comes to the tourism industry, this interpretation requires further clarification (Bosovska, 2015). We define a tourist cluster as a geographically localized, interconnected and mutually reinforcing group of market players ready to delegate management functions to a co-ordination centre to achieve a synergistic effect through entrepreneurial initiatives promoting territorial development. A green tourism cluster may be considered a group of enterprises and organizations (institutional feature) that agree to interact and conduct business (structural feature) coherently with other cluster members (system characteristic). They operate within a specific tourist destination (geographic feature) and share technologies (technological feature) while combining resources through vertical and horizontal co-operation (functional feature). Clusters jointly use regional natural and recreational resources to promote territorial development (green tourism feature). They develop management systems (organizational and management features) and form behavioural patterns (behavioural features). Clusters are flexible alliances (evolutionary feature) that promote innovativeness among participants (innovative feature), ensuring their competitiveness and goal achievement (strategic feature) through a synergistic effect (qualitative feature).

Rural Green Tourist Clusters: A New Approach

We propose a five-block model of green tourism clusters (Fig. 25.1) that outlines the attributes that drive co-operation, target orientations and initiatives among enterprises. At the core of a cluster is the sector of green tourism production, while supplying and additional services sectors form a cluster basis. The drivers of cluster activity include the availability of assets, knowledge and competencies, which provide the necessary conditions and resources for cluster co-operation. The environment determines a tourist destination's attractiveness based on certain

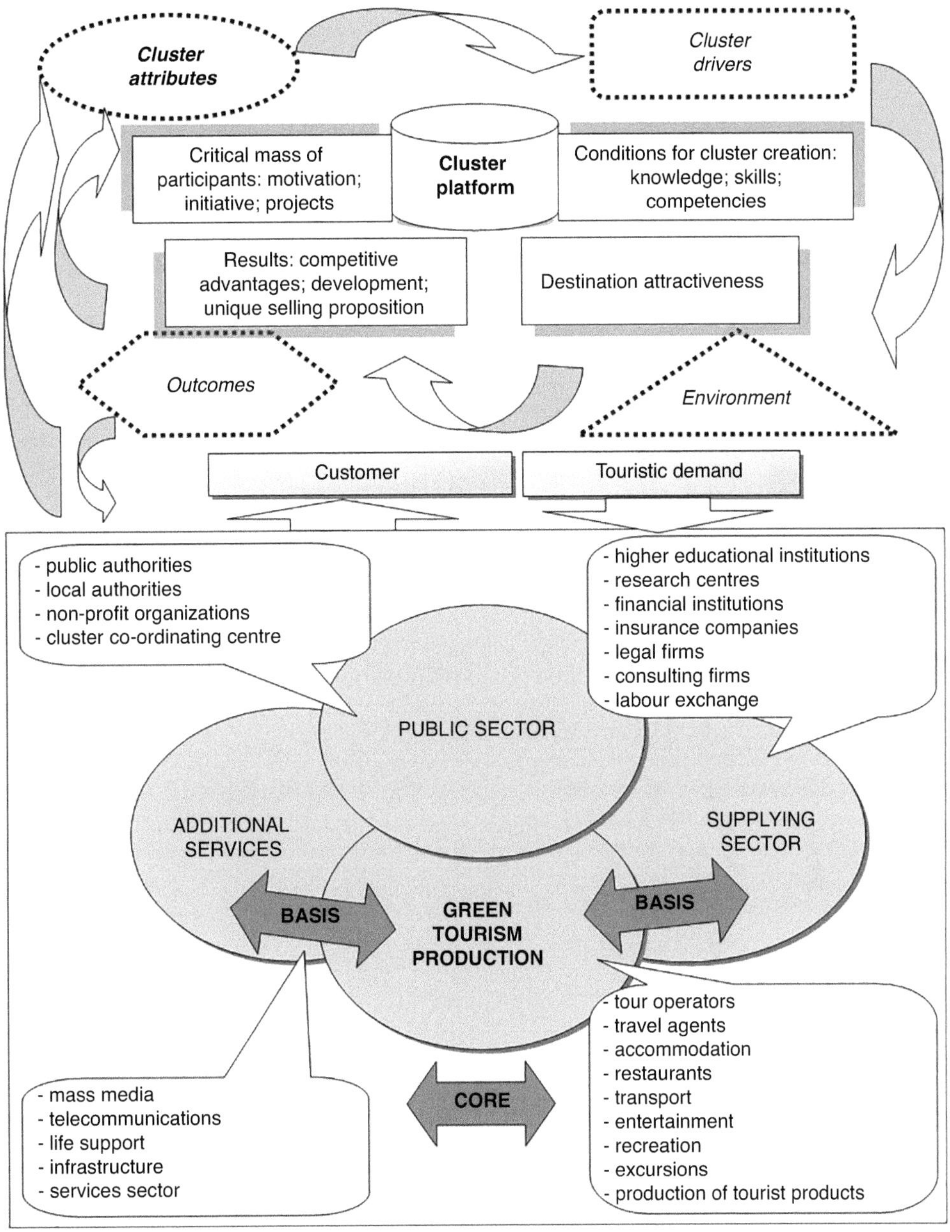

Fig. 25.1. Model for developing green tourism clusters.

unique natural and recreational resources, tourist and transport infrastructure, investment programmes for regional development, institutional support of cluster interaction and availability of qualified personnel. The outcomes of cluster co-operation reflect the benefits and risks for members and regional development. Finally, the cluster platform comprises knowledge, competencies, trust, interests, strategic advantages, business processes and innovation.

Research indicates that effective cluster development relies heavily on the interaction between its members. Successful cluster development must balance the expectations of all participants with their strategic capabilities within the given limitations. Relationships within the cluster can evolve and revolutionize, leading to a cyclical pattern and a specific vector (spiral), which can be formalized and based on a systematic approach (Pylypenko, 2007; Bosovska *et al.*, 2019). Four basic strategies can be employed to develop green tourism clusters, as illustrated in Table 25.1.

In the context of cluster development, stabilization in business activities refers to a situation where cluster members strive to maintain their economic independence while sharing resources carefully, which can create additional opportunities and reduce risks. This type of cluster tends to have a high level of competition, but it can offer short-term stability for its members. However, the lack of co-operation can hinder further development. The strategic vector of this type of cluster involves implementing short-term initiatives and existing projects of participants.

On the other hand, a business expansion strategy may be employed when cluster members have a positive business reputation and enjoy close, long-term relationships. This strategy involves expanding cluster products, setting new goals, preserving formed images and values, and intensifying participant responsibility for possible risks and results of joint cluster activities. In this case, enterprises are motivated to co-operate, which can stimulate long-term development and ensures project expansion for the cluster.

In cluster development, absolute competitive advantages can be achieved when the growth in business activity leads to positive economic effects through the introduction of unique green tourism products. This can be achieved through innovations, higher consumer satisfaction, optimal cluster structure and high organizational competencies, which reduce costs through economies of scale. With these advantages, cluster members may achieve a monopolistic position in the market. However, the alliance may quickly collapse because of inconsistent ties between members and the need to co-ordinate business activities and power distribution.

Alternatively, each cluster member can utilize its individual and organizational competencies to gain relative competitive advantages. Cluster effectiveness also depends on the nature and duration of the relationships (symbiosis) between the members. In this case, companies perceive each other as strategic partners rather than competitors, generating added value together. The cluster represents a system built through the co-operation of many participants, aimed at effective long-term development and gaining new competitive advantages in the business environment.

Rural Green Tourism in Ukraine

Countryside recreation has been popular in Ukraine since the late 19th century, particularly among intellectuals and members of youth organizations. In the 1960s, tourist and health complexes emerged in the traditionally popular summer recreation areas, such as Morshyn, Truskavets and Svalyava, in the west of Ukraine. However, these areas were often overcrowded during peak seasons and lacked available accommodation. In 1996, the Union for Promotion of Rural Green Tourism in Ukraine was founded

Table 25.1. Matrix of strategies for green tourism clusters.

		Growth in business effectiveness	
		Low	High
Intensification of connections between cluster members	Low (independence)	Stabilization in business activities	Benefitting from absolute competitive advantages
	High (co-operation)	Business expansion strategy (introducing new products)	Benefitting from relative competitive advantages

and introduced the 'rural green tourism' domestic brand for promotion abroad. The organization has been a member of Eurogites since 2003 (Zbarsky and Gribova, 2020).

Green and rural tourism are the most popular forms of nature-based tourism in Ukraine, with high potential due to the country's natural and cultural heritage. Ukraine has thousands of unique historical monuments and beautiful landscapes with cultural value comparable to famous landmarks such as the Egyptian pyramids or Sumerian cities. These include hillforts of Tryplian culture, Scythian burial mounds, priceless treasures of Kyiv Rus, medieval castles and wonderful aristocratic parks. The country also boasts 15 unique protected areas, 55 national natural parks and 71 regional landscape parks, as shown in Fig. 25.2 and Table 25.2.

Ukraine's natural, historical and cultural resources have the potential to attract significant interest from domestic and foreign tourists. The country's state resort and recreational areas cover approximately 9.1 million ha, representing around 15% of the territory. Additionally, rural regions of Ukraine with high agricultural development pose rich tourist and recreational potential, as shown in Fig. 25.3.

As of 2022, Ukraine was ranked 60th in the Environmental Performance Index (EPI, 2022). It was placed in the first-in-the-world position in the agricultural index, indicating the best environmental performance of agricultural production. Additionally, Ukraine's climate is favourable compared to many other countries worldwide, ranking 26th in the EPI rating. The country has abundant water resources, fisheries and marine protected areas, making it suitable for developing ecotourism. Despite its significant natural, ethnographic, historical and cultural potential, the development of rural green tourism in Ukraine remains low. Only an estimated 1500–2000 homesteads provide rural tourism services, mainly in the west of the country. The country lacks an organizational and legislative framework for this sector; there is no financial or statistical reporting, and very few investments have been made in this area.

Rural Green Tourist Potential of the Kharkiv Region

Kharkiv is the second-largest Ukrainian city, located in the east of the country near the Russian border (Fig. 25.4). Since the Russian invasion on 24 February 2022, the region has suffered significant damage, with over 550 cultural and historical monuments destroyed as of December 2022, mainly in the Kharkiv and Donetsk regions (UACulture, n.d.). The Postcards from Ukraine project has documented the destruction of 113 architectural objects in the area, including university facilities, museums, Kharkiv Freedom Square, the Opera and Ballet Theatre, the Church of the Transfiguration of Our Lord and the Holocaust Memorial Complex. Although the Kharkiv region has suffered considerable damage, its tourism potential may still be utilized during the postwar reconstruction, given the strong interest in historic sites and landmarks.

The region's land provides ample opportunity for green and rural tourism, including its forests, green zones of cities and suburban areas, and water resources (Kharkiv Regional State Administration, 2021). The region boasts 91,300 ha of water, accounting for 2.9% of its territory, with 46,300 ha occupied by reservoirs and ponds. The Kharkiv region is predominantly located within the basins of the Siverskyi Donets river, covering 21,930 km^2 (69.8% of the region's territory), and the Dnipro river, covering 9,470 km^2 (30.2% of the region's territory). The region's forests span a total area of 431,873 ha. These forests are crucial in ecological functions such as recreation, water and nature protection. While the forest coverage of the region ranks 15th in Ukraine, they are distributed fragmentarily across the area (Green Ukraine, 2020).

Table 25.3 indicates that 84,709.15 ha have been designated for green and rural tourism within the Kharkiv region's protected areas, constituting 2.70% of the region's total area. Aside from its natural resources, the region also boasts a rich cultural heritage, including architectural, archaeological, historical and monumental art monuments, with 9561 units – approximately 5% of Ukraine's historical and cultural potential.

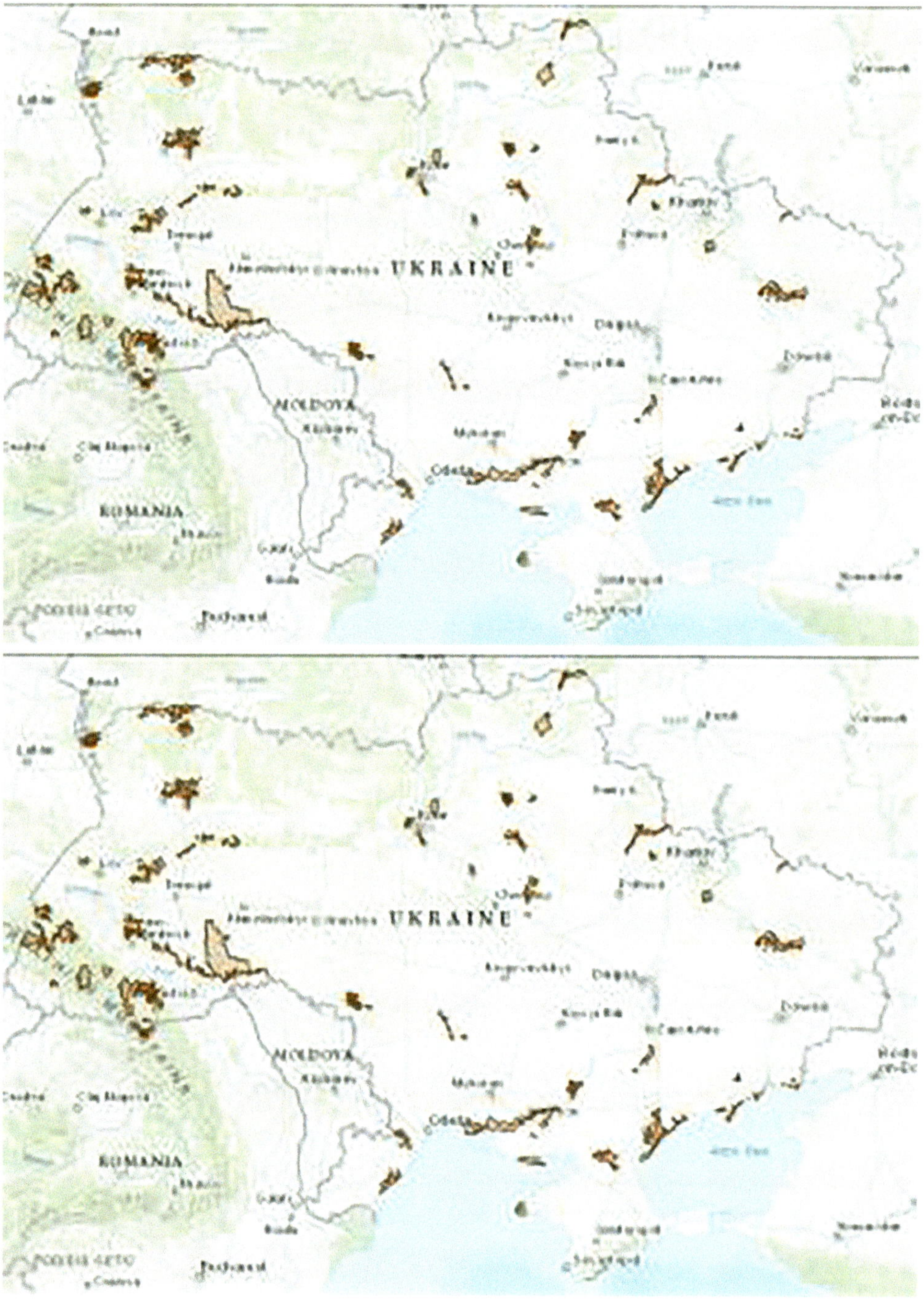

Fig. 25.2. Ukraine's national natural parks (a) and regional landscape parks (b). (From Grachev, n.d. -a,-b)

Table 25.2. Distribution of national natural parks and regional landscape parks in Ukraine. (From Ministry of Environmental Protection and Natural Resources, 2021.)

	National natural parks		Regional landscape parks	
Region	Number	Area (ha)	Number	Area (ha)
Vinnytsia	1	20,203.40	4	18,468.38
Volynsk	3	121,767.84	0	0
Dnipropetrovsk	0	0	5	14,499.67
Donetsk	2	61,326.50	6	28,927.31
Zhytomyr	0	0	0	0
Zakarpattia	3	99,680.50	2	14,961.96
Zaporizhzhia	2	94,882.92	1	1025.00
Ivano-Frankivsk	5	120,339.70	3	38,417.00
Kyivska	2	17,206.72	4	17,206.70
Kirovohradsk	0	0	2	78,758.05
Luhansk	1	7269.00	1	14,011.00
Lviv	5	79,585.50	5	56,540.68
Mykolayivska	2	41,497.48	5	39,345.20
Odesa	2	49,176.10	2	15,320.00
Poltava	2	22,792.62	5	53,056.45
Rivne	2	30,767.11	3	58,708.00
Sumy	2	39,575.20	1	98,857.90
Ternopilsk	2	18,681.48	3	42,997.00
Kharkiv	3	22,690.00	7	20,544.33
Khersonsk	5	166,692.74	0	0
Khmelnytska	2	270,078.70	1	16,915.30
Cherkassy	2	11,227.23	1	5562.50
Chernivtsi	3	27,801.60	2	36,473.30
Chernihivska	3	41,988.50	3	85,045.35
City of Kyiv	1	10,988.14	4	2651.60
Sevastopol	0	0	1	83.90
Total	**55**	**1,376,218.98**	**71**	**758,376.58**

Efforts to promote green and rural tourism in the region necessitate addressing critical problems that hinder the sector's development. Inadequate social infrastructure and engineering lines, including issues with heat and water supply, inadequate household facilities and lack of transportation networks are among the primary restraining factors. Additionally, the low standard of living in rural areas reduces their potential attractiveness to both domestic and foreign tourists. Imperfect tourism legislation in green services, low quality of services, and a lack of information and marketing support for rural green tourism further compound the issues. Finally, the lack of a clear strategy for developing green tourism at the regional and state levels also negatively impacts the sector's growth, as depicted in Table 25.4.

A cluster-based approach can be effectively utilized to address the issues mentioned above. Cluster systems generate synergistic effects by

Fig. 25.3. Ukrainian landscape. (From Polishchuk, 2022.)

Fig. 25.4. Kharkiv region bioclimatic potential. (From Reshetchenko *et al.*, 2018.)

Table 25.3. Protected areas in the Kharkiv region. (From Green Ukraine, 2020.)

	Of national value		Of local value		Total	
Category	Number	Area (,000 ha)	Number	Area (,000 ha)	Number	Area (,000 ha)
National parks	3	22,690	-	-	3	22,690
Regional landscape parks	-	-	7	20,544.3	7	20,544.3
Reserves	3	1038	166	36,921.6	169	37,959.6
including landscape	-	-	13	26,043.2	13	26,043.2
forest	-	-	9	3207.1	9	3207.1
botanical	1	185	51	3169.99	52	3354.99
zoological	2	853	5	1292.4	7	2145.4
ornithological	-	-	7	787.9	7	787.9
entomological	-	-	63	594.8	63	594.8
hydrological	-	-	17	1811.6	17	1811.6
geological	-	-	1	14.6	1	14.6
Nature memorials	-	-	44	645.9	44	645.9
including complex	-	-	2	176.3	2	176.3
botanical	-	-	38	455.2	38	455.2
hydrological	-	-	4	14.4	4	14.4
Reserve tracts	-	-	9	2537.2	9	2537.2
Botanical gardens	1	41.9	1	13.25	2	55.15
Dendrological parks	1	22.8	1	51.5	2	74.3
Monuments of garden and park art	4	169.9	1	10.8	5	180.7
Zoological parks	1	22	-	-	1	22
Total	13	23,984.6	229	60,724.5	242	84,709.15

fostering the development of new nature and culture-based destinations (Fig. 25.5). This is crucial for the postwar reconstruction of Ukrainian regions and the country as a whole.

Perspectives for Touristic Clusters in the Kharkiv Region After the War

After the war, the Kharkiv region of Ukraine can develop ecotourism by following a four-step plan that utilizes cluster principles and instruments (see Table 25.5). This plan will help address the issues listed in Table 25.4. The first step involves collaborating with local authorities to manage rural green tourism, which can be facilitated through tourist clusters that bring together various rural tourism businesses. Introducing tourist service databases and rural tourism development projects in different regions and rural communities is necessary. Additionally, the regional authorities in Kharkiv should invest in creating a legislative framework, establishing new institutions and promoting green tourism programmes at all levels.

The second step of the postwar reconstruction strategy for rural tourism in Ukraine involves developing and implementing quality standards. The aim is to increase the attractiveness of rural tourism by introducing standards that will certify, categorize and catalogue rural

Fig. 25.5. Sharivskyi Palace and Dvorichanskyi National Nature Park. (From Polishchuk, 2019; Kletionkin, 2014, 2021a, b.)

Table 25.4. Factors inhibiting the development of rural green tourism in the Kharkiv region.

Factor	Description	Consequences
Lack of support from authorities	Lack of rural green tourism law creates uncertain legal status and low interest from local and state authorities. No active advertising or co-operation with uncertified tourist actors. No government programmes or funding for rural green tourism. Insufficient competencies and co-ordination among responsible ministries.	Unclear legal status results in misunderstandings with controlling authorities. Low interest from authorities reduces funding and resources. Lack of government programmes and funding leads to insufficient promotion and lack of investment in rural green tourism.
Lack of promotion for green and rural tourism	Lack of integrated advertising, tourist information centres, catalogues, guidebooks and databases. No reservation system and poor co-ordination of marketing activities.	Lack of awareness and information about rural tourism. Difficulty in planning and booking rural tourism activities.
Low comfort level of existing rural estates	Lack of centralized water supply and sewage, low-quality drinking water, lack of information and standardization for owners and no unified state system or database.	Poor living conditions for tourists. Lack of trust in the quality of accommodation. Difficulty in planning and booking rural tourism activities.
Poor rural engineering infrastructure	Poor land improvement, waste management, highway conditions and transportation links.	Difficulty in accessing rural estates. Poor living conditions for tourists. Difficulty in planning and booking rural tourism activities.
Narrow product line and low-quality rural tourist services	Lack of quality standard for accommodation, poor entertainment options, no transfer to rural tourism facilities, no public associations or clusters, low professional level, narrow range of additional services and lack of foreign-language proficiency.	Low satisfaction and retention rate among tourists. Negative word-of-mouth and reviews. Difficulty in attracting new customers. Difficulty in expanding and diversifying rural tourism offerings.

estates. This will encourage rural tourism sector players to improve the quality of their accommodation and additional services.

The third step focuses on upgrading the competencies of rural estate owners in the tourism business through advanced training, seminars, conferences, and the publication of educational literature, such as handbooks.

The fourth and final step aims to enhance rural tourism competitiveness in the Kharkiv region by utilizing various marketing instruments, including promotional activities, opening tourist information centres, participating in local and international exhibitions, publishing catalogues, maps and guidebooks, and using informative web pages and social media.

Taking inspiration from the experience of other countries that have faced armed conflicts, it may be worthwhile to integrate rural green and military tourism into Ukraine's postwar reconstruction strategy. For example, after the end of the war in Croatia, a broad public relations campaign was conducted to promote previously unknown regions, resulting in tourism contributing 15% to the country's GDP (Goulding and Domic, 2009; Arnaud, 2016; Zarubina *et al.*, 2022). Bosnia and Herzegovina experienced an annual increase in tourist flows

Table 25.5. Implementation steps for developing rural green tourism in the Kharkiv region after the war.

Step	Description
1. Partnership and programme development	Expand the Coordination Council for the Development of Rural Territories by including representatives from non-governmental organizations and rural tourism entities.
	Develop and submit a rural tourism development programme for approval to Ukraine's Verkhovna Rada (Ukrainian Parliament), based on the concept of rural tourism development in Ukraine, specifically for the Kharkiv region.
	Develop and approve local programmes to promote rural tourism during postwar reconstruction.
	Develop and approve rural green tourism programmes at the village council level.
	Test selected rural tourism models in terms of costs and profits.
2. Quality standards and promotion	Develop uniform quality standards for rural tourism estates to ensure consistency and quality.
	Create a rural tourism database to promote rural tourism to potential tourists.
	Clarify the accommodation certification rules for all rural tourism representatives.
	Develop an integrated system to categorize rural tourism estates based on quality, type and other relevant factors.
	Create a catalogue of certified and categorized rural tourism estates.
3. Competency development for estate owners	Publish a handbook on organizational forms of rural tourism entities to guide estate owners.
	Introduce ongoing seminars on rural tourism to develop rural areas for estate owners and local executive bodies.
	Publish a manual on rural tourism business for estate owners.
	Organize conferences to share experiences and discuss rural tourism problems.
	Conduct training and seminars for estate owners and residents willing to engage in rural tourism.
4. Marketing and integration	Create a database of additional services for tourists in rural areas to promote rural estate services.
	Develop a website for rural tourism to provide information to potential tourists.
	Open rural tourism information centres based on clusters.
	Promote the Kharkiv calendar of events for the tourist season to attract tourists to the region.
	Integrate rural green and military tourism into the Kharkiv region's overall tourism promotion strategy to increase rural tourism's attractiveness.

of 24% following the end of the conflict in 1995 (Causevic, 2010; Aussems, 2016; Kassouha, 2019). Despite the ongoing conflict with Palestine, Israel's tourism industry accounts for 6.2% of the economy (Brin, 2006; Parfinenko, 2015; Strömbom, 2019). Georgia also attracted investments in the tourism industry amid its armed conflict with Russia (Arghutashvili, 2018; Romanova, 2018).

Drawing on the experience of countries that have undergone armed conflicts, it may be beneficial to integrate military and memory tourism with rural green tourism in the postwar reconstruction strategy for Ukraine (Godis and Nilsson, 2018). Military tourism can highlight the country's military glory, including defence structures, battlesites, museums, memorials, monuments, and burials related to military operations. This can demonstrate the negative consequences of war and the resilience of the Ukrainian people who defended their independence in the most significant war of the early 21st century. Croatia, Bosnia and Herzegovina, Israel and Georgia have all experienced increased tourist flows and investments in their tourism industries following armed conflicts. Therefore, Ukraine can also benefit from this strategy to promote its rural and military tourism.

Conclusions and Discussion

Implementing the cluster principles has the potential to facilitate the establishment of tourist clusters in Ukraine, which can enhance the overall quality of tourism offerings and promote the development of specific regions and destinations for green tourism. To effectively implement cluster initiatives in the economy of green tourism, it is crucial to comprehend the conceptual framework for cluster interaction and formulate appropriate mechanisms for implementation. Further research on cluster system management in tourism, particularly during the postwar reconstruction phase in Ukraine, is necessary.

Despite the numerous challenges facing Ukraine's tourism industry, including the full-scale military invasion, we believe that green tourism can provide opportunities for revitalizing the front-line and formerly occupied regions. Integrating military and green tourism can foster regional development, related economic sectors and agriculture, thereby leveraging the country's natural, historical, cultural and human resources to ensure sustainable economic and social development after the war.

References

Arghutashvili, V. (2018) The role of tourism in economic development of Georgia. *Journal of International Management Studies* 18(2), 59–64. DOI: 10.18374/JIMS-18-2.6.

Arnaud, F. (2016) Memorial policies and restoration of Croatian tourism two decades after the war in former Yugoslavia. *Journal of Tourism and Cultural Change* 14(3), 270–290. DOI: 10.1080/14766825.2016.1169348.

Aussems, E. (2016) Cross-community tourism in Bosnia and Herzegovina: a path to reconciliation? *Journal of Tourism and Cultural Change* 14(3), 240–254. DOI: 10.1080/14766825.2016.1169347.

Boiko, M., Bosovska, M., Vedmid, N., Melnychenko, S. and Okhrimenko, A. (2017) Development of the tourism cluster. *Problems and Perspectives in Management* 15(4), 134–149. DOI: 10.21511/ppm.15(4).2017.12.

Bosovska, M. (2015) Consolidation of tourism sector economic entities based on clusterization. *Agrosvit* 3–4, 47–54.

Bosovska, M., Vedmid, N. and Kulyk, M. (2019) Tourist enterprises' integration processes diagnostics and structuring. *Herald of Knute* (20). DOI: 10.32782/2524-0072/2019-20-19.

Brandenburger, A.M. and Nalebuff, B.J. (2011) *Co-opetition*. Crown Business.

Brin, E. (2006) Politically-oriented tourism in Jerusalem. *Tourist Studies* 6(3), 215–243. DOI: 10.1177/1468797607076672.

Causevic, S. (2010) Tourism which erases borders: an introspection into Bosnia and Herzegovina. In: Moufakkir, O. and Kelly, I. (eds) *Tourism, Progress and Peace*. CAB International, Wallingford, UK, pp. 48–64. DOI: 10.1079/9781845936778.0000.

Drevinh, S. (2008) The institutional framework of inter-regional innovation systems: the role of innovation clusters. *Problems of Modern Economy [Problemy Suchasnoi Ekonomiky]* 3(27). Available at: http://www.m-economy.ru/art.php?nArtId=2067

EPI (2022) Environmental Performance Index: Ukraine. Yale Center for Environmental Law & Policy, The Center for International Earth Science Information Network at Columbia University's Earth Institute. Available at: https://epi.yale.edu/epi-results/2020/country/ukr (accessed 29 October 2023).

Godis, N. and Nilsson, J.H. (2018) Memory tourism in a contested landscape: exploring identity discourses in Lviv, Ukraine. *Current Issues in Tourism* 21(15), 1690–1709. DOI: 10.1080/13683500.2016.1216529.

Gorina, G.O., Barabanova, V.V., Bohatyryova, G.A., Nikolaichuk, O.A. and Romanukha, A.M. (2020) Clustering of regional tourism service markets according to indicators of the functioning of subjects of tourism activity. *Journal of Geology, Geography and Geoecology* 29(4), 684–692. DOI: 10.15421/112061.

Goulding, C. and Domic, D. (2009) Heritage, identity and ideological manipulation: the case of Croatia. *Annals of Tourism Research* 36(1), 85–102. DOI: 10.1016/j.annals.2008.10.004.

Grachev, A. (n.d.-a) Natural national parks. Ukrainian protected areas. Nature of Ukraine. Available at: https://pzf.land.kiev.ua/pzf3.html (accessed 29 October 2023).

Grachev, A. (n.d.-b) Regional landscape parks. Ukrainian protected areas. Nature of Ukraine. Available at: https://pzf.land.kiev.ua/pzf4.html (accessed 29 October 2023).

Green Ukraine (2020) Development of green and rural tourism in Ukraine. Available at: https://www.hneu.edu.ua/wp-content/uploads/2020/03/21_Zelena-Ukrayina.pdf (accessed 29 October 2023).

Kassouha, Z.A. (2019) Post-conflict tourist landscapes: between the heritage of conflict and the hybridization of tourism activity. *Open Edition Journals* 15. DOI: 10.4000/viatourism.3984.

Kharkiv Regional State Administration (2021) Ecological passport of Kharkiv region in 2020. Available at: https://kharkivoda.gov.ua/content/documents/1110/110928/Attaches/ekologichniy_pasport_harkivskoyi_oblasti_za_2020_rik.pdf?sv (accessed 29 October 2023).

Kletionkin, V. (2014) Dvorichanskyi National Nature Park: zerynthia Polyxena (copyright Volodymyr Kletionkin). Available at: https://www.facebook.com/photo/?fbid=409390335863364&set=a.715559048579823 (accessed 29 October 2023).

Kletionkin, V. (2021a) Dvorichanskyi National Nature Park: chalky mountain (copyright Volodymyr Kletionkin). Available at: https://m.facebook.com/story.php?story_fbid=pfbid0XNsfrYKgNMqdoXNjd761c2qDusZ7TejsjB81bb6CY44XsDMgovGnMWdSYEj4VosMl&id=100003772028357 (accessed 29 October 2023).

Kletionkin, V. (2021b) *Dvorichanskyi National Nature Park: bombyliidae*. [photograph]. Facebook. Available at: https://www.facebook.com/photo/?fbid=2382819535187091&set=a.715559048579823 (copyright permission of Volodymyr Kletionkin)

Marshall, A. (1964) *L'Europe solidaire*. Éditions Cujas, Paris.

Melnychenko, S., Bosovska, M. and Okhrimenko, A. (2021) The formation of a nation tourism brand of Ukraine. *Baltic Journal of Economic Studies* 7(2), 161–169. DOI: 10.30525/2256-0742/2021-7-2-161-169.

Ministry of Environmental Protection and Natural Resources (2021) Administrative-territorial units analyze the protected areas of the natural reserve fund in Ukraine. Available at: https://mepr.gov.ua/news/37155.html (accessed 29 October 2023).

Parfinenko, A. (2015) International tourism in Ukraine: geopolitical aspects of a global phenomenon. *Current Issues of International Relations [Aktualni problemy mizhnarodnykh vidnosyn]* 126(1), 12–23.

Polishchuk, K. (2019) Sharivskyi Palace (copyright Kateryna Polischuk). Available at: https://m.facebook.com/story.php?story_fbid=pfbid02zkTQVZdDUfVTRgD1QhuCsobamSM4V2wveB9LgvJZ8E2Jqn4Papwfsao9jbJCf7Kl&id=1411519114 (accessed 29 October 2023).

Polishchuk, K. (2022) Ukrainian landscape (copyright Kateryna Polischuk). Available at: https://m.facebook.com/story.php?story_fbid=pfbid02Hha2pfkRacnvUAN8s4AMk248UrnqaPLHVwW6bSQkzkxAVbQZaZGrR4BJQovLANVpl&id=1411519114 (accessed 29 October 2023).

Porter, M.E. (1980) *Competitive Strategy*. The Free Press, New York.

Pylypenko, A.A. (2007) *Strategic Integration of Enterprises: The Mechanism of Management and Modeling of Development*. INGEC (International Green Energy Cluster).

Reshetchenko, S., Popovych, N., Shulika, B., Porvan, A. and Cherkashyna, N. (2018) Evaluation of the environmental status of agricultural resources in the territory of Ukraine under conditions of climate change. *Technology Audit and Production Reserves* 3(41), 21–32. DOI: 10.15587/2312-8372.2018.134890.

Roelandt, T.J.A. and den Hertog, P. (1999) Summary report of the focus group on clusters. Paper prepared for the OECD. Available at: http://www.oecd.org/dataoecd/56/47/2369025.pdf (accessed 29 October 2023).

Romanova, A. (2018) Management of tourism industry development amidst armed conflicts. *Modern Economics* 9(1), 93–104. DOI: 10.31521/modecon.V9(2018)-12.

Sokolenko, S.I. (2004) *Clusters in a Global Economy [Klastery v hlobal'nij ekonomitsi]*. Lohos.

Strömbom, L. (2019) Exploring prospects for agonistic encounters in conflict zones: investigating dual narrative tourism in Israel/Palestine. *Alternatives* 44(2–4), 75–93. DOI: 10.1177/0304375419857421.

Sushchenko, O.A. (2009) Formation and development of regional clusters under global competition. In: *Proceedings of the International Scientific Conference Pereyaslav Rada: Its Historical Significance and Prospects for the Development of the East Slavic Civilization, Kharkiv, Ukraine*, NTU KPI, pp. 199–204.

UACulture (n.d.) Map of cultural losses. Available at: https://uaculture.org/culture-loss-en/ (accessed 29 October 2023).

Wolfgang, P. (1999) Human behavior: a factor in applied economics. *Perspective Research [Perspektyvni doslidzhennia]* 2, 3–16.

Zarubina, A., Sira, E. and Demchuk, L. (2022) Peculiarities of tourism in a state of war. *Economics and Society [Ekonomika Ta Suspilstvo]* 41. DOI: 10.32782/2524-0072/2022-41-14.

Zbarsky, V.K. and Gribova, D.V. (2020) Rural green tourism as a form of small entrepreneurship of the rural population. Scientific works collection of Dmytro Motorny Tavria state Agrotechnological University (economic sciences) [Zbirnyk naukovykh prats TDATU imeni Dmytra Motornoho (ekonomichni nauky)] 1(41), 125–131. DOI: 10.31388/2519-884X-2020-41-125-131.

26 Conclusions

Nature-based Tourism and Wellbeing explores diverse viewpoints and presents different research topics and methods on a growing sector of global tourism. It presents a rich and multi-faceted array of perspectives, delving into a plethora of compelling questions in many distinct tourism destinations. This comprehensive exploration also leaves open questions on themes and phenomena that warrant further scholarly investigation.

There are nature-based tourism (NBT) research questions that need in-depth treatment. Among them is the question of under what conditions is NBT truly sustainable. The answer, obviously, is complex and multi-faceted and depends on different socio-ecological contexts, levels of tourism and types of recreational activities, and associated infrastructure.

Some tourists rent private jets, emitting enormous amounts of CO_2, to visit natural destinations. Furthermore, many visitors to natural areas show little concern for their impact, disregarding basic behavioural rules, like staying on designated trails, respecting wildlife, using proper waste disposal facilities and other 'leave-no-trace' principles. The challenge lies in establishing a virtuous cycle between destination plans and rules, tourists' behaviour, and the impact of NBT on local communities and livelihoods.

From the supply side, the book shows that significant lessons can be derived from many successful experiences in protected areas that successfully balance nature conservation, authentic and enriching tourism experiences, and support for local livelihoods, cultures and economic development. These protected areas could offer useful examples of sustainable practices in tourism management that need to be taken to scale for larger territorial public administrations to adopt.

Additionally, and from the demand side, protected areas play a critical role in educating tourists about sustainable behaviour, fostering respect for nature through environmental education and interpretation. This line of action would also need to be expanded over wider territorial contexts than those of the protected areas, to change tourist behaviour beyond specific natural destinations.

Achieving sustainable and responsible NBT requires collaboration among different public- and private-sector actors and appropriate governance regimes, another fruitful area for comparative research on appropriate institutional roles and functions. By drawing inspiration from good practices, such as some protected areas, and by promoting education for sustainability, destinations can pave the way for a better future in NBT, preserving natural wonders for generations to come. At the same time, direct involvement of local communities, including traditional and indigenous peoples, as actors and beneficiaries of well-designed NBT destinations, is a global imperative and worthy of additional research on tools and techniques for achieving lasting benefits to local communities from NBT development.

Some further aspects to consider are the relationship between the sustainable development goals and NBT, and between NBT and the Global Biodiversity Strategy, including the global goal of conserving 30% of the world's land and sea area in protected areas and through other effective conservation measures by 2030. This is a daunting goal as, currently, less than 17% of the world's terrestrial surface and less than 10% of its seas have some measure of protection. Undoubtedly, NBT in both existing and to-be-created parks and reserves will depend heavily on making these areas user-friendly for an increasing number of visitors and to help fund their management. How to take NBT to scale

DOI: 10.1079/9781800621411.0026

to make good practices the norm and not the exception in protected areas and other nature-based destinations will be a huge task in the coming years and a rich environment for applied research on what works in sustainable NBT.

Climate change and the challenges it presents to sustainable NBT should also occupy a pre-eminent position in discussions about NBT due to its profound implications for this sector. It is evident that destinations worldwide will need to swiftly decarbonize and review their NBT offerings, and even their landscape management strategies, in response to climate change. Additionally, due to climate change effects, the demand for NBT is expected to evolve, particularly during certain months, with a potential shift towards what can be termed 'survival' or 'climate refugee' tourism, characterized by masses of travellers seeking refuge from extreme temperatures in cooler nature-based locations, such as coasts or mountain forests. These phenomena warrant in-depth and extensive research to gain comprehensive insights.

Undoubtedly, NBT stands as a phenomenon that demands continuous scientific study and exploration, engaging tourism operators, conservationists and researchers, especially in reaching consensus on ways and means of bolstering climate change resilience and reducing the global carbon footprint of the sector.

In conclusion, this book provides a compelling basis for further research in the domain of NBT. The urgent need to delve into specific aspects is evident as we confront the challenges posed by the dual crises of climate change and loss of biodiversity, the evolving demands of tourists, and the growing recognition of the role of NBT in promoting human physical and mental health and contributing to achieving all 17 global sustainable development goals. Thus, maintaining a dynamic and sustained scholarly engagement on the topic of NBT is indispensable for advancing our understanding of both positive and negative impacts of NBT and steering it in a fully sustainable direction.

Index

Note: page numbers in *italic* denote tables and figures.